A Course in Model Theory

This concise introduction to model theory begins with standard notions and takes the reader through to more advanced topics such as stability, simplicity and Hrushovski constructions. The authors introduce the classic results, as well as more recent developments in this vibrant area of mathematical logic. Concrete mathematical examples are included throughout to make the concepts easier to follow. The book also contains over 200 exercises, many with solutions, making the book a useful resource for graduate students as well as researchers.

KATRIN TENT is Professor of Mathematics at the Universität Münster, Germany.

MARTIN ZIEGLER is Professor of Mathematics at the Albert-Ludwigs-Universität Freiburg, Germany.

LECTURE NOTES IN LOGIC

A Publication for The Association for Symbolic Logic

This series serves researchers, teachers, and students in the field of symbolic logic, broadly interpreted. The aim of the series is to bring publications to the logic community with the least possible delay and to provide rapid dissemination of the latest research. Scientific quality is the overriding criterion by which submissions are evaluated.

More information, including a list of the books in the series, can be found at http://www.aslonline.org/books-lnl.html

A Course in Model Theory

KATRIN TENT
Universität Münster

MARTIN ZIEGLER
Universität Freiburg

CAMBRIDGE
UNIVERSITY PRESS

University Printing House, Cambridge CB2 8BS, United Kingdom

Published in the United States of America by Cambridge University Press, New York

Cambridge University Press is part of the University of Cambridge.

It furthers the University's mission by disseminating knowledge in the pursuit of education, learning and research at the highest international levels of excellence.

www.cambridge.org
Information on this title: www.cambridge.org/9780521763240

First published 2012

A catalogue record for this publication is available from the British Library

ISBN 978-0-521-76324-0 Hardback

CONTENTS

PREFACE

This book aims to be an introduction to model theory which can be used without any background in logic. We start from scratch, introducing first-order logic, structures, languages etc. but move on fairly quickly to the fundamental results in model theory and stability theory. We also decided to cover simple theories and Hrushovski constructions, which over the last decade have developed into an important subject. We try to give the necessary background in algebra, combinatorics and set theory either in the course of the text or in the corresponding section of the appendices. The exercises form an integral part of the book. Some of them are used later on, others complement the text or present aspects of the theory that we felt should not be completely ignored. For the most important exercises (and the more difficult ones) we include (hints for) solutions at the end of the book. Those exercises which will be used in the text have their solution marked with an asterisk.

The book falls into four parts. The first three chapters introduce the basics as would be contained in a course giving a general introduction to model theory. This first part ends with Chapter 4 which introduces and explores the notion of a type, the topology on the space of types and a way to make sure that a certain type will not be realized in a model to be constructed. The chapter ends with Fraïssé's amalgamation method, a simple but powerful tool for constructing models.

Chapter 5 is devoted to Morley's famous theorem that a theory with a unique model in *some* uncountable cardinality has a unique model in *every* uncountable cardinality. To prove this theorem, we describe the analysis of uncountably categorical theories due to Baldwin and Lachlan in terms of strongly minimal sets. These are in some sense the easiest examples of stable theories and serve as an introduction to the topic. This chapter forms a unit with Chapter 6 in which the Morley rank is studied in a bit more detail.

For the route to more general stable theories we decided to go via simplicity. The notion of a simple theory was introduced by Shelah in [56]. Such theories allow for a notion of independence which is presented in Chapter 7. Fundamental examples such as pseudo-finite fields make simple theories an important generalisation of the stable ones. We specialise this notion of independence in Chapter 8 to characterise forking in stable theories.

In Chapters 9 and 10 we go back to more classical topics of stability theory such as existence and uniqueness of prime extensions and their analysis in the uncountably categorical case due to Hrushovski. We end the exposition by explaining a variant of Hrushovski's construction of a strongly minimal set.

Model theory does not exist independently of set theory or other areas of mathematics. Many proofs require a knowledge of certain principles of infinite combinatorics which we were hesitant to assume as universally known. Similarly, to study theories of fields we felt it necessary to explain a certain amount of algebra. In the three appendices we try to give enough background about set theory and algebra to be able to follow the exposition in the text.

Other books, some general introductions, others emphasising particular aspects of the theory, that we recommend for further reading include those by Pillay [44] and [42], by Marker [39], Buechler [12], Hodges [24], Poizat [45] [46], Casanovas [14], Wagner [60] and of course Shelah [54]. We refer the reader to these books also for their excellent accounts of the historical background on the material we present.

We would like to thank Manuel Bleichner, Juan-Diego Caycedo, Philipp Doebler, Heinz-Dieter Ebbinghaus, Antongiulio Fornasiero, Nina Frohn, Zaniar Ghadernezhad, John Goodrick, Guntram Hainke, Immanuel Halupczok, Franziska Jahnke, Leander Jehl, Itay Kaplan, Magnus Kollmann, Alexander Kraut, Moritz Müller, Alexandra Omar Aziz, Amador Martin Pizarro, Sebastian Rombach, Lars Scheele and Nina Schwarze for carefully reading earlier versions of the manuscript and Bijan Afshordel for suggesting Exercises 1.1.2 and 5.4.1. We also thank Andreas Baudisch for trying out the book in a seminar and Bernhard Herwig, who translated early parts of the lecture notes from which parts of this book evolved.

Chapter 1

THE BASICS

1.1. Structures

In this section we start at the very beginning, by introducing the prerequisites for the objects of study. We deal with first-order logic and its structures. To this end we first introduce the languages. These will be chosen in different ways for the different mathematical structures that one wants to study.

DEFINITION 1.1.1. A *language* L is a set of constants, function symbols and relation symbols[1].

Function symbols and relation symbols have an arity ≥ 1. One can think of constants as 0-ary function symbols[2]. This allows us to omit the constant symbol case in many proofs.

The language *per se* has no inherent meaning. However, the choice of language will reflect the nature of the intended objects. Here are some standard examples:

$\mathrm{L}_{\emptyset}$	$= \emptyset$	The empty language.
$\mathrm{L}_{\mathrm{AbG}}$	$= \{\underline{0}, +, -\}$	The language of abelian groups.
$\mathrm{L}_{\mathrm{Ring}}$	$= \mathrm{L}_{\mathrm{AbG}} \cup \{\underline{1}, \cdot\}$	The language of rings.
$\mathrm{L}_{\mathrm{Group}}$	$= \{\underline{e}, \circ, {}^{-1}\}$	The language of groups.
$\mathrm{L}_{\mathrm{Order}}$	$= \{<\}$	The language of orders.
$\mathrm{L}_{\mathrm{ORing}}$	$= \mathrm{L}_{\mathrm{Ring}} \cup \mathrm{L}_{\mathrm{Order}}$	The language of ordered rings.
$\mathrm{L}_{\mathrm{Numbers}}$	$= \{\underline{0}, \mathrm{S}, +, \cdot, <\}$	The language of the natural numbers.
$\mathrm{L}_{\mathrm{Set}}$	$= \{\varepsilon\}$	The language of set theory.

The symbols are

constants:	$\underline{0}$, $\underline{1}$, $\underline{e}$
unary function symbols:	$-$, ${}^{-1}$, S
binary function symbols:	$+$, $\cdot$, $\circ$
binary relation symbols:	$<$, ε.

[1]We also use *predicate* for *relation symbol*.

[2]By an unfortunate convention 0-ary relation symbols are not considered.

The languages obtain their meaning only when interpreted in an appropriate structure:

DEFINITION 1.1.2. Let L be a language. An *L-structure* is a pair $\mathfrak{A} = (A, (Z^{\mathfrak{A}})_{Z \in L})$, where

A	is a non-empty set, the *domain* or *universe* of $\mathfrak{A}$,
$Z^{\mathfrak{A}} \in A$	if Z is a constant,
$Z^{\mathfrak{A}} : A^n \longrightarrow A$	if Z is an n-ary function symbol, and
$Z^{\mathfrak{A}} \subseteq A^n$	if Z is an n-ary relation symbol.

We call $Z^{\mathfrak{A}}$ the *$\mathfrak{A}$-interpretation* of Z.

The requirement on A to be not empty is merely a (sometimes annoying) convention. The *cardinality* of a structure is the cardinality of its universe. We write $|\mathfrak{A}|$ or $|A|$ for the cardinality of $\mathfrak{A}$.

DEFINITION 1.1.3. Let $\mathfrak{A}$ and $\mathfrak{B}$ be L-structures. A map $h\colon A \to B$ is called a *homomorphism* if for all $a_1, \dots, a_n \in A$

$$\begin{aligned} h(c^{\mathfrak{A}}) &= c^{\mathfrak{B}} \\ h(f^{\mathfrak{A}}(a_1, \dots, a_n)) &= f^{\mathfrak{B}}(h(a_1), \dots, h(a_n)) \\ R^{\mathfrak{A}}(a_1, \dots, a_n) &\Rightarrow R^{\mathfrak{B}}(h(a_1), \dots, h(a_n)) \end{aligned}$$

for all constants c, n-ary function symbols f and relation symbols R from L. We denote this by

$$h\colon \mathfrak{A} \to \mathfrak{B}.$$

If in addition h is injective and

$$R^{\mathfrak{A}}(a_1, \dots, a_n) \Leftrightarrow R^{\mathfrak{B}}(h(a_1), \dots, h(a_n))$$

for all $a_1, \dots, a_n \in A$, then h is called an (isomorphic) *embedding*. An *isomorphism* is a surjective embedding. We denote isomorphisms by

$$h : \mathfrak{A} \xrightarrow{\sim} \mathfrak{B}.$$

If there is an isomorphism between $\mathfrak{A}$ and $\mathfrak{B}$, the two structures are called *isomorphic* and we write

$$\mathfrak{A} \cong \mathfrak{B}.$$

It is easy to see that being isomorphic is an equivalence relation between structures and that bijections can be used to transfer the structures between sets.

DEFINITION 1.1.4. An *automorphism* of $\mathfrak{A}$ is an isomorphism $\mathfrak{A} \xrightarrow{\sim} \mathfrak{A}$. The set of automorphisms $\mathrm{Aut}(\mathfrak{A})$ forms a group under composition.

DEFINITION 1.1.5. We call $\mathfrak{A}$ a *substructure* of $\mathfrak{B}$ if $A \subseteq B$ and if the inclusion map is an embedding from $\mathfrak{A}$ to $\mathfrak{B}$. We denote this by

$$\mathfrak{A} \subseteq \mathfrak{B}.$$

We say $\mathfrak{B}$ is an *extension* of $\mathfrak{A}$ if $\mathfrak{A}$ is a substructure of $\mathfrak{B}$.

REMARK 1.1.6. If $\mathfrak{B}$ is an L-structure and A a non-empty subset of B, then A is the universe of a (uniquely determined) substructure $\mathfrak{A}$ if and only if A contains all $c^{\mathfrak{B}}$ and A is closed under all operations $f^{\mathfrak{B}}$. In particular, if L does not contain any constants or function symbols, then any non-empty subset of an L-structure is again an L-structure. Also, if $h\colon \mathfrak{A} \to \mathfrak{B}$, then $h(A)$ is the universe of a substructure of $\mathfrak{B}$.

It is also clear that for any family $\mathfrak{A}_i$ of substructures of $\mathfrak{B}$, the intersection of the A_i is either empty or a substructure of $\mathfrak{B}$. Therefore, if S is any non-empty subset of $\mathfrak{B}$, then there exists a smallest substructure $\mathfrak{A} = \langle S\rangle^{\mathfrak{B}}$ which contains S. We call $\mathfrak{A}$ the substructure *generated* by S. If S is finite, then $\mathfrak{A}$ is said to be *finitely generated.*

If L contains a constant, then the intersection of all substructures of $\mathfrak{B}$ is not empty as it contains the $\mathfrak{B}$-interpretation of this constant. Thus $\mathfrak{B}$ has a smallest substructure $\langle\emptyset\rangle^{\mathfrak{B}}$. If L has no constants, we set $\langle\emptyset\rangle^{\mathfrak{B}} = \emptyset$

LEMMA 1.1.7. *If $\mathfrak{A}$ is generated by S, then every homomorphism $h\colon \mathfrak{A} \to \mathfrak{B}$ is determined by its values on S.*

PROOF. If $h'\colon \mathfrak{A} \to \mathfrak{B}$ is another homomorphism, then $C = \{b \mid h(b) = h'(b)\}$ is either empty or a substructure. If h and h' coincide on S, then S is a subset of C, and therefore $C = A$. ⊣

LEMMA 1.1.8. *Let $h : \mathfrak{A} \xrightarrow{\sim} \mathfrak{A}'$ be an isomorphism and $\mathfrak{B}$ an extension of $\mathfrak{A}$. Then there exists an extension $\mathfrak{B}'$ of $\mathfrak{A}'$ and an isomorphism $g : \mathfrak{B} \xrightarrow{\sim} \mathfrak{B}'$ extending h.*

PROOF. First extend the bijection $h\colon A \to A'$ to a bijection $g\colon B \to B'$ and use g to define an L-structure on B'. ⊣

DEFINITION 1.1.9. Let $(I, \leq)$ be a *directed partial order*. This means that for all $i, j \in I$ there exists a $k \in I$ such that $i \leq k$ and $j \leq k$. A family $(\mathfrak{A}_i)_{i\in I}$ of L-structures is called *directed* if

$$i \leq j \Rightarrow \mathfrak{A}_i \subseteq \mathfrak{A}_j.$$

If I is linearly ordered, we call $(\mathfrak{A}_i)_{i\in I}$ a *chain*. If, for example, a structure $\mathfrak{A}_1$ is isomorphic to a substructure $\mathfrak{A}_0$ of itself,

$$h_0 : \mathfrak{A}_0 \xrightarrow{\sim} \mathfrak{A}_1,$$

then Lemma 1.1.8 gives an extension

$$h_1 : \mathfrak{A}_1 \xrightarrow{\sim} \mathfrak{A}_2.$$

Continuing in this way, we obtain a chain $\mathfrak{A}_0 \subseteq \mathfrak{A}_1 \subseteq \mathfrak{A}_2 \subseteq \cdots$ and an increasing sequence $h_i : \mathfrak{A}_i \xrightarrow{\sim} \mathfrak{A}_{i+1}$ of isomorphisms.

LEMMA 1.1.10. *Let $(\mathfrak{A}_i)_{i \in I}$ be a directed family of L-structures. Then $A = \bigcup_{i \in I} A_i$ is the universe of a (uniquely determined) L-structure*

$$\mathfrak{A} = \bigcup_{i \in I} \mathfrak{A}_i,$$

which is an extension of all $\mathfrak{A}_i$.

PROOF. Let R be an n-ary relation symbol and $a_1, \ldots, a_n \in A$. As I is directed, there exists $k \in I$ such that all a_i are in A_k. We define (and this is the only possibility)

$$R^{\mathfrak{A}}(a_1, \ldots, a_n) \Leftrightarrow R^{\mathfrak{A}_k}(a_1, \ldots, a_n).$$

Constants and function symbols are treated similarly. ⊣

A subset K of L is called a sublanguage. An L-structure becomes a K-structure, the *reduct*, by simply forgetting the interpretations of the symbols from $L \setminus K$:

$$\mathfrak{A} \upharpoonright K = \left(A, (Z^{\mathfrak{A}})_{Z \in K}\right).$$

Conversely we call $\mathfrak{A}$ an *expansion* of $\mathfrak{A} \upharpoonright K$. Here are some examples: Let $\mathfrak{A}$ be an L-structure.

a) Let R be an n-ary relation on A. We introduce a new relation symbol $\underline{\mathrm{R}}$ and we denote by

$$(\mathfrak{A}, R)$$

the expansion of $\mathfrak{A}$ to an $L \cup \{\underline{\mathrm{R}}\}$-structure in which $\underline{\mathrm{R}}$ is interpreted by R.

b) For given elements $a_1, \ldots, a_n$ we may introduce new constants $\underline{\mathrm{a}}_1, \ldots, \underline{\mathrm{a}}_n$ and consider the $L \cup \{\underline{\mathrm{a}}_1, \ldots, \underline{\mathrm{a}}_n\}$-structure

$$(\mathfrak{A}, a_1, \ldots, a_n).$$

c) Let B be a subset of A. By considering every element of B as a new constant, we obtain the new language

$$L(B) = L \cup B$$

and the $L(B)$-structure

$$\mathfrak{A}_B = (\mathfrak{A}, b)_{b \in B}.$$

Note that $\mathrm{Aut}(\mathfrak{A}_B)$ is the group of automorphisms of $\mathfrak{A}$ fixing B elementwise. We denote this group by $\mathrm{Aut}(\mathfrak{A}/B)$.

Similarly, if C is a set of new constants, we write $L(C)$ for the language $L \cup C$.

Many-sorted structures. Without much effort, the concepts introduced here can be extended to many-sorted languages and structures, which we shall need to consider later on.

Let S be a set, which we call the set of sorts. An S-sorted language L is given by a set of constants for each sort in S, and typed function and relation symbols which carry the information about their arity and the sorts of their domain and range. More precisely, for any tuple $(s_1, \dots, s_n)$ and $(s_1, \dots, s_n, t)$ there is a set of relation symbols and function symbols, respectively. An S-sorted structure is a pair $\mathfrak{A} = (A, (Z^{\mathfrak{A}})_{Z \in L})$, where

A	is a family $(A_s)_{s \in S}$ of non-empty sets.
$Z^{\mathfrak{A}} \in A_s$	if Z is a constant of sort $s \in S$,
$Z^{\mathfrak{A}} : A_{s_1} \times \cdots \times A_{s_n} \longrightarrow A_t$	if Z is a function symbol of type $(s_1, \dots, s_n, t)$,
$Z^{\mathfrak{A}} \subseteq A_{s_1} \times \cdots \times A_{s_n}$	if Z is a relation symbol of type $(s_1, \dots, s_n)$.

It should be clear how to define homomorphisms between many-sorted structures $\mathfrak{A}$ and $\mathfrak{B}$: they are given by maps taking A_s to B_s for $s \in S$ and behaving as before with respect to constants, function and relation symbols.

EXAMPLE. Consider the two-sorted language $\mathrm{L}_{\mathrm{Perm}}$ for permutation groups with a sort x for the set and a sort g for the group. The constants and function symbols for $\mathrm{L}_{\mathrm{Perm}}$ are those of $\mathrm{L}_{\mathrm{Group}}$ restricted to the sort g and an additional function symbol φ of type (x, g, x). Thus, an $\mathrm{L}_{\mathrm{Perm}}$-structure (X, G) is given by a set X and an $\mathrm{L}_{\mathrm{Group}}$-structure G together with a function $X \times G \longrightarrow X$.

EXERCISE 1.1.1 (Direct products). Let $\mathfrak{A}_1, \mathfrak{A}_2$ be L-structures. Define an L-structure $\mathfrak{A}_1 \times \mathfrak{A}_2$ with universe $A_1 \times A_2$ such that the natural epimorphisms $\pi_i : \mathfrak{A}_1 \times \mathfrak{A}_2 \longrightarrow \mathfrak{A}_i$ for $i = 1, 2$ satisfy the following universal property: given any L-structure $\mathfrak{D}$ and homomorphisms $\varphi_i : \mathfrak{D} \longrightarrow \mathfrak{A}_i, i = 1, 2$ there is a unique homomorphism $\psi : \mathfrak{D} \longrightarrow \mathfrak{A}_1 \times \mathfrak{A}_2$ such that $\pi_i \circ \psi = \varphi_i, i = 1, 2$, i.e., this is the product in the category of L-structures with homomorphisms.

EXERCISE 1.1.2. Let $f : \mathfrak{A} \to \mathfrak{A}$ be an embedding. Then there is an extension $\mathfrak{A} \subseteq \mathfrak{B}$ and an extension of f to an automorphism g of $\mathfrak{B}$. We can find B as the union of the chain $A \subseteq g^{-1}(A) \subseteq g^{-2}(A) \subseteq \cdots$. The pair $(\mathfrak{B}, g)$ is uniquely determined by that property.

1.2. Language

Starting from the inventory of the languages defined in Section 1.1 we now describe the grammar which allows us to build well-formed terms and formulas which will again be interpreted in the according structures.

DEFINITION 1.2.1. An *L-term* is a word (sequence of symbols) built from constants, the function symbols of L and the *variables* $v_0, v_1, \dots$ according to the following rules:

1. Every variable v_i and every constant c is an L-term.
2. If f is an n-ary function symbol and $t_1, \dots, t_n$ are L-terms, then $f t_1 \dots t_n$ is also an L-term.

The number of occurrences of function symbols in a term is called its *complexity*. This will be used in induction arguments.

We often write $f(t_1, \dots, t_n)$ instead of $f t_1 \dots t_n$ for better readability and use the usual conventions for some particular function symbols. For example

$$(x + y) \cdot (z + w)$$

stands for

$$\cdot + xy + zw$$

and $(x \circ y)^{-1}$ for ${}^{-1} \circ xy$.

Let $\mathfrak{A}$ be an L-structure and $\vec{b} = (b_0, b_1, \dots)$ a sequence of elements which we consider as *assignments* to the variables $v_0, v_1, \dots$. If we replace in t each variable v_i by a_i, the term t determines an element $t^{\mathfrak{A}}[\vec{b}]$ of $\mathfrak{A}$ in an obvious way:

DEFINITION 1.2.2. For an L-term t, an L-structure $\mathfrak{A}$ and an assignment $\vec{b}$ we define the interpretation $t^{\mathfrak{A}}[\vec{b}]$ by

$$v_i^{\mathfrak{A}}[\vec{b}] = b_i$$
$$c^{\mathfrak{A}}[\vec{b}] = c^{\mathfrak{A}}$$
$$f t_1 \dots t_n^{\mathfrak{A}}[\vec{b}] = f^{\mathfrak{A}}(t_1^{\mathfrak{A}}[\vec{b}], \dots, t_n^{\mathfrak{A}}[\vec{b}]).$$

This (recursive) definition is possible because every term has a unique decomposition into its constituents: if $f t_1 \dots t_n = f t'_1 \dots t'_n$, then $t_1 = t'_1, \dots, t_n = t'_n$. This as well as the following lemma are easy to prove using induction on the complexity of the terms involved.

LEMMA 1.2.3. *The interpretation $t^{\mathfrak{A}}[\vec{b}]$ depends on b_i only if v_i occurs in t.*

If the variables $x_1, \dots, x_n$ are pairwise distinct[3] and if no other variables occur in t, we write

$$t = t(x_1, \dots, x_n).$$

According to the previous lemma, if $\vec{b}$ is an assignment for the variables which assigns a_i to x_i, we can define

$$t^{\mathfrak{A}}[a_1, \dots, a_n] = t^{\mathfrak{A}}[\vec{b}].$$

If $t_1, \dots, t_n$ are terms, we can substitute $t_1, \dots, t_n$ for the variables $x_1, \dots, x_n$. The resulting term is denoted by

$$t(t_1, \dots, t_n).$$

[3] Remember that $x_i \in \{v_0, v_1, \dots\}$.

One easily proves:

LEMMA 1.2.4 (Substitution Lemma).

$$t(t_1,\dots,t_n)^{\mathfrak{A}}[\vec{b}] = t^{\mathfrak{A}}\left[t_1^{\mathfrak{A}}[\vec{b}],\dots,t_n^{\mathfrak{A}}[\vec{b}]\right]. \qquad \dashv$$

If we expand $\mathfrak{A}$ to the $L(A)$-structure $\mathfrak{A}_A$, we get as a special case

$$t(a_1,\dots,a_n)^{\mathfrak{A}_A} = t^{\mathfrak{A}}[a_1,\dots,a_n].$$

LEMMA 1.2.5. *Let $h\colon \mathfrak{A} \to \mathfrak{B}$ be a homomorphism and $t(x_1,\dots,x_n)$ a term. Then we have for all $a_1,\dots,a_n$ from A*

$$t^{\mathfrak{B}}[h(a_1),\dots,h(a_n)] = h\left(t^{\mathfrak{A}}[a_1,\dots,a_n]\right).$$

PROOF. Induction on the complexity of t. $\dashv$

LEMMA 1.2.6. *Let S be a subset of the L-structure $\mathfrak{A}$. Then*

$$\langle S\rangle^{\mathfrak{A}} = \left\{t^{\mathfrak{A}}[s_1,\dots,s_n] \mid t(x_1,\dots,x_n)\ L\text{-term},\ s_1,\dots,s_n \in S\right\}.$$

PROOF. We may assume that S is not empty or that L contains a constant since otherwise both sides of the equation are empty. It follows from Lemma 1.2.5 that the universe of a substructure is closed under interpretations of terms $t^{\mathfrak{A}}[-,\dots,-]$. Thus the right hand side is contained in $\langle S\rangle^{\mathfrak{A}}$. For the converse we have to show that the right hand side is closed under the operations $f^{\mathfrak{A}}$. The assertion now follows using Remark 1.1.6. $\dashv$

A *constant term* is a term without variables. As a special case of Lemma 1.2.6 we thus have

$$\langle\emptyset\rangle^{\mathfrak{A}} = \left\{t^{\mathfrak{A}} \mid t \text{ constant } L\text{-term}\right\}.$$

The previous lemma implies:

COROLLARY 1.2.7. $|\langle S\rangle^{\mathfrak{A}}| \le \max(|S|,|L|,\aleph_0)$.

PROOF. There are at most $\max(|L|,\aleph_0)$ many L-terms and for every term t at most $\max(|S|,\aleph_0)$ many assignments of elements of S to the variables of t. $\dashv$

We still need to define *L-formulas*. These are sequences of symbols which are built from the symbols of L, the parentheses "(" and ")" as auxiliary symbols and the following *logical symbols*:

variables	$v_0, v_1, \dots$
equality symbol	$\doteq$
negation symbol	$\neg$
conjunction symbol	$\wedge$
existential quantifier	$\exists$

DEFINITION 1.2.8. *L-formulas* are

1. $t_1 \doteq t_2$ where t_1, t_2 are L-terms,
2. $Rt_1\dots t_n$ where R is an n-ary relation symbol from L and $t_1,\dots,t_n$ are L-terms,

3. $\neg\psi$ where ψ is an L-formula,
4. $(\psi_1 \wedge \psi_2)$ where ψ_1 and ψ_2 are L-formulas,
5. $\exists x\ \psi$ where ψ is an L-formula and x a variable.

Formulas of the form $t_1 \doteq t_2$ or $Rt_1 \ldots t_n$ are called *atomic*.

As with terms, we define the complexity of a formula as the number of occurrences of $\neg$, $\exists$ and $\wedge$. This allows us to do induction on (the complexity of) formulas.

We use the following abbreviations:

$$(\psi_1 \vee \psi_2) = \neg(\neg\psi_1 \wedge \neg\psi_2)$$
$$(\psi_1 \rightarrow \psi_2) = \neg(\psi_1 \wedge \neg\psi_2)$$
$$(\psi_1 \leftrightarrow \psi_2) = ((\psi_1 \rightarrow \psi_2) \wedge (\psi_2 \rightarrow \psi_1))$$
$$\forall x\ \psi = \neg\exists x \neg\psi$$

for disjunction, implication, equivalence and universal quantifier.

Sometimes we write t_1Rt_2 for Rt_1t_2, $\exists x_1 \ldots x_n$ for $\exists x_1 \ldots \exists x_n$ and $\forall x_1 \ldots x_n$ for $\forall x_1 \ldots \forall x_n$. To improve readability we might use superfluous parentheses. On the other hand we might omit parentheses with an implicit understanding of the *binding strength* of logical symbols: $\neg$, $\exists$, and $\forall$ bind more strongly than $\wedge$ which in turn binds more strongly than $\vee$. Finally $\rightarrow$ and $\leftrightarrow$ have the least binding strength. For example $\neg\psi_1 \wedge \psi_2 \rightarrow \psi_3$ is understood to mean $((\neg\psi_1 \wedge \psi_2) \rightarrow \psi_3)$.

Given an L-structure $\mathfrak{A}$ and an L-formula $\varphi(x_1, \ldots x_n)$ it should now be clear what it means for φ to *hold* for $\vec{b}$. Here is the formal definition.

DEFINITION 1.2.9. Let $\mathfrak{A}$ be an L-structure. For L-formulas φ and all assignments $\vec{b}$ we define the relation

$$\mathfrak{A} \models \varphi[\vec{b}]$$

recursively over φ:

$$\mathfrak{A} \models t_1 \doteq t_2\,[\vec{b}] \Leftrightarrow t_1^{\mathfrak{A}}[\vec{b}] = t_2^{\mathfrak{A}}[\vec{b}]$$
$$\mathfrak{A} \models Rt_1 \ldots t_n\,[\vec{b}] \Leftrightarrow R^{\mathfrak{A}}\left(t_1^{\mathfrak{A}}[\vec{b}], \ldots, t_n^{\mathfrak{A}}[\vec{b}]\right)$$
$$\mathfrak{A} \models \neg\psi\,[\vec{b}] \Leftrightarrow \mathfrak{A} \not\models \psi\,[\vec{b}]$$
$$\mathfrak{A} \models (\psi_1 \wedge \psi_2)\,[\vec{b}] \Leftrightarrow \mathfrak{A} \models \psi_1\,[\vec{b}] \text{ and } \mathfrak{A} \models \psi_2\,[\vec{b}]$$
$$\mathfrak{A} \models \exists x\psi\,[\vec{b}] \Leftrightarrow \text{ there exists } a \in A \text{ such that } \mathfrak{A} \models \psi\left[\vec{b}\frac{a}{x}\right].$$

Here we use the notation

$$\vec{b}\frac{a}{x} = (b_0, \ldots, b_{i-1}, a, b_{i+1,\ldots}) \quad \text{if } x = v_i.$$

If $\mathfrak{A} \models \varphi[\vec{b}]$ holds we say φ *holds in* $\mathfrak{A}$ *for* $\vec{b}$ or $\vec{b}$ *satisfies* φ *(in* $\mathfrak{A}$*)*.

For this definition to work one has to check that every formula has a *unique decomposition*[4] into subformulas: if $Rt_1 \dots t_n = Rt'_1 \dots t'_n$, then $t_1 = t'_1, \dots, t_n = t'_n$; and $(\psi_1 \wedge \psi_2) = (\psi'_1 \wedge \psi'_2)$ implies $\psi_1 = \psi'_1$ and $\psi_2 = \psi'_2$.

It should be clear that our abbreviations have the intended meaning, e.g.,

$$\mathfrak{A} \models (\psi_1 \to \psi_2)[\vec{b}] \text{ if and only if } (\mathfrak{A} \models \psi_1[\vec{b}] \text{ implies } \mathfrak{A} \models \psi_2[\vec{b}]).$$

Whether φ holds in $\mathfrak{A}$ for $\vec{b}$ depends only on the *free* variables of φ:

DEFINITION 1.2.10. The variable x occurs *freely* in the formula φ if it occurs at a place which is not within the scope of a quantifier $\exists x$. Otherwise its occurrence is called *bound*. Here is the formal definition (recursive in φ):

$$x \text{ free in } t_1 \doteq t_2 \Leftrightarrow x \text{ occurs in } t_1 \text{ or in } t_2.$$
$$x \text{ free in } Rt_1 \dots t_n \Leftrightarrow x \text{ occurs in one of the } t_i.$$
$$x \text{ free in } \neg\psi \Leftrightarrow x \text{ free in } \psi.$$
$$x \text{ free in } (\psi_1 \wedge \psi_2) \Leftrightarrow x \text{ free in } \psi_1 \text{ or } x \text{ free in } \psi_2.$$
$$x \text{ free in } \exists y\, \psi \Leftrightarrow x \neq y \text{ and } x \text{ free in } \psi.$$

For example the variable v_0 does not occur freely in $\forall v_0(\exists v_1 R(v_0, v_1) \wedge P(v_1))$; v_1 occurs both freely and bound[5].

LEMMA 1.2.11. *Suppose $\vec{b}$ and $\vec{c}$ agree on all variables which are free in φ. Then*

$$\mathfrak{A} \models \varphi[\vec{b}] \Leftrightarrow \mathfrak{A} \models \varphi[\vec{c}].$$

PROOF. By induction on the complexity of φ. ⊣

If we write a formula in the form $\varphi(x_1, \dots, x_n)$, we mean:

- the x_i are pairwise distinct,
- all free variables in φ are among $\{x_1, \dots, x_n\}$.

If furthermore $a_1, \dots, a_n$ are elements of the structure $\mathfrak{A}$, we define

$$\mathfrak{A} \models \varphi[a_1, \dots, a_n]$$

by $\mathfrak{A} \models \varphi[\vec{b}]$, where $\vec{b}$ is an assignment satisfying $\vec{b}(x_i) = a_i$. Because of Lemma 1.2.11 this is well defined.

Thus $\varphi(x_1, \dots, x_n)$ defines an n-ary relation

$$\varphi(\mathfrak{A}) = \{\overline{a} \mid \mathfrak{A} \models \varphi[\overline{a}]\}$$

on A, the *realisation set* of φ. Such realisation sets are called 0*-definable subsets* of A^n, or 0-definable relations.

Let B be a subset of A. A B*-definable* subset of $\mathfrak{A}$ is a set of the form $\varphi(\mathfrak{A})$ for an $L(B)$-formula $\varphi(x)$. We also say that φ (and $\varphi(\mathfrak{A})$) are defined *over*

[4]It is precisely because of this uniqueness that we introduced brackets when defining formulas.

[5]However, we usually make sure that no variable occurs both freely and bound. This can be done by renaming the free occurrence with an unused variable.

B and that the set $\varphi(\mathfrak{A})$ is defined by φ. Often we don't explicitly specify a parameter set B and just talk about *definable* subsets. A 0-definable set is definable over the empty set. We call two formulas *equivalent* if in every structure they define the same set.

DEFINITION 1.2.12. A formula φ without free variables is called a *sentence*. We write $\mathfrak{A} \models \varphi$ if $\mathfrak{A} \models \varphi[\vec{b}]$ for some (all) $\vec{b}$.

In that case $\mathfrak{A}$ is called a *model* of φ. We also say φ *holds in* $\mathfrak{A}$. If Σ is a set of sentences, then $\mathfrak{A}$ is a model of Σ if all sentences of Σ hold in $\mathfrak{A}$. We denote this by

$$\mathfrak{A} \models \Sigma.$$

Let $\varphi = \varphi(x_1, \ldots, x_n)$ and let $t_1, \ldots, t_n$ be terms. The formula

$$\varphi(t_1, \ldots, t_n)$$

is the formula obtained by first renaming all bound variables by variables which do not occur in the t_i and then replacing every free occurrence of x_i by t_i.

LEMMA 1.2.13 (Substitution lemma).

$$\mathfrak{A} \models \varphi(t_1, \ldots, t_n)[\vec{b}] \iff \mathfrak{A} \models \varphi\left[t_1^{\mathfrak{A}}[\vec{b}], \ldots, t_n^{\mathfrak{A}}[\vec{b}]\right].$$

PROOF. The proof is an easy induction on φ. ⊣

Also note this (trivial) special case:

$$\mathfrak{A}_A \models \varphi(a_1, \ldots, a_n) \iff \mathfrak{A} \models \varphi[a_1, \ldots, a_n].$$

Henceforth we often suppress the assignment (and the subscript) and simply write

$$\mathfrak{A} \models \varphi(a_1, \ldots, a_n).$$

Atomic formulas and their negations are called *basic*. Formulas without quantifiers (or: *quantifier-free* formulas) are Boolean combinations of basic formulas, i.e., they can be built from basic formulas by successively applying $\neg$ and $\wedge$. The conjunction of formulas π_i is denoted by $\bigwedge_{i<m} \pi_i$ and $\bigvee_{i<m} \pi_i$ denotes their disjunction. By convention $\bigwedge_{i<1} \pi_i = \bigvee_{i<1} \pi_i = \pi_0$. It is convenient to allow the empty conjunction and the empty disjunction. For that we introduce two new formulas: the formula $\top$, which is always true, and the formula $\bot$, which is always false. We define

$$\bigwedge_{i<0} \pi_i = \top$$

$$\bigvee_{i<0} \pi_i = \bot$$

A formula is in *negation normal form* if it is built from basic formulas using $\wedge, \vee, \exists, \forall$.

LEMMA 1.2.14. *Every formula can be transformed into an equivalent formula which is in negation normal form.*

PROOF. Let $\sim$ denote equivalence of formulas. We consider formulas which are built using $\wedge, \vee, \exists, \forall$ and $\neg$ and move the negation symbols in front of atomic formulas using

$$\begin{aligned}
\neg(\varphi \wedge \psi) &\sim (\neg\varphi \vee \neg\psi)\\
\neg(\varphi \vee \psi) &\sim (\neg\varphi \wedge \neg\psi)\\
\neg\exists x\varphi &\sim \forall x\neg\varphi\\
\neg\forall x\varphi &\sim \exists x\neg\varphi\\
\neg\neg\varphi &\sim \varphi.
\end{aligned}$$ ⊣

DEFINITION 1.2.15. A formula in negation normal form which does not contain any existential quantifier is called *universal*. Formulas in negation normal form without universal quantifiers are called *existential*.

Clearly an isomorphism $h\colon \mathfrak{A} \to \mathfrak{B}$ preserves the validity of every formula:

$$\mathfrak{A} \models \varphi[a_1, \dots, a_n] \iff \mathfrak{B} \models \varphi[h(a_1), \dots, h(a_n)].$$

Embeddings preserve the validity of existential formulas:

LEMMA 1.2.16. *Let* $h\colon \mathfrak{A} \to \mathfrak{B}$ *be an embedding. Then for all existential formulas* $\varphi(x_1, \dots, x_n)$ *and all* $a_1, \dots, a_n \in A$ *we have*

$$\mathfrak{A} \models \varphi[a_1, \dots, a_n] \implies \mathfrak{B} \models \varphi[h(a_1), \dots, h(a_n)].$$

For universal φ, *the dual holds:*

$$\mathfrak{B} \models \varphi[h(a_1), \dots, h(a_n)] \implies \mathfrak{A} \models \varphi[a_1, \dots, a_n].$$

PROOF. By an easy induction on φ: for basic formulas the assertion follows from the definition of an embedding and Lemma 1.2.5. The inductive step is trivial for the cases $\wedge$ and $\vee$. Let finally $\varphi(\overline{x})$ be $\exists y\, \psi(\overline{x}, y)$. If $\mathfrak{A} \models \varphi[\overline{a}]$, there exists an $a \in A$ such that $\mathfrak{A} \models \psi[\overline{a}, a]$. By induction we have $\mathfrak{B} \models \psi[h(\overline{a}), h(a)]$. Thus $\mathfrak{B} \models \varphi[h(\overline{a})]$. ⊣

Let $\mathfrak{A}$ be an L-structure. The *atomic diagram* of $\mathfrak{A}$ is

$$\mathrm{Diag}(\mathfrak{A}) = \{\varphi \text{ basic } L(A)\text{-sentence} \mid \mathfrak{A}_A \models \varphi\},$$

the set of all basic sentences with parameters from A which hold in $\mathfrak{A}$.

LEMMA 1.2.17. *The models of* Diag($\mathfrak{A}$) *are precisely those structures* $(\mathfrak{B}, h(a))_{a\in A}$ *for embeddings* $h\colon \mathfrak{A} \to \mathfrak{B}$.

PROOF. The structures $(\mathfrak{B}, h(a))_{a \in A}$ are models of the atomic diagram by Lemma 1.2.16. For the converse note that a map h is an embedding if and only if it preserves the validity of all formulas of the form

$$\begin{aligned}&(\neg)\, x_1 \doteq x_2\\&c \doteq x_1\\&f(x_1,\dots,x_n) \doteq x_0\\&(\neg)\, R(x_1,\dots,x_n).\end{aligned} \qquad \dashv$$

Many-sorted languages. In a many-sorted language with sorts in S, terms and formulas are built with respect to the sorts. For each sort $s \in S$ we have variables $v_0^s, v_1^s, \dots$ from which we build the following terms of sort s.

Every variable v_i^s is an L-term of sort s.
Every constant c of sort s is an L-term of sort s.
If f is a function symbol of type $(s_1,\dots,s_n,s)$ and t_i is an L-term of sort s_i for $i = 1,\dots,n$, then $f t_1 \dots t_n$ is an L-term of sort s.

We construct L-formulas as before with the following adjustments:

$t_1 \doteq t_2$	where t_1, t_2 are L-terms of the same sort,
$Rt_1 \dots t_n$	where R is a relation symbol from L of type $(s_1,\dots,s_n)$ and t_i is an L-term of sort s_i,
$\exists x\ \psi$	where ψ is an L-formula and x a variable (of some sort s).

It should be clear how to extend the definitions of this section to the many-sorted situation and that the results presented here continue to hold without change. In what follows we will not deal separately with many-sorted languages until we meet them again in Section 8.4.

EXERCISE 1.2.1. Let L be a language and P be a new n-ary relation symbol. Let $\varphi = \varphi(P)$ be an $L(P) = L \cup \{P\}$-sentence and $\pi(x_1,\dots,x_n)$ an L-formula. Now replace every occurrence of P in φ by π. More precisely, every subformula of the form $Pt_1 \dots t_n$ is replaced by $\pi(t_1 \dots t_n)$. We denote the resulting L-formula by $\varphi(\pi)$. Show that

$$\mathfrak{A} \models \varphi(\pi) \text{ if and only if } (\mathfrak{A}, \pi(\mathfrak{A})) \models \varphi(P).$$

EXERCISE 1.2.2. Every quantifier-free formula is equivalent to a formula of the form

$$\bigwedge_{i<m} \bigvee_{j<m_i} \pi_{ij}$$

and to a formula of the form

$$\bigvee_{i<m} \bigwedge_{j<m_i} \pi_{ij}$$

where the π_{ij} are basic formulas. The first form is called the *conjunctive* normal form; the second, the *disjunctive* normal form.

EXERCISE 1.2.3. Every formula is equivalent to a formula in prenex normal form:

$$Q_1 x_1 \ldots Q_n x_n \varphi.$$

The Q_i are quantifiers ($\exists$ or $\forall$) and φ is quantifier-free.

EXERCISE 1.2.4 (Ultraproducts and Łos's Theorem). A *filter* on a set I is a non-empty set $\mathcal{F} \subseteq \mathfrak{P}(I)$ which does not contain the empty set and is closed under intersections and supersets, i.e., for $A, B \in \mathcal{F}$, we have $A \cap B \in \mathcal{F}$ and if $A \in \mathcal{F}$ and $A \subseteq C \subseteq I$ we have $C \in \mathcal{F}$. A filter $\mathcal{F}$ is called an *ultrafilter* if for every $A \in \mathfrak{P}$ we have $A \in \mathcal{F}$ or $I \setminus A \in \mathcal{F}$. (By Zorn's Lemma, any filter can be extended to an ultrafilter.)

For a family $(\mathfrak{A}_i \mid i \in I)$ of L-structures and $\mathcal{F}$ an ultrafilter on I we define the *ultraproduct* $\Pi_{i \in I} \mathfrak{A}_i / \mathcal{F}$ as follows. On the Cartesian product $\Pi_{i \in I} \mathfrak{A}_i$, the ultrafilter $\mathcal{F}$ defines an equivalence relation $\sim_{\mathcal{F}}$ by

$$(a_i)_{i \in I} \sim_{\mathcal{F}} (b_i)_{i \in I} \Leftrightarrow \{i \in I \mid a_i = b_i\} \in \mathcal{F}.$$

On the set of equivalence classes $(a_i)_{\mathcal{F}}$ we define an L-structure $\Pi_{i \in I} \mathfrak{A}_i / \mathcal{F}$.

- For constants $c \in L$, put $c^{\Pi_{\mathcal{F}} \mathfrak{A}_i} = (c^{\mathfrak{A}_i})_{\mathcal{F}}$.
- For n-ary function symbols $f \in L$ put

$$f^{\Pi_{\mathcal{F}} \mathfrak{A}_i}((a_i^1)_{\mathcal{F}}, \ldots, (a_i^n)_{\mathcal{F}})) = (f^{\mathfrak{A}_i}(a_i^1, \ldots, a_i^n))_{\mathcal{F}}.$$

- For n-ary relation symbols $R \in L$ put

$$R^{\Pi_{\mathcal{F}} \mathfrak{A}_i}((a_i^1)_{\mathcal{F}}, \ldots, (a_i^n)_{\mathcal{F}})) \Leftrightarrow \{i \in I \mid R^{\mathfrak{A}_i}(a_i^1, \ldots, a_i^n)\} \in \mathcal{F}.$$

1. Show that the ultraproduct $\Pi_{i \in I} \mathfrak{A}_i / \mathcal{F}$ is well-defined.
2. Prove Łos's Theorem: for any L-formula φ we have

$$\Pi_{i \in I} \mathfrak{A}_i / \mathcal{F} \models \varphi((a_i^1)_{\mathcal{F}}, \ldots, (a_i^n)_{\mathcal{F}}) \Leftrightarrow \{i \in I \mid \mathfrak{A}_i \models \varphi(a_i^1, \ldots, a_i^n)\} \in \mathcal{F}.$$

1.3. Theories

Having defined a language, we can now take a closer look at which formulas hold in a given structure. Conversely, we can start with a set of sentences and consider those structures in which they hold. In this way, these sentences serve as a set of *axioms* for a theory.

DEFINITION 1.3.1. An *L-theory T* is a set of L-sentences.

A theory which has a model is a *consistent* theory. More generally, we call a set Σ of L-formulas *consistent* if there is an L-structure $\mathfrak{A}$ and an assignment $\vec{b}$ such that $\mathfrak{A} \models \varphi[\vec{b}]$ for all $\varphi \in \Sigma$. We say that Σ is *consistent with T* if $T \cup \Sigma$ is consistent.

LEMMA 1.3.2. *Let T be an L-theory and L' be an extension of L. Then T is consistent as an L-theory if and only if T is consistent as a L'-theory.*

PROOF. This follows from the (trivial) fact, that every L-structure is expandable to an L'-structure. ⊣

EXAMPLE. To keep algebraic expressions readable we will write 0 and 1 for the symbols $\underline{0}$ and $\underline{1}$ in the following examples. We will omit the dot for the multiplication and brackets if they are implied by the order of operations rule.

AbG, the theory of abelian groups, has the axioms:

- $\forall x, y, z\ (x + y) + z \doteq x + (y + z)$
- $\forall x\ 0 + x \doteq x$
- $\forall x\ (-x) + x \doteq 0$
- $\forall x, y\ x + y \doteq y + x$.

Ring, the theory of commutative rings:

- AbG
- $\forall x, y, z\ (xy)z \doteq x(yz)$
- $\forall x\ 1x \doteq x$
- $\forall x, y\ xy \doteq yx$
- $\forall x, y, z\ x(y + z) \doteq xy + xz$.

Field, the theory of fields:

- Ring
- $\neg 0 \doteq 1$
- $\forall x\ (\neg x \doteq 0 \rightarrow \exists y\ xy \doteq 1)$.

DEFINITION 1.3.3. If a sentence φ holds in all models of T, we say that φ *follows from* T (or that T *proves* φ) and write[6]

$$T \vdash \varphi.$$

By Lemma 1.3.2 this relation is independent of the language. Sentences φ which follow from the empty theory $\emptyset$ are called *valid*. We denote this by $\vdash \varphi$.

The most important properties of $\vdash$ are:

LEMMA 1.3.4. 1. *If $T \vdash \varphi$ and $T \vdash (\varphi \rightarrow \psi)$, then $T \vdash \psi$.*

2. *If $T \vdash \varphi(c_1, \ldots, c_n)$ and the constants $c_1, \ldots, c_n$ occur neither in T nor in $\varphi(x_1, \ldots, x_n)$, then $T \vdash \forall x_1 \ldots x_n \varphi(x_1, \ldots, x_n)$.*

PROOF. We prove 2. Let $L' = L \setminus \{c_1, \ldots, c_n\}$. If the L'-structure $\mathfrak{A}$ is a model of T and $a_1, \ldots, a_n$ are arbitrary elements, then $(\mathfrak{A}, a_1, \ldots, a_n) \models \varphi(c_1, \ldots, c_n)$. That means $\mathfrak{A} \models \forall x_1 \ldots x_n \varphi(x_1, \ldots, x_n)$. Thus $T \vdash \forall x_1 \ldots x_n\ \varphi(x_1, \ldots, x_n)$. ⊣

We generalise this relation to theories S: we write $T \vdash S$ if all models of T are models of S. S and T are called *equivalent*, $S \equiv T$, if S and T have the same models.

---

[6]Note that sometimes this relation is denoted by $T \models \varphi$ to distinguish this notion from the more syntactic notion of *logical inference*, see [57, section 2.6].

DEFINITION 1.3.5. A consistent L-theory T is called *complete* if for all L-sentences φ

$$T \vdash \varphi \text{ or } T \vdash \neg\varphi.$$

This notion clearly depends on L. If T is complete and L' is an extension of L, then T will in general not be complete as an L'-theory.

DEFINITION 1.3.6. For a complete theory T we define

$$|T| = \max(|L|, \aleph_0):$$

$|T|$ is exactly the number of L–formulas. This will be explained in the proof of Corollary 2.1.3.

The typical (and, as we will see below, only) example of a complete theory is the theory of a structure $\mathfrak{A}$

$$\mathrm{Th}(\mathfrak{A}) = \{\varphi \mid \mathfrak{A} \models \varphi\}.$$

LEMMA 1.3.7. *A consistent theory is complete if and only if it is maximal consistent, i.e., if it is equivalent to every consistent extension.*

PROOF. We call φ independent from T if neither φ nor $\neg\varphi$ follows from T. So φ is independent from T exactly when $T \cup \{\varphi\}$ is a proper (i.e., not equivalent) consistent extension of T. From this the lemma follows directly. ⊣

DEFINITION 1.3.8. Two L-structures $\mathfrak{A}$ and $\mathfrak{B}$ are called *elementary equivalent*,

$$\mathfrak{A} \equiv \mathfrak{B},$$

if they have the same theory; that is, if for all L-sentences φ

$$\mathfrak{A} \models \varphi \iff \mathfrak{B} \models \varphi.$$

Isomorphic structures are always elementarily equivalent. The converse holds only for finite structures, see Exercise 1.3.3 and Theorem 2.3.1.

LEMMA 1.3.9. *Let T be a consistent theory. Then the following are equivalent:*

a) *T is complete.*
b) *All models of T are elementarily equivalent.*
c) *There exists a structure $\mathfrak{A}$ with $T \equiv \mathrm{Th}(\mathfrak{A})$.*

PROOF. a) ⇒ c): Let $\mathfrak{A}$ be a model of T. If φ holds in $\mathfrak{A}$, then $T \not\vdash \neg\varphi$ and thus $T \vdash \varphi$. So $T \equiv \mathrm{Th}(\mathfrak{A})$ holds.

c) ⇒ b): If $\mathfrak{B} \models T$, then $\mathfrak{B} \models \mathrm{Th}(\mathfrak{A})$ and therefore $\mathfrak{B} \equiv \mathfrak{A}$. Note that $\equiv$ is an equivalence relation.

b) ⇒ a): Let $\mathfrak{A}$ be a model of T. If φ holds in $\mathfrak{A}$, then φ holds in all models of T, i.e., $T \vdash \varphi$. Otherwise, $\neg\varphi$ holds in $\mathfrak{A}$ and we have $T \vdash \neg\varphi$. ⊣

From now on, when we fix a complete theory it is generally assumed to have an infinite model. In many cases, the results will still be true for the complete theory of a finite model, often for trivial reasons since in this case the model is unique up to isomorphism (see Exercise 1.3.3).

A class of L-structures forms an *elementary class* if it is the class of models of some L-theory T. By the previous examples, the class of all abelian groups (commutative rings, fields, respectively) is an elementary class as is the subclass of elementary abelian p-groups for some prime p. However, the class of all finite abelian p-groups does not form an elementary class since by Łos's Theorem elementary classes are closed under ultraproducts (see Exercises 1.3.4, 2.1.2 and 1.2.4).

EXERCISE 1.3.1. Write down the axioms for the theory DLO of dense linear orders without endpoint in the language L_{Order} of orders and the axioms for the theory ACF of algebraically closed fields in L_{Ring}. Is ACF complete?

EXERCISE 1.3.2.
1. For a prime number p, let $\mathbb{Z}_{p^\infty}$ denote the p-Prüfer group, i.e., the group of all p^k-th roots of unity for all $k \in \mathbb{N}$. Show that the groups $\mathbb{Z}_{p^\infty}^m$ and $\mathbb{Z}_{p^\infty}^n$ are not elementarily equivalent for $m \neq n$.
2. Show that $\mathbb{Z}^n \not\equiv \mathbb{Z}^m$ in the language of groups if $n \neq m$.

EXERCISE 1.3.3. Show that if $\mathfrak{A}$ is a finite L-structure and $\mathfrak{B}$ is elementarily equivalent to $\mathfrak{A}$, then they are isomorphic. (Show this first for finite L.)

EXERCISE 1.3.4. Use ultraproducts to show that the class of all finite groups (all torsion groups, all nilpotent groups, respectively) does not form an elementary class.

EXERCISE 1.3.5. Two structures $\mathfrak{A}$ and $\mathfrak{B}$ are *partially isomorphic* if there is a non-empty set $\mathcal{I}$ of isomorphisms between substructures of $\mathfrak{A}$ and $\mathfrak{B}$ with the *back-and-forth* property:
1. For every $f \in \mathcal{I}$ and $a \in A$ there is an extension of f in $\mathcal{I}$ with a in its domain.
2. For every $f \in \mathcal{I}$ and $b \in B$ there is an extension of f in $\mathcal{I}$ with b in its image.

Show that partially isomorphic structures are elementarily equivalent.

Chapter 2

ELEMENTARY EXTENSIONS AND COMPACTNESS

2.1. Elementary substructures

As in other fields of mathematics, we need to compare structures and consider maps from one structure to another. For this to make sense we consider a fixed language L. Maps and extensions are then required to respect this language.

Let $\mathfrak{A}$ and $\mathfrak{B}$ be two L-structures. A map $h\colon A \to B$ is called *elementary* if it preserves the validity of arbitrary formulas $\varphi(x_1, \dots, x_n)$[1]. More precisely, for all $a_1, \dots, a_n \in A$ we have:

$$\mathfrak{A} \models \varphi[a_1, \dots, a_n] \iff \mathfrak{B} \models \varphi[h(a_1), \dots, h(a_n)].$$

In particular, h preserves quantifier-free formulas and is therefore an embedding. Hence h is also called elementary embedding. We write

$$h : \mathfrak{A} \overset{\prec}{\longrightarrow} \mathfrak{B}.$$

The following lemma is clear.

LEMMA 2.1.1. *The models of* $\mathrm{Th}(\mathfrak{A}_A)$ *are exactly the structures of the form* $(\mathfrak{B}, h(a))_{a\in A}$ *for elementary embeddings* $h : \mathfrak{A} \overset{\prec}{\longrightarrow} \mathfrak{B}$. ⊣

We call $\mathrm{Th}(\mathfrak{A}_A)$ the *elementary diagram* of $\mathfrak{A}$.

A substructure $\mathfrak{A}$ of $\mathfrak{B}$ is called *elementary* if the inclusion map is elementary, i.e., if

$$\mathfrak{A} \models \varphi[a_1, \dots, a_n] \iff \mathfrak{B} \models \varphi[a_1, \dots, a_n]$$

for all $a_1, \dots, a_n \in A$. In this case we write

$$\mathfrak{A} \prec \mathfrak{B}$$

and $\mathfrak{B}$ is called an elementary extension of $\mathfrak{A}$.

---

[1]This only means that formulas which hold in $\mathfrak{A}$ also hold in $\mathfrak{B}$. But taking negations, the converse follows.

THEOREM 2.1.2 (Tarski's Test). *Let $\mathfrak{B}$ be an L-structure and A a subset of B. Then A is the universe of an elementary substructure if and only if every $L(A)$-formula $\varphi(x)$ which is satisfiable in $\mathfrak{B}$ can be satisfied by an element of A.*

PROOF. If $\mathfrak{A} \prec \mathfrak{B}$ and $\mathfrak{B} \models \exists x\varphi(x)$, we also have $\mathfrak{A} \models \exists x\varphi(x)$ and there exists $a \in A$ such that $\mathfrak{A} \models \varphi(a)$. Thus $\mathfrak{B} \models \varphi(a)$.

Conversely, suppose that the condition of Tarski's test is satisfied. First we show that A is the universe of a substructure $\mathfrak{A}$. The $L(A)$-formula $x \doteq x$ is satisfiable in $\mathfrak{B}$, so A is not empty. If $f \in L$ is an n-ary function symbol $(n \geq 0)$ and $a_1, \ldots, a_n$ is from A, we consider the formula

$$\varphi(x) = f(a_1, \ldots, a_n) \doteq x.$$

Since $\varphi(x)$ is always satisfied by an element of A, it follows that A is closed under $f^{\mathfrak{B}}$.

Now we show, by induction on ψ, that

$$\mathfrak{A} \models \psi \iff \mathfrak{B} \models \psi$$

for all $L(A)$-sentences ψ. This is clear for atomic sentences. The induction steps for $\psi = \neg\varphi$ and $\psi = (\varphi_1 \wedge \varphi_2)$ are trivial.

It remains to consider the case $\psi = \exists x\varphi(x)$. If ψ holds in $\mathfrak{A}$, there exists $a \in A$ such that $\mathfrak{A} \models \varphi(a)$. The induction hypothesis yields $\mathfrak{B} \models \varphi(a)$, thus $\mathfrak{B} \models \psi$. For the converse suppose ψ holds in $\mathfrak{B}$. Then $\varphi(x)$ is satisfiable in $\mathfrak{B}$ and by Tarski's test we find $a \in A$ such that $\mathfrak{B} \models \varphi(a)$. By induction $\mathfrak{A} \models \varphi(a)$ and $\mathfrak{A} \models \psi$ holds. ⊣

We use Tarski's Test to construct small elementary substructures.

COROLLARY 2.1.3. *Suppose S is a subset of the L-structure $\mathfrak{B}$. Then $\mathfrak{B}$ has an elementary substructure $\mathfrak{A}$ containing S and of cardinality at most*

$$\max(|S|, |L|, \aleph_0).$$

PROOF. We construct A as the union of an ascending sequence $S_0 \subseteq S_1 \subseteq \cdots$ of subsets of B. We start with $S_0 = S$. If S_i is already defined, we choose an element $a_\varphi \in B$ for every $L(S_i)$-formula $\varphi(x)$ which is satisfiable in $\mathfrak{B}$ and define S_{i+1} to be S_i together with these a_φ. It is clear that A is the universe of an elementary substructure. It remains to prove the bound on the cardinality of A.

An L-formula is a finite sequence of symbols from L, auxiliary symbols and logical symbols. These are $|L| + \aleph_0 = \max(|L|, \aleph_0)$ many symbols and therefore there are exactly $\max(|L|, \aleph_0)$ many L-formulas (see Corollary A.3.4).

Let $\kappa = \max(|S|, |L|, \aleph_0)$. There are κ many $L(S)$-formulas: therefore $|S_1| \leq \kappa$. Inductively it follows for every i that $|S_i| \leq \kappa$. Finally we have $|A| \leq \kappa \cdot \aleph_0 = \kappa$. ⊣

A directed family $(\mathfrak{A}_i)_{i \in I}$ of structures is *elementary* if $\mathfrak{A}_i \prec \mathfrak{A}_j$ for all $i \leq j$. The following lemma is mainly applied to elementary chains, hence its name.

THEOREM 2.1.4 (Tarski's Chain Lemma). *The union of an elementary directed family is an elementary extension of all its members.*

PROOF. Let $\mathfrak{A} = \bigcup_{i \in I} (\mathfrak{A}_i)_{i \in I}$. We prove by induction on $\varphi(\overline{x})$ that for all i and $\overline{a} \in \mathfrak{A}_i$

$$\mathfrak{A}_i \models \varphi(\overline{a}) \iff \mathfrak{A} \models \varphi(\overline{a}).$$

If φ is atomic, nothing is to be proved. If φ is a negation or a conjunction, the claim follows directly from the induction hypothesis.

If $\varphi(\overline{x}) = \exists y \psi(\overline{x}, y)$, then $\varphi(\overline{a})$ holds in $\mathfrak{A}$ exactly if there exists $b \in A$ with $\mathfrak{A} \models \psi(\overline{a}, b)$. As the family is directed, there always exists a $j \geq i$ with $b \in A_j$. By the induction hypothesis we have $\mathfrak{A} \models \psi(\overline{a}, b) \iff \mathfrak{A}_j \models \psi(\overline{a}, b)$. Thus $\varphi(\overline{a})$ holds in $\mathfrak{A}$ exactly if it holds in an $\mathfrak{A}_j$ $(j \geq i)$. Now the claim follows from $\mathfrak{A}_i \prec \mathfrak{A}_j$. ⊣

EXERCISE 2.1.1. Let $\mathfrak{A}$ be an L-structure and $(\mathfrak{A}_i)_{i \in I}$ a chain of elementary substructures of $\mathfrak{A}$. Show that $\bigcup_{i \in I} A_i$ is an elementary substructure of $\mathfrak{A}$.

EXERCISE 2.1.2. Consider a class $\mathcal{C}$ of L-structures. Prove:

1. Let $\text{Th}(\mathcal{C}) = \{\varphi \mid \mathfrak{A} \models \varphi \text{ for all } \mathfrak{A} \in \mathcal{C}\}$ be the *theory of* $\mathcal{C}$. Then $\mathfrak{M}$ is a model of $\text{Th}(\mathcal{C})$ if and only if $\mathfrak{M}$ is elementarily equivalent to an ultraproduct of elements of $\mathcal{C}$.
2. Show that $\mathcal{C}$ is an elementary class if and only if $\mathcal{C}$ is closed under ultraproducts and elementary equivalence.
3. Assume that $\mathcal{C}$ is a class of finite structures containing only finitely many structures of size n for each $n \in \omega$. Then the infinite models of $\text{Th}(\mathcal{C})$ are exactly the models of

$$\text{Th}_{\text{a}}(\mathcal{C}) = \{\varphi \mid \mathfrak{A} \models \varphi \text{ for all but finitely many } \mathfrak{A} \in \mathcal{C}\}.$$

2.2. The Compactness Theorem

In this section we prove the Compactness Theorem, one of the fundamental results in first-order logic. It states that a first-order theory has a model if every finite part of it does. Its name is motivated by the results in Section 4.2 which associate to each theory a certain compact topological space.

We call a theory T *finitely satisfiable* if every finite subset of T is consistent.

THEOREM 2.2.1 (Compactness Theorem). *Finitely satisfiable theories are consistent.*

Let L be a language and C a set of new constants. An $L(C)$-theory T' is called a *Henkin theory* if for every $L(C)$-formula $\varphi(x)$ there is a constant $c \in C$ such that

$$\exists x \varphi(x) \rightarrow \varphi(c) \in T'.$$

The elements of C are called *Henkin constants* of T'.

Let us call an L-theory T *finitely complete* if it is finitely satisfiable and if every L-sentence φ satisfies $\varphi \in T$ or $\neg\varphi \in T$. This terminology is only preliminary (and not standard): by the Compactness Theorem a theory is equivalent to a finitely complete one if and only if it is complete.

The Compactness Theorem follows from the following two lemmas.

LEMMA 2.2.2. *Every finitely satisfiable L-theory T can be extended to a finitely complete Henkin theory T^*.*

Note that conversely the lemma follows directly from the Compactness Theorem. Choose a model $\mathfrak{A}$ of T. Then $\mathrm{Th}(\mathfrak{A}_A)$ is a finitely complete Henkin theory with A as a set of Henkin constants.

PROOF. We define an increasing sequence $\emptyset = C_0 \subseteq C_1 \subseteq \cdots$ of new constants by assigning to every $L(C_i)$-formula $\varphi(x)$ a constant $c_{\varphi(x)}$ and

$$C_{i+1} = \{c_{\varphi(x)} \mid \varphi(x)\ L(C_i)\text{-formula}\}.$$

Let C be the union of the C_i and T^H the set of all Henkin axioms

$$\exists x\varphi(x) \rightarrow \varphi(c_{\varphi(x)})$$

for $L(C)$-formulas $\varphi(x)$. It is easy to see that one can expand every L-structure to a model of T^H. Hence $T \cup T^H$ is a finitely satisfiable Henkin theory. Using the fact that the union of a chain of finitely satisfiable theories is also finitely satisfiable, we can apply Zorn's Lemma and get a maximal finitely satisfiable $L(C)$-theory T^* which contains $T \cup T^H$. As in Lemma 1.3.7 we show that T^* is finitely complete: if neither φ nor $\neg\varphi$ belongs to T^*, neither $T^* \cup \{\varphi\}$ nor $T^* \cup \{\neg\varphi\}$ would be finitely satisfiable. Hence there would be a finite subset Δ of T^* which would be consistent neither with φ nor with $\neg\varphi$. Then Δ itself would be inconsistent and T^* would not be finitely satisfiable. This proves the lemma. ⊣

LEMMA 2.2.3. *Every finitely complete Henkin theory T^* has a model $\mathfrak{A}$ (unique up to isomorphism)* consisting of constants; *i.e.*,

$$(\mathfrak{A}, a_c)_{c\in C} \models T^*$$

with $A = \{a_c \mid c \in C\}$.

PROOF. Let us first note that since T^* is finitely complete, every sentence which follows from a finite subset of T^* belongs to T^*. Otherwise the negation of that sentence would belong to T^*, but would also be inconsistent together with a finite part of T^*.

We define for $c, d \in C$

$$c \simeq d \iff c \doteq d \in T^*.$$

As $c \doteq c$ is valid, and $d \doteq c$ follows from $c \doteq d$, and $c \doteq e$ follows from $c \doteq d$ and $d \doteq e$, we have that $\simeq$ is an equivalence relation. We denote the equivalence class of c by a_c and set

$$A = \{a_c \mid c \in C\}.$$

We expand A to an L-structure $\mathfrak{A}$ by defining

$$R^{\mathfrak{A}}(a_{c_1}, \dots, a_{c_n}) \iff R(c_1, \dots, c_n) \in T^* \tag{2.1}$$

$$f^{\mathfrak{A}}(a_{c_1}, \dots, a_{c_n}) = a_{c_0} \iff f(c_1, \dots, c_n) \doteq c_0 \in T^* \tag{2.2}$$

for relation symbols R and function symbols f ($n \geq 0$-ary) from L.

We have to show that this is well-defined. For (2.1) we have to show that

$$a_{c_1} = a_{d_1}, \dots, a_{c_n} = a_{d_n}, \ R(c_1, \dots, c_n) \in T^*$$

implies $R(d_1, \dots, d_n) \in T^*$. But clearly $R(d_1, \dots, d_n)$ holds in any structure satisfying

$$c_1 \doteq d_1, \dots, c_n \doteq d_n, \ R(c_1, \dots, c_n).$$

Similarly for (2.2) we first notice that $a_{c_0} = a_{d_0}$ follows from

$$a_{c_1} = a_{d_1}, \dots, a_{c_n} = a_{d_n}, \ f(c_1, \dots, c_n) \doteq c_0 \in T^*, \ f(d_1, \dots, d_n) \doteq d_0 \in T^*.$$

For (2.2) we also have to show that for all $c_1, \dots, c_n$ there exists c_0 with $f(c_1, \dots, c_n) \doteq c_0 \in T^*$. As T^* is a Henkin theory, there exists c_0 with

$$\exists x f(c_1, \dots, c_n) \doteq x \to f(c_1, \dots, c_n) \doteq c_0 \in T^*.$$

Now the valid sentence $\exists x f(c_1, \dots, c_n) \doteq x$ belongs to T^*, so $f(c_1, \dots, c_n) \doteq c_0$ belongs to T^*. This shows that everything is well defined.

Let $\mathfrak{A}^*$ be the $L(C)$-structure $(\mathfrak{A}, a_c)_{c \in C}$. We show by induction on the complexity of φ that for every $L(C)$-sentence φ

$$\mathfrak{A}^* \models \varphi \iff \varphi \in T^*.$$

There are four cases:

a) φ is atomic. If φ has the form $c \doteq d$ or $R(c_1, \dots, c_n)$, the statement follows from the construction of $\mathfrak{A}^*$. Other atomic sentences contain function symbols f or constants (which we consider in this proof as 0-ary function symbols) from L. We inductively reduce the number of such occurrences and apply the previous case. Suppose φ contains a function symbol from L. Then φ can be written as

$$\varphi = \psi(f(c_1, \dots, c_n))$$

for a function symbol $f \in L$ and an $L(C)$-formula $\psi(x)$. Choose c_0 satisfying $f(c_1, \dots, c_n) \doteq c_0 \in T^*$. By construction $f(c_1, \dots, c_n) \doteq c_0$ holds in $\mathfrak{A}^*$. Thus $\mathfrak{A}^* \models \varphi \iff \mathfrak{A}^* \models \psi(c_0)$ and $\varphi \in T^* \iff \psi(c_0) \in T^*$. From the induction hypothesis on the number of occurrences we have $\mathfrak{A}^* \models \psi(c_0) \iff \psi(c_0) \in T^*$. This suffices.

b) $\varphi = \neg\psi$. As T^* is finitely complete, $\varphi \in T^* \iff \psi \notin T^*$ holds and by the induction hypothesis we have

$$\mathfrak{A}^* \models \varphi \iff \mathfrak{A}^* \not\models \psi \iff \psi \notin T^* \iff \varphi \in T^*.$$

c) $\varphi = (\psi_1 \wedge \psi_2)$. As T^* contains all sentences which follow from a finite subset of T^*, φ belongs to T^* if and only if ψ_1 and ψ_2 belong to T^*. Thus

$$\mathfrak{A}^* \models \varphi \iff \mathfrak{A}^* \models \psi_i \; (i = 1, 2) \iff \psi_i \in T^* \; (i = 1, 2) \iff \varphi \in T^*.$$

d) $\varphi = \exists x \psi(x)$. We have

$$\mathfrak{A}^* \models \varphi \Leftrightarrow \mathfrak{A}^* \models \psi(c) \text{ for some } c \in C \Leftrightarrow \psi(c) \in T^* \text{ for some } c \in C \Leftrightarrow \varphi \in T^*.$$

The second equivalence is the induction hypothesis and for the third we argue as follows: if $\varphi \in T^*$, we choose c satisfying $\varphi \to \psi(c) \in T^*$. As $\varphi \in T^*$ we also have $\psi(c) \in T^*$. ⊣

COROLLARY 2.2.4. *We have $T \vdash \varphi$ if and only if $\Delta \vdash \varphi$ for a finite subset Δ of T.*

PROOF. The formula φ follows from T if and only if $T \cup \{\neg\varphi\}$ is inconsistent. ⊣

COROLLARY 2.2.5. *A set of formulas $\Sigma(x_1, \ldots, x_n)$ is consistent with T if and only if every finite subset of Σ is consistent with T.*

PROOF. Introduce new constants $c_1, \ldots, c_n$. Then Σ is consistent with T if and only if $T \cup \Sigma(c_1, \ldots, c_n)$ is consistent. Now apply the Compactness Theorem. ⊣

DEFINITION 2.2.6. Let $\mathfrak{A}$ be an L-structure and $B \subseteq A$. Then $a \in A$ *realises* a set of $L(B)$-formulas $\Sigma(x)$ (containing at most the free variable x), if a satisfies all formulas from $\Sigma(x)$. We write

$$\mathfrak{A} \models \Sigma(a).$$

We call $\Sigma(x)$ *finitely satisfiable* in $\mathfrak{A}$ if every finite subset of Σ is realised in $\mathfrak{A}$.

LEMMA 2.2.7. *The set $\Sigma(x)$ is finitely satisfiable in $\mathfrak{A}$ if and only if there is an elementary extension of $\mathfrak{A}$ in which $\Sigma(x)$ is realised.*

PROOF. By Lemma 2.1.1, Σ is realised in an elementary extension of $\mathfrak{A}$ if and only if Σ is consistent with $\mathrm{Th}(\mathfrak{A}_A)$. So the lemma follows from the easy observation that a finite set of $L(A)$-formulas is consistent with $\mathrm{Th}(\mathfrak{A}_A)$ if and only if it is realised in $\mathfrak{A}$. ⊣

DEFINITION 2.2.8. Let $\mathfrak{A}$ be an L-structure and B a subset of A. A set $p(x)$ of $L(B)$-formulas is a *type* over B if $p(x)$ is maximal finitely satisfiable in $\mathfrak{A}$. We call B the *domain* of p. Let

$$\mathrm{S}(B) = \mathrm{S}^{\mathfrak{A}}(B)$$

denote the set of types over B.

Every element a of $\mathfrak{A}$ determines a type

$$\mathrm{tp}(a/B) = \mathrm{tp}^{\mathfrak{A}}(a/B) = \{\varphi(x) \mid \mathfrak{A} \models \varphi(a),\ \varphi \text{ an } L(B)\text{-formula}\}.$$

So an element a realises the type $p \in \mathrm{S}(B)$ exactly if $p = \mathrm{tp}(a/B)$. Note that if $\mathfrak{A}'$ is an elementary extension of $\mathfrak{A}$, then

$$\mathrm{S}^{\mathfrak{A}}(B) = \mathrm{S}^{\mathfrak{A}'}(B) \quad \text{and} \quad \mathrm{tp}^{\mathfrak{A}'}(a/B) = \mathrm{tp}^{\mathfrak{A}}(a/B).$$

We will use the notation $\mathrm{tp}(a)$ for $\mathrm{tp}(a/\emptyset)$.

Similarly, maximal finitely satisfiable sets of formulas in $x_1, \ldots, x_n$ are called *n-types* and

$$\mathrm{S}_n(B) = \mathrm{S}_n^{\mathfrak{A}}(B)$$

denotes the set of n-types over B. For an n-tuple $\overline{a}$ from $\mathfrak{A}$, there is an obvious definition of $\mathrm{tp}^{\mathfrak{A}}(\overline{a}/B) \in \mathrm{S}_n^{\mathfrak{A}}(B)$. Very much in the same way, we can define the type $\mathrm{tp}(C/B)$ of an arbitrary set C over B. This will be convenient in later chapters. In order to do this properly we allow free variables x_c indexed by $c \in C$ and define

$$\mathrm{tp}(C/B) = \{\varphi(x_{c_1}, \ldots, x_{c_n}) \mid \mathfrak{A} \models \varphi(c_1, \ldots, c_n),\ \varphi \text{ an } L(B)\text{-formula}\}.$$

Many theorems which we will formulate for 1-types will hold, with the same proofs, for n-types and often for types with infinitely many variables.

COROLLARY 2.2.9. *Every structure $\mathfrak{A}$ has an elementary extension $\mathfrak{B}$ in which all types over A are realised.*

PROOF. We choose for every $p \in \mathrm{S}(A)$ a new constant c_p. We have to find a model of

$$\mathrm{Th}(\mathfrak{A}_A) \cup \bigcup_{p \in \mathrm{S}(A)} p(c_p).$$

It is easy to see that this theory is finitely satisfiable using that every p is finitely satisfiable in $\mathfrak{A}$. The Compactness Theorem now shows that the model exists.

We give a second proof which only uses Lemma 2.2.7. Let $(p_\alpha)_{\alpha<\lambda}$ be an enumeration of $\mathrm{S}(A)$, where λ is an ordinal number (see Section A.2). Using Theorem A.2.2, we construct an elementary chain

$$\mathfrak{A} = \mathfrak{A}_0 \prec \mathfrak{A}_1 \prec \cdots \prec \mathfrak{A}_\beta \prec \cdots \quad (\beta \leq \lambda)$$

such that each p_α is realised in $\mathfrak{A}_{\alpha+1}$.

Let us suppose that the elementary chain $(\mathfrak{A}_{\alpha'})_{\alpha'<\beta}$ is already constructed. If β is a limit ordinal, we let $\mathfrak{A}_\beta = \bigcup_{\alpha<\beta} \mathfrak{A}_\alpha$.[2] The longer chain $(\mathfrak{A}_{\alpha'})_{\alpha'\leq\beta}$ is elementary because of Lemma 2.1.4. If $\beta = \alpha + 1$ we first note that p_α is also finitely satisfiable in $\mathfrak{A}_\alpha$. Therefore we can realise p_α in a suitable elementary extension $\mathfrak{A}_\beta \succ \mathfrak{A}_\alpha$. Then $\mathfrak{B} = \mathfrak{A}_\lambda$ is the model we were looking for. ⊣

[2] We call a chain $(\mathfrak{A}_\alpha)$ indexed by ordinal numbers *continuous* if $\mathfrak{A}_\beta = \bigcup_{\alpha<\beta} \mathfrak{A}_\alpha$ for all limit ordinals β.

EXERCISE 2.2.1. Prove the Compactness Theorem using ultraproducts (see Exercise 1.2.4).

EXERCISE 2.2.2. A class C of L-structure is *finitely axiomatisable* if it is the class of models of a finite theory. Show that C is finitely axiomatisable if and only if both C and its complement form an elementary class.

EXERCISE 2.2.3. Show that the class of connected graphs is not an elementary class. A graph (V, R) is a set V with a symmetric, irreflexive binary relation. It is *connected* if for any $x, y \in V$ there is a sequence of elements $x_0 = x, \ldots, x_k = y$ such that $(x_{i-1}, x_i) \in R$ for $i = 1, \ldots, n$.

EXERCISE 2.2.4. Let $\mathfrak{A} = (\mathbb{R}, 0, <, f^{\mathfrak{A}})$, where f is a unary function symbol. Call an element $x \in \mathfrak{A}^* \succ \mathfrak{A}$ *infinitesimal* if $-\frac{1}{n} < x < \frac{1}{n}$ for all positive natural numbers n. Show that if $f^{\mathfrak{A}}(0) = 0$, then $f^{\mathfrak{A}}$ is continuous in 0 if and only if for any elementary extension $\mathfrak{A}^*$ of $\mathfrak{A}$ the map $f^{\mathfrak{A}^*}$ takes infinitesimal elements to infinitesimal elements.

EXERCISE 2.2.5. Let T be an L_{Ring}-theory containing Field. Show that:

1. If T has models of arbitrary large characteristic, then it has a model of characteristic 0.
2. The theory of fields of characteristic 0 is not finitely axiomatisable.

2.3. The Löwenheim–Skolem Theorem

One of the consequences of the Compactness Theorem is the fact that a first-order theory cannot pin down the size of an infinite structure. This is the content of the following theorem.

THEOREM 2.3.1 (Löwenheim–Skolem). *Let $\mathfrak{B}$ be an L-structure, S a subset of B and κ an infinite cardinal.*

1. *If*

$$\max(|S|, |L|) \leq \kappa \leq |\mathfrak{B}|,$$

then $\mathfrak{B}$ has an elementary substructure of cardinality κ containing S.

2. *If $\mathfrak{B}$ is infinite and*

$$\max(|\mathfrak{B}|, |L|) \leq \kappa,$$

then $\mathfrak{B}$ has an elementary extension of cardinality κ.

PROOF. 1: Choose a set $S \subseteq S' \subseteq B$ of cardinality κ and apply Corollary 2.1.3.

2: We first construct an elementary extension $\mathfrak{B}'$ of cardinality at least κ. Choose a set C of new constants of cardinality κ. As $\mathfrak{B}$ is infinite, the theory

$$\mathrm{Th}(\mathfrak{B}_B) \cup \{\neg c \doteq d \mid c, d \in C,\ c \neq d\}$$

is finitely satisfiable (even in $\mathfrak{B}$: just interpret the finitely many new constants in a finite subset by elements of B). By Lemma 2.1.1, any model $(\mathfrak{B}'_B, b_c)_{c \in C}$ is an elementary extension of $\mathfrak{B}$ with κ many different elements (b_c).

Finally we apply the first part of the theorem to $\mathfrak{B}'$ and $S = B$. ⊣

Note that in Theorem 2.3.1(1) the assumption $\kappa \geq \max(|S|, |L|)$ is certainly necessary in general.

COROLLARY 2.3.2. *A theory which has an infinite model has a model in every cardinality* $\kappa \geq \max(|L|, \aleph_0)$. ⊣

Thus, no theory with an infinite model can describe this model up to isomorphism. So the best we can hope for is a unique model for a given cardinality.

DEFINITION 2.3.3 (preliminary, see 2.3.5). Let κ be an infinite cardinal. A theory T is called *κ-categorical* if all models of T of cardinality κ are isomorphic.

THEOREM 2.3.4 (Vaught's Test). *A κ-categorical theory T is complete if the following conditions are satisfied*:

a) *T is consistent,*
b) *T has no finite model,*
c) $|L| \leq \kappa$.

PROOF. We have to show that all models $\mathfrak{A}$ and $\mathfrak{B}$ of T are elementarily equivalent. As $\mathfrak{A}$ and $\mathfrak{B}$ are infinite, $\mathrm{Th}(\mathfrak{A})$ and $\mathrm{Th}(\mathfrak{B})$ have models $\mathfrak{A}'$ and $\mathfrak{B}'$ of cardinality κ. By assumption $\mathfrak{A}'$ and $\mathfrak{B}'$ are isomorphic, and it follows that

$$\mathfrak{A} \equiv \mathfrak{A}' \equiv \mathfrak{B}' \equiv \mathfrak{B}.$$ ⊣

EXAMPLES.

1. (Theorem of Cantor, see Exercise 1.3.1) The theory DLO of dense linear orders without endpoints is $\aleph_0$-categorical and by Vaught's test complete. To see this let $\mathfrak{A}$ and $\mathfrak{B}$ be countable dense linear orders without endpoints, and let $A = \{a_i \mid i \in \omega\}$, $B = \{b_i \mid i \in \omega\}$. We inductively define sequences $(c_i)_{i<\omega}$, $(d_i)_{i<\omega}$ exhausting A and B, respectively, and such that the assignment $c_i \mapsto d_i$ is the required isomorphism. Assume that $(c_i)_{i<m}$, $(d_i)_{i<m}$ have been defined so that $c_i \mapsto d_i, i < m$ is an order isomorphism. If $m = 2k$, let $c_m = a_j$ where a_j is the element with minimal index in $\{a_i \mid i \in \omega\}$ not occurring in $(c_i)_{i<m}$. Since $\mathfrak{B}$ is a dense linear order without endpoints, there is some element $d_m \in \{b_i \mid i \in \omega\}$ such that $(c_i)_{i \leq m}$ and $(d_i)_{i \leq m}$ are order isomorphic. If $m = 2k + 1$ we interchange the roles of $\mathfrak{A}$ and $\mathfrak{B}$.
2. For any prime p or $p = 0$, the theory ACF_p of algebraically closed fields of characteristic p is κ-categorical for any $\kappa > \aleph_0$. Any two algebraically closed fields of the same characteristic and of cardinality $\kappa > \aleph_0$ have transcendence bases (over the algebraic closure of the prime

field) of cardinality κ, see Corollary A.3.4 and Section C.1. Any bijection between these transcendence bases induces an isomorphism of the fields. It follows that ACF_p is complete.

Considering Theorem 2.3.4 we strengthen our definition.

DEFINITION 2.3.5. Let κ be an infinite cardinal. A theory T is called κ-*categorical* if it is complete, $|T| \leq \kappa$ and, up to isomorphism, has exactly one model of cardinality κ.

EXERCISE 2.3.1.

1. Two functions $f, g : \mathbb{N} \to \mathbb{N}$ are *almost disjoint* if $f(n) \neq g(n)$ for almost all n. Show that there are $2^{\aleph_0}$-many almost disjoint functions from $\mathbb{N}$ to $\mathbb{N}$. (Hint: For every real r choose a sequence of rational numbers which converges to r.)
2. Let $\mathcal{F}$ be the set of all functions $\mathbb{N} \to \mathbb{N}$. Show that $(\mathbb{N}, <, f)_{f \in \mathcal{F}}$ has no countable proper elementary extension.
3. Let $\mathbb{Q}$ be the ordered field of rational numbers. For every real r introduce two predicates P_r, R_r for $\{q \in \mathbb{Q} \mid q < r\}$ and $\{q \in \mathbb{Q} \mid r \leq q\}$. Show that $(\mathbb{Q}, P_r, Q_r)_{r \in \mathbb{R}_\varphi}$ has no countable proper elementary extension.

EXERCISE 2.3.2.

1. The theory of K-vector spaces $\mathsf{Mod}(K)$ (see p. 38) is κ-categorical for $\kappa > |K|$.
2. Is ACF_p $\aleph_0$-categorical?

EXERCISE 2.3.3. Show that an $\forall\exists$-sentence, which holds in all finite fields, is true in all algebraically closed fields.

Chapter 3

QUANTIFIER ELIMINATION

3.1. Preservation theorems

In general, it can be difficult to tell which sentences belong to a given theory or which extensions are consistent. Therefore it is helpful to know that in certain theories one can restrict attention to sentences of a certain class, e.g., quantifier-free or, say, existential formulas. We consider a fixed language L and first prove some separation results.

LEMMA 3.1.1 (Separation Lemma). *Let T_1 and T_2 be two theories. Assume $\mathcal{H}$ is a set of sentences which is closed under $\wedge$ and $\vee$ and contains $\top$ and $\bot$ (true and false). Then the following are equivalent:*

a) *There is a sentence $\varphi \in \mathcal{H}$ which* separates *T_1 from T_2. This means*

$$T_1 \vdash \varphi \text{ and } T_2 \vdash \neg\varphi.$$

b) *All models $\mathfrak{A}_1$ of T_1 can be separated from all models $\mathfrak{A}_2$ of T_2 by a sentence $\varphi \in \mathcal{H}$. This means*

$$\mathfrak{A}_1 \models \varphi \text{ and } \mathfrak{A}_2 \models \neg\varphi.$$

PROOF. a) $\Rightarrow$ b): If φ separates T_1 from T_2, it separates all models of T_1 from all models of T_2.

b) $\Rightarrow$ a): For any model $\mathfrak{A}_1$ of T_1 let $\mathcal{H}_{\mathfrak{A}_1}$ be the set of all sentences from $\mathcal{H}$ which are true in $\mathfrak{A}_1$. b) implies that $\mathcal{H}_{\mathfrak{A}_1}$ and T_2 cannot have a common model. By the Compactness Theorem there is a finite conjunction $\varphi_{\mathfrak{A}_1}$ of sentences from $\mathcal{H}_{\mathfrak{A}_1}$ inconsistent with T_2. Clearly,

$$T_1 \cup \{\neg\varphi_{\mathfrak{A}_1} \mid \mathfrak{A}_1 \models T_1\}$$

is inconsistent since any model $\mathfrak{A}_1$ of T_1 satisfies $\varphi_{\mathfrak{A}_1}$. Again by compactness T_1 implies a disjunction φ of finitely many of the $\varphi_{\mathfrak{A}_1}$. This formula φ is in $\mathcal{H}$ and separates T_1 from T_2. $\dashv$

For structures $\mathfrak{A}, \mathfrak{B}$ and a map $f : A \to B$ preserving all formulas from a set of formulas Δ, we use the notation

$$f : \mathfrak{A} \to_\Delta \mathfrak{B}.$$

We also write

$$\mathfrak{A} \Rrightarrow_\Delta \mathfrak{B}$$

to express that all sentences from Δ true in $\mathfrak{A}$ are also true in $\mathfrak{B}$.

LEMMA 3.1.2. *Let T be a theory, $\mathfrak{A}$ a structure and Δ a set of formulas, closed under existential quantification, conjunction and substitution of variables. Then the following are equivalent:*

a) *All sentences $\varphi \in \Delta$ which are true in $\mathfrak{A}$ are consistent with T.*
b) *There is a model $\mathfrak{B} \models T$ and a map $f : \mathfrak{A} \to_\Delta \mathfrak{B}$.*

PROOF. b) $\Rightarrow$ a): Assume $f : \mathfrak{A} \to_\Delta \mathfrak{B} \models T$. If $\varphi \in \Delta$ is true in $\mathfrak{A}$ it is also true in $\mathfrak{B}$ and therefore is consistent with T.

a) $\Rightarrow$ b): Consider $\mathrm{Th}_\Delta(\mathfrak{A}_A)$, the set of all sentences $\delta(\overline{a})$, $(\delta(\overline{x}) \in \Delta)$, which are true in $\mathfrak{A}_A$. The models $(\mathfrak{B}, f(a)_{a\in A})$ of this theory correspond to maps $f : \mathfrak{A} \to_\Delta \mathfrak{B}$. This means that we have to find a model of $T \cup \mathrm{Th}_\Delta(\mathfrak{A}_A)$. To show finite satisfiability it is enough to show that $T \cup D$ is consistent for every finite subset D of $\mathrm{Th}_\Delta(\mathfrak{A}_A)$. Let $\delta(\overline{a})$ be the conjunction of the elements of D. Then $\mathfrak{A}$ is a model of $\varphi = \exists\overline{x}\ \delta(\overline{x})$, so by assumption T has a model $\mathfrak{B}$ which is also a model of φ. This means that there is a tuple $\overline{b}$ such that $(\mathfrak{B}, \overline{b}) \models \delta(\overline{a})$. ⊣

Note that Lemma 3.1.2 applied to $T = \mathrm{Th}(\mathfrak{B})$ shows that $\mathfrak{A} \Rrightarrow_\Delta \mathfrak{B}$ if and only if there exists a map f and a structure $\mathfrak{B}' \equiv \mathfrak{B}$ such that $f : \mathfrak{A} \to_\Delta \mathfrak{B}'$.

THEOREM 3.1.3. *Let T_1 and T_2 be two theories. Then the following are equivalent:*

a) *There is a universal sentence which separates T_1 from T_2.*
b) *No model of T_2 is a substructure of a model of T_1.*

PROOF. a) $\Rightarrow$ b): Let φ be a universal sentence which separates T_1 from T_2. Let $\mathfrak{A}_1$ be a model of T_1 and $\mathfrak{A}_2$ a substructure of $\mathfrak{A}_1$. Since $\mathfrak{A}_1$ is a model of φ, then by Lemma 1.2.16 $\mathfrak{A}_2$ is also model of φ. Therefore $\mathfrak{A}_2$ cannot be a model of T_2.

b) $\Rightarrow$ a): If T_1 and T_2 cannot be separated by a universal sentence, then they have models A_1 and A_2 which cannot be separated by a universal sentence. This can be denoted by

$$\mathfrak{A}_2 \Rrightarrow_\exists \mathfrak{A}_1.$$

Now Lemma 3.1.2 implies that $\mathfrak{A}_2$ has an extension $\mathfrak{A}_1' \equiv \mathfrak{A}_1$. Then $\mathfrak{A}_1'$ is again a model of T_1 contradicting b). ⊣

DEFINITION 3.1.4. For any L-theory T, the formulas $\varphi(\overline{x})$, $\psi(\overline{x})$ are said to be *equivalent* modulo T (or relative to T) if $T \vdash \forall\overline{x}(\varphi(\overline{x}) \leftrightarrow \psi(\bar{x}))$.

COROLLARY 3.1.5. *Let T be a theory.*

1. *Consider a formula $\varphi(x_1, \ldots, x_n)$. The following are equivalent:*
   a) *$\varphi(x_1, \ldots, x_n)$ is, modulo T, equivalent to a universal formula.*

b) *If $\mathfrak{A} \subseteq \mathfrak{B}$ are models of T and $a_1, \ldots, a_n \in A$, then $\mathfrak{B} \models \varphi(a_1, \ldots, a_n)$ implies $\mathfrak{A} \models \varphi(a_1, \ldots, a_n)$.*

2. *We say that a theory which consists of universal sentences is* universal. *Then T is equivalent to a universal theory if and only if all substructures of models of T are again models of T.*

PROOF. 1): Assume b). We extend L by an n-tuple $\overline{c}$ of new constants $c_1, \ldots, c_n$ and consider the theory

$$T_1 = T \cup \{\varphi(\overline{c})\} \quad \text{and} \quad T_2 = T \cup \{\neg\varphi(\overline{c})\}.$$

Then b) says that substructures of models of T_1 cannot be models of T_2. By Theorem 3.1.3, T_1 and T_2 can be separated by a universal $L(\overline{c})$-sentence $\psi(\overline{c})$. By Lemma 1.3.4(2), $T_1 \vdash \psi(\overline{c})$ implies

$$T \vdash \forall \overline{x}\, (\varphi(\overline{x}) \to \psi(\overline{x}))$$

and from $T_2 \vdash \neg\psi(\overline{c})$ we see

$$T \vdash \forall \overline{x}\, (\neg\varphi(\overline{x}) \to \neg\psi(\overline{x})).$$

2): It is clear that substructures of models of a universal theory are models again. Now suppose that a theory T has this property. Let φ be an axiom of T. If $\mathfrak{A}$ is a substructure of $\mathfrak{B}$, it is not possible for $\mathfrak{B}$ to be a model of T and for $\mathfrak{A}$ to be a model of $\neg\varphi$ at the same time. By 3.1.3 there is a universal sentence ψ with $T \vdash \psi$ and $\neg\varphi \vdash \neg\psi$. Hence all axioms of T follow from

$$T_\forall = \{\psi \mid T \vdash \psi,\ \psi \text{ universal}\}. \qquad \dashv$$

An $\forall\exists$-formula is of the form

$$\forall x_1 \ldots x_n \psi$$

where ψ is existential (see p. 11). The following is clear.

LEMMA 3.1.6. *Suppose φ is an $\forall\exists$-sentence, $(\mathfrak{A}_i)_{i \in I}$ is a directed family of models of φ and $\mathfrak{B}$ the union of the $\mathfrak{A}_i$. Then $\mathfrak{B}$ is also a model of φ.*

PROOF. Write

$$\varphi = \forall \overline{x}\, \psi(\overline{x}),$$

where ψ is existential. For any $\overline{a} \in B$ there is an A_i containing $\overline{a}$. Since $\mathfrak{A}_i \models \varphi$, clearly $\psi(\overline{a})$ holds in $\mathfrak{A}_i$. As $\psi(\overline{a})$ is existential it must also hold in $\mathfrak{B}$. $\dashv$

DEFINITION 3.1.7. We call a theory T *inductive* if the union of any directed family of models of T is again a model.

THEOREM 3.1.8. *Let T_1 and T_2 be two theories. Then the following are equivalent:*

a) *There is an $\forall\exists$-sentence which separates T_1 from T_2.*

b) *No model of T_2 is the union of a chain (or of a directed family) of models of T_1.*

PROOF. a) $\Rightarrow$ b): Assume φ is a $\forall\exists$-sentence which separates T_1 from T_2, $(\mathfrak{A}_i)_{i\in I}$ is a directed family of models of T_1 and $\mathfrak{B}$ the union of the $\mathfrak{A}_i$. Since the $\mathfrak{A}_i$ are models of φ, by Lemma 3.1.6 $\mathfrak{B}$ is also a model of φ. Since $\mathfrak{B}\models\varphi$, $\mathfrak{B}$ cannot be a model of T_2.

b) $\Rightarrow$ a): If a) is not true, T_1 and T_2 have models which cannot be separated by an $\forall\exists$-sentence. Since $\exists\forall$-formulas are equivalent to negated $\forall\exists$-formulas, we have

$$\mathfrak{B}^0 \Rightarrow_{\exists\forall} \mathfrak{A}.$$

By Lemma 3.1.2 there is a map

$$f:\mathfrak{B}^0 \to_{\forall} \mathfrak{A}^0$$

with $\mathfrak{A}^0\equiv\mathfrak{A}$. We can assume that $\mathfrak{B}^0\subseteq\mathfrak{A}^0$ and f is the inclusion map. Then

$$\mathfrak{A}^0_B \Rightarrow_{\exists} \mathfrak{B}^0_B.$$

Applying Lemma 3.1.2 again, we obtain an extension $\mathfrak{B}^1_B$ of $\mathfrak{A}^0_B$ with $\mathfrak{B}^1_B\equiv\mathfrak{B}^0_B$, i.e., $\mathfrak{B}^0\prec\mathfrak{B}^1$.

$$\underbrace{\mathfrak{B}^0 \subseteq \mathfrak{A}^0 \subseteq \mathfrak{B}^1}_{\prec}$$

The same procedure applied to $\mathfrak{A}$ and $\mathfrak{B}^1$ gives us two extensions $\mathfrak{A}^1\subseteq\mathfrak{B}^2$ with $\mathfrak{A}^1\equiv\mathfrak{A}$ and $\mathfrak{B}^1\prec\mathfrak{B}^2$. This results in an infinite chain

$$\underbrace{\mathfrak{B}^0 \subseteq \mathfrak{A}^0 \subseteq \mathfrak{B}^1}_{\prec}\underbrace{\subseteq \mathfrak{A}^1 \subseteq \mathfrak{B}^2}_{\prec}\underbrace{\subseteq \cdots}_{\prec}$$

with $\mathfrak{A}^i\equiv\mathfrak{A}$ and $\mathfrak{B}^i\prec\mathfrak{B}^{i+1}$. Let $\mathfrak{B}$ be the union of the $\mathfrak{A}^i$. Since $\mathfrak{B}$ is also the union of the elementary chain of the $\mathfrak{B}^i$, it is an elementary extension of $\mathfrak{B}^0$ and hence a model of T_2. But the $\mathfrak{A}^i$ are models of T_1, so b) does not hold. $\dashv$

COROLLARY 3.1.9. *Let T be a theory.*

1. *For each sentence φ the following are equivalent:*
   a) *φ is, modulo T, equivalent to an $\forall\exists$-sentence.*
   b) *If*

$$\mathfrak{A}^0\subseteq\mathfrak{A}^1\subseteq\cdots$$

and their union $\mathfrak{B}$ are models of T, then φ holds in $\mathfrak{B}$ if it is true in all the $\mathfrak{A}^i$.

2. *T is inductive if and only if it can be axiomatised by $\forall\exists$-sentences.*

PROOF. 1): Theorem 3.1.8 shows that $\forall\exists$-formulas are preserved by unions of chains. Hence a) $\Rightarrow$ b). For the converse consider the theories

$$T_1 = T \cup \{\varphi\} \quad \text{and} \quad T_2 = T \cup \{\neg\varphi\}.$$

Part b) says that the union of a chain of models of T_1 cannot be a model of T_2. By Theorem 3.1.8 we can separate T_1 and T_2 by an $\forall\exists$-sentence ψ. Now $T_1 \vdash \psi$ implies $T \vdash \varphi \to \psi$ and $T_2 \vdash \neg\psi$ implies $T \vdash \neg\varphi \to \neg\psi$.

2): Clearly $\forall\exists$-axiomatised theories are inductive. For the converse assume that T is inductive and φ an axiom of T. If $\mathfrak{B}$ is a union of models of T, it cannot be a model of $\neg\varphi$. By Theorem 3.1.8 there is an $\forall\exists$-sentence ψ with $T \vdash \psi$ and $\neg\varphi \vdash \neg\psi$. Hence all axioms of T follow from

$$T_{\forall\exists} = \{\psi \mid T \vdash \psi,\ \psi\ \forall\exists\text{-formula}\}. \qquad \dashv$$

EXERCISE 3.1.1. Let X be a topological space, Y_1 and Y_2 quasi-compact[1] subsets, and $\mathcal{H}$ a set of clopen subsets. Then the following are equivalent:

a) There is a positive Boolean combination B of elements from $\mathcal{H}$ such that $Y_1 \subseteq B$ and $Y_2 \cap B = \emptyset$.
b) For all $y_1 \in Y_1$ and $y_2 \in Y_2$ there is an $H \in \mathcal{H}$ such that $y_1 \in H$ and $y_2 \notin H$.

(This, in fact, is a generalisation of the Separation Lemma 3.1.1.)

3.2. Quantifier elimination

Having quantifier elimination in a reasonable language is a property which makes a theory 'tame'. In this section we will collect some criteria and extensions of this concept. They will be applied in Section 3.3 to show that a number of interesting theories have quantifier elimination in the appropriate language.

DEFINITION 3.2.1. A theory T has *quantifier elimination* if every L-formula $\varphi(x_1, \ldots, x_n)$ in the theory is equivalent modulo T to some quantifier-free formula $\rho(x_1, \ldots, x_n)$.

For $n = 0$ this means that modulo T every sentence is equivalent to a quantifier-free sentence. If L has no constants, $\top$ and $\bot$ are the only quantifier-free sentences. Then T is either inconsistent or complete.

Note that it is easy to transform any theory T into a theory with quantifier elimination if one is willing to expand the language: just enlarge L by adding an n-place relation symbol R_φ for every L-formula $\varphi(x_1, \ldots, x_n)$ and T by adding all axioms

$$\forall x_1, \ldots, x_n (R_\varphi(x_1, \ldots, x_n) \leftrightarrow \varphi(x_1, \ldots, x_n)).$$

---

[1] That is, compact but not necessarily Hausdorff.

The resulting theory, the *Morleyisation* T^{m} of T, has quantifier elimination. Many other properties of a theory are not affected by Morleyisation. So T is complete if and only if T^{m} is; similarly for κ-categoricity and other properties we will define in later chapters.

A *prime structure* of T is a structure which embeds into all models of T. The following is clear.

LEMMA 3.2.2. *A consistent theory T with quantifier elimination which possesses a prime structure is complete.* ⊣

DEFINITION 3.2.3. A *simple existential formula* has the form

$$\varphi = \exists y\ \rho$$

for a quantifier-free formula ρ. If ρ is a conjunction of basic formulas, φ is called *primitive existential*.

LEMMA 3.2.4. *The theory T has quantifier elimination if and only if every primitive existential formula is, modulo T, equivalent to a quantifier-free formula.*

PROOF. We can write every simple existential formula in the form $\exists y \bigvee_{i<n} \rho_i$ for ρ_i which are conjunctions of basic formulas. This shows that every simple existential formula is equivalent to a disjunction of primitive existential formulas, namely to $\bigvee_{i<n}(\exists y\ \rho_i)$. We can therefore assume that every simple existential formula is, modulo T, equivalent to a quantifier-free formula.

We are now able to eliminate the quantifiers in arbitrary formulas in prenex normal form (see Exercise 1.2.3)

$$Q_1 x_1 \ldots Q_n x_n \rho.$$

If $Q_n = \exists$, we choose a quantifier-free formula ρ_0 which, modulo T, is equivalent to $\exists x_n \rho$. Then we proceed with the formula $Q_1 x_1 \ldots Q_{n-1} x_{n-1} \rho_0$. If $Q_n = \forall$, we find a quantifier-free ρ_1 which is, modulo T, equivalent to $\exists x_n \neg \rho$ and proceed with $Q_1 x_1 \ldots Q_{n-1} x_{n-1} \neg \rho_1$. ⊣

The following theorem gives useful criteria for quantifier elimination.

THEOREM 3.2.5. *For a theory T the following are equivalent:*

a) *T has quantifier elimination.*
b) *For all models $\mathfrak{M}^1$ and $\mathfrak{M}^2$ of T with a common substructure $\mathfrak{A}$ we have*

$$\mathfrak{M}^1_A \equiv \mathfrak{M}^2_A.$$

c) *For all models $\mathfrak{M}^1$ and $\mathfrak{M}^2$ of T with a common substructure $\mathfrak{A}$ and for all primitive existential formulas $\varphi(x_1, \ldots, x_n)$ and parameters $a_1, \ldots, a_n$ from A we have*

$$\mathfrak{M}^1 \models \varphi(a_1, \ldots, a_n) \Rightarrow \mathfrak{M}^2 \models \varphi(a_1, \ldots, a_n).$$

If L has no constants, $\mathfrak{A}$ is allowed to be the empty "structure".

PROOF. a) $\Rightarrow$ b): Let $\varphi(\overline{a})$ be an $L(A)$-sentence which holds in $\mathfrak{M}^1$. Choose a quantifier-free $\rho(\overline{x})$ which is, modulo T, equivalent to $\varphi(\overline{x})$. Then

$$\begin{array}{lllllll} \mathfrak{M}^1 & \models & \varphi(\overline{a}) & \Rightarrow & \mathfrak{M}^1 & \models & \rho(\overline{a}) \\ & & & \Rightarrow & \mathfrak{A} & \models & \rho(\overline{a}) \Rightarrow \\ \mathfrak{M}^2 & \models & \rho(\overline{a}) & \Rightarrow & \mathfrak{M}^2 & \models & \varphi(\overline{a}). \end{array}$$

b) $\Rightarrow$ c): Clear.

c) $\Rightarrow$ a): Let $\varphi(\overline{x})$ be a primitive existential formula. In order to show that $\varphi(\overline{x})$ is equivalent, modulo T, to a quantifier-free formula $\rho(\overline{x})$ we extend L by an n-tuple $\overline{c}$ of new constants $c_1, \ldots, c_n$. We have to show that we can separate $T \cup \{\varphi(\overline{c})\}$ and $T \cup \{\neg\varphi(\overline{c})\}$ by a quantifier free sentence $\rho(\overline{c})$. We apply the Separation Lemma. Let $\mathfrak{M}^1$ and $\mathfrak{M}^2$ be two models of T with two distinguished n-tuples $\overline{a}^1$ and $\overline{a}^2$. Suppose that $(\mathfrak{M}^1, \overline{a}^1)$ and $(\mathfrak{M}^2, \overline{a}^2)$ satisfy the same quantifier-free $L(\overline{c})$-sentences. We have to show that

$$\mathfrak{M}^1 \models \varphi(\overline{a}^1) \Rightarrow \mathfrak{M}^2 \models \varphi(\overline{a}^2). \tag{3.1}$$

Consider the substructures $\mathfrak{A}^i = \langle \overline{a}^i \rangle^{\mathfrak{M}^i}$, generated by $\overline{a}^i$. If we can show that there is an isomorphism

$$f : \mathfrak{A}^1 \to \mathfrak{A}^2$$

taking $\overline{a}^1$ to $\overline{a}^2$, we may assume that $\mathfrak{A}^1 = \mathfrak{A}^2 = \mathfrak{A}$ and $\overline{a}^1 = \overline{a}^2 = \overline{a}$. Then (3.1) follows directly from c).

Every element of $\mathfrak{A}^1$ has the form $t^{\mathfrak{M}^1}[\overline{a}^1]$ for an L-term $t(\overline{x})$, (see 1.2.6). The isomorphism f to be constructed must satisfy

$$f\,(t^{\mathfrak{M}^1}[\overline{a}^1]) = t^{\mathfrak{M}^2}[\overline{a}^2].$$

We now *define* f by this equation, and we have to check that f is well defined and injective. Assume

$$s^{\mathfrak{M}^1}[\overline{a}^1] = t^{\mathfrak{M}^1}[\overline{a}^1].$$

Then $s(\overline{c}) \doteq t(\overline{c})$ holds in $(\mathfrak{M}^1, \overline{a}^1)$, and by our assumption also in $(\mathfrak{M}^2, \overline{a}^2)$, which means

$$s^{\mathfrak{M}^2}[\overline{a}^2] = t^{\mathfrak{M}^2}[\overline{a}^2].$$

This shows that f is well defined. Swapping the two sides yields injectivity.

That f is surjective is clear. It remains to show that f commutes with the interpretation of the relation symbols. Now

$$\mathfrak{M}^1 \models R\,[t_1^{\mathfrak{M}^1}[\overline{a}^1], \ldots, t_m^{\mathfrak{M}^1}[\overline{a}^1]],$$

is equivalent to $(\mathfrak{M}^1, \overline{a}^1) \models R(t_1(\overline{c}), \ldots, t_m(\overline{c}))$, which is equivalent to $(\mathfrak{M}^2, \overline{a}^2) \models R(t_1(\overline{c}), \ldots, t_m(\overline{c}))$, which in turn is equivalent to

$$\mathfrak{M}^2 \models R\,[t_1^{\mathfrak{M}^2}[\overline{a}^2], \ldots, t_m^{\mathfrak{M}^2}[\overline{a}^2]]. \qquad \dashv$$

Note that part b) of Theorem 3.2.5 is saying that T is *substructure complete*; i.e., for any model $\mathfrak{M} \models T$ and substructure $\mathfrak{A} \subseteq \mathfrak{M}$ the theory $T \cup \mathrm{Diag}(\mathfrak{A})$ is complete.

DEFINITION 3.2.6. We call T *model complete* if for all models $\mathfrak{M}^1$ and $\mathfrak{M}^2$ of T

$$\mathfrak{M}^1 \subseteq \mathfrak{M}^2 \;\Rightarrow\; \mathfrak{M}^1 \prec \mathfrak{M}^2.$$

Note that T is model complete if and only if for any $\mathfrak{M} \models T$ the theory $T \cup \mathrm{Diag}(\mathfrak{M})$ is complete.

Clearly, by 3.2.5(b) applied to $\mathfrak{A} = \mathfrak{M}^1$ all theories with quantifier elimination are model complete.

LEMMA 3.2.7 (Robinson's Test). *Let T be a theory. Then the following are equivalent*:

a) *T is model complete.*

b) *For all models $\mathfrak{M}^1 \subseteq \mathfrak{M}^2$ of T and all existential sentences φ from $L(M^1)$*

$$\mathfrak{M}^2 \models \varphi \Rightarrow \mathfrak{M}^1 \models \varphi.$$

c) *Each formula is, modulo T, equivalent to a universal formula.*

PROOF. a) $\Rightarrow$ b) is trivial.

a) $\Leftrightarrow$ c) follows from 3.1.5(1).

b) implies that every existential formula is, modulo T, equivalent to a universal formula. As in the proof of 3.2.4 this implies c). ⊣

If $\mathfrak{M}^1 \subseteq \mathfrak{M}^2$ satisfies b), we call $\mathfrak{M}^1$ *existentially closed* in $\mathfrak{M}^2$. We denote this by

$$\mathfrak{M}^1 \prec_1 \mathfrak{M}^2.$$

DEFINITION 3.2.8. Let T be a theory. A theory T^* is a *model companion* of T if the following three conditions are satisfied.

a) Each model of T can be extended to a model of T^*.

b) Each model of T^* can be extended to a model of T.

c) T^* is model complete.

THEOREM 3.2.9. *A theory T has, up to equivalence, at most one model companion T^*.*

PROOF. If T^+ is another model companion of T, every model of T^+ is contained in a model of T^* and conversely. Let $\mathfrak{A}_0$ be a model of T^+. Then $\mathfrak{A}_0$ can be embedded in a model $\mathfrak{B}_0$ of T^*. In turn $\mathfrak{B}_0$ is contained in a model $\mathfrak{A}_1$ of T^+. In this way we find two elementary chains, $(\mathfrak{A}_i)$ and $(\mathfrak{B}_i)$, which have a common union $\mathfrak{C}$. Then $\mathfrak{A}_0 \prec \mathfrak{C}$ and $\mathfrak{B}_0 \prec \mathfrak{C}$ implies $\mathfrak{A}_0 \equiv \mathfrak{B}_0$. Thus $\mathfrak{A}_0$ is a model of T^*. Interchanging T^* and T^+ yields that every model of T^* is a model of T^+. ⊣

Digression: existentially closed structures and the Kaiser hull. Let T be an L-theory. It follows from 3.1.2 that the models of $T_\forall$ are the substructures of models of T. The conditions a) and b) in the definition of "model companion" can therefore be expressed as

$$T_\forall = T^*_\forall.$$

Hence the model companion of a theory T depends only on $T_\forall$.

DEFINITION 3.2.10. An L-structure $\mathfrak{A}$ is called *T-existentially closed* (or T-ec), if

a) $\mathfrak{A}$ can be embedded in a model of T.
b) $\mathfrak{A}$ is existentially closed in every extension which is a model of T.

A structure $\mathfrak{A}$ is T-ec exactly if it is $T_\forall$-ec. This is clear for condition a) since every model $\mathfrak{B}$ of $T_\forall$ can be embedded in a model $\mathfrak{M}$ of T. For b) this follows from the fact that $\mathfrak{A} \subseteq \mathfrak{B} \subseteq \mathfrak{M}$ and $\mathfrak{A} \prec_1 \mathfrak{M}$ implies $\mathfrak{A} \prec_1 \mathfrak{B}$.

LEMMA 3.2.11. *Every model of a theory T can be embedded in a T-ec structure.*

PROOF. Let $\mathfrak{A}$ be a model of $T_\forall$. We choose an enumeration $(\varphi_\alpha)_{\alpha<\kappa}$ of all existential $L(A)$-sentences and construct an ascending chain $(\mathfrak{A}_\alpha)_{\alpha\leq\kappa}$ of models of $T_\forall$. We begin with $\mathfrak{A}_0 = \mathfrak{A}$. Let $\mathfrak{A}_\alpha$ be constructed. If φ_α holds in an extension of $\mathfrak{A}_\alpha$ which is a model of T, we let $\mathfrak{A}_{\alpha+1}$ be such a model. Otherwise we set $\mathfrak{A}_{\alpha+1} = \mathfrak{A}_\alpha$. For limit ordinals λ we define $\mathfrak{A}_\lambda$ to be the union of all $\mathfrak{A}_\alpha$, $(\alpha < \lambda)$. Note that $\mathfrak{A}_\lambda$ is again a model of $T_\forall$.

The structure $\mathfrak{A}^1 = \mathfrak{A}_\kappa$ has the following property: every existential $L(A)$-sentence which holds in an extension of $\mathfrak{A}^1$ that is a model of T holds in $\mathfrak{A}^1$. Now, in the same manner, we construct $\mathfrak{A}^2$ from $\mathfrak{A}^1$, etc. The union $\mathfrak{M}$ of the chain $\mathfrak{A}^0 \subseteq \mathfrak{A}^1 \subseteq \mathfrak{A}^2 \subseteq \cdots$ is the desired T-ec structure. ⊣

The structure $\mathfrak{M}$ constructed in the proof can be very big. On the other hand, it is easily seen that every elementary substructure $\mathfrak{N}$ of a T-ec structure $\mathfrak{M}$ is again T-ec. To this end let $\mathfrak{N} \subseteq \mathfrak{A}$ be a model of T. Since $\mathfrak{M}_N \Rightarrow_\exists \mathfrak{A}_N$, there is an embedding of $\mathfrak{M}$ in an elementary extension $\mathfrak{B}$ of $\mathfrak{A}$ which is the identity on N. Since $\mathfrak{M}$ is existentially closed in $\mathfrak{B}$, it follows that $\mathfrak{N}$ is existentially closed in $\mathfrak{B}$ and therefore also in $\mathfrak{A}$.

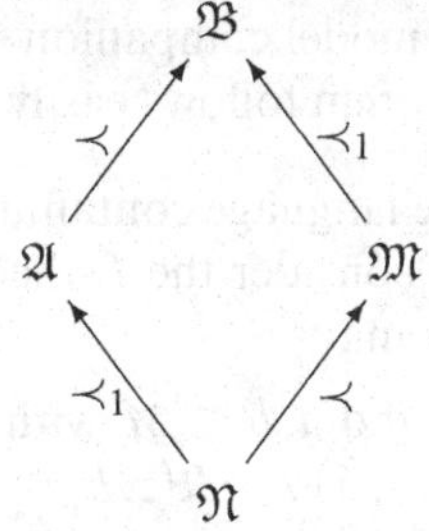

LEMMA 3.2.12 ([32]). *Let T be a theory. Then there is a biggest inductive theory T^{KH} with $T_\forall = T^{\mathrm{KH}}_\forall$. We call T^{KH} the* Kaiser hull *of T.*

PROOF. Let T^1 and T^2 be two inductive theories with $T^1_\forall = T^2_\forall = T_\forall$. We have to show that $(T^1 \cup T^2)_\forall = T_\forall$. Let $\mathfrak{M}$ be a model of T. As in the proof of 3.2.9 we extend $\mathfrak{M}$ by a chain $\mathfrak{A}_0 \subseteq \mathfrak{B}_0 \subseteq \mathfrak{A}_1 \subseteq \mathfrak{B}_1 \subseteq \cdots$ of models of T^1 and T^2. The union of this chain is a model of $T^1 \cup T^2$. ⊣

LEMMA 3.2.13. *The Kaiser hull T^{KH} is the $\forall\exists$-part of the theory of all T-ec structures.*

PROOF. Let T^* be the $\forall\exists$-part of the theory of all T-ec structures. Since T-ec structures are models of $T_\forall$, we have $T_\forall \subseteq T^*_\forall$. It follows from 3.2.11 that $T^*_\forall \subseteq T_\forall$. Hence T^* is contained in the Kaiser hull.

It remains to show that every T-ec structure $\mathfrak{M}$ is a model of the Kaiser hull. Choose a model $\mathfrak{N}$ of T^{KH} which contains $\mathfrak{M}$. Then $\mathfrak{M} \prec_1 \mathfrak{N}$. This implies $\mathfrak{N} \Rightarrow_{\forall\exists} \mathfrak{M}$ and therefore $\mathfrak{M} \models T^{\mathrm{KH}}$. ⊣

The previous lemma implies immediately that T-ec structures are models of $T_{\forall\exists}$.

THEOREM 3.2.14. *For any theory T the following are equivalent:*

a) *T has a model companion T^*.*
b) *All models of T^{KH} are T-ec.*
c) *The T-ec structures form an elementary class.*

If T^* exists, we have

$$T^* = T^{\mathrm{KH}} = \text{theory of all } T\text{-ec structures.}$$

PROOF. a) ⇒ b): Let T^* be the model companion of T. As a model complete theory, T^* is inductive. So T^* is contained in the Kaiser hull and it suffices to show that every model $\mathfrak{M}$ of T^* is T-ec. Let $\mathfrak{A}$ be a model of T which extends $\mathfrak{M}$. $\mathfrak{A}$ can be embedded in a model $\mathfrak{N}$ of T^*. Now $\mathfrak{M} \prec \mathfrak{N}$ implies $\mathfrak{M} \prec_1 \mathfrak{A}$.

b) ⇒ c): By the last lemma all T-ec structures are models of T^{KH}. Thus b) implies that T-ec structures are exactly the models of T^{KH}.

c) ⇒ a): Assume that the T-ec structures are exactly the models of the theory T^+. By 3.2.11 we have $T_\forall = T^+_\forall$. Criterion 3.2.7 implies that T^+ is model complete. So T^+ is the model companion of T.

The last assertion of the theorem follows easily from the proof. ⊣

EXERCISE 3.2.1. Let L be the language containing a unary function f and a binary relation symbol R and consider the L-theory $T = \{\forall x \forall y (R(x, y) \to (R(x, f(y))\}$. Show the following

1. For any T- structure $\mathfrak{M}$ and $a, b \in M$ with $b \notin \{a, f^{\mathfrak{M}}(a), (f^{\mathfrak{M}})^2(a), \dots\}$ we have $\mathfrak{M} \models \exists z (R(z, a) \wedge \neg R(z, b))$.

2. Let $\mathfrak{M}$ be a model of T and a an element of M such that $\{a, f^{\mathfrak{M}}(a), (f^{\mathfrak{M}})^2(a), \dots\}$ is infinite. Then in an elementary extension $\mathfrak{M}'$ there is an element b with $\mathfrak{M}' \models \forall z(R(z,a) \to R(z,b))$.
3. The class of T-ec structures is not elementary, so T does not have a model companion.

EXERCISE 3.2.2. Prove:

1. If T is inductive and has infinite models, then it has existentially closed models in every cardinality $\kappa \geq |T|$.
2. If T has an infinite model which is not existentially closed, it has such a model in every cardinality $\kappa \geq |T|$.
3. Lindström's Theorem: Every inductive κ-categorical theory is model complete.

End of digression.

EXERCISE 3.2.3. A theory T with quantifier elimination is axiomatisable by sentences of the form

$$\forall x_1 \dots x_n \psi$$

where ψ is primitive existential formula.

3.3. Examples

In this section we present a number of theories with quantifier elimination, or at least elimination down to some well-understood formulas, among them the theories of vector spaces and of algebraically, differentially, and real closed fields. Since such theories are comparatively easy to understand, they form a core inventory of the working model theorist. One notable omission is the theory of valued fields. Their model theory can be found in [15] and in [48] and [23].

3.3.1. Infinite sets. The models of the theory Infset of *infinite sets* are all infinite sets without additional structure. The language $L_\emptyset$ is empty, the axioms are (for $n = 1, 2, \dots$)

- $\exists x_0 \dots x_{n-1} \bigwedge_{i<j<n} \neg x_i \doteq x_j$

THEOREM 3.3.1. *The theory* Infset *of infinite sets has quantifier elimination and is complete.*

PROOF. Clear. ⊣

3.3.2. Dense linear orderings.

THEOREM 3.3.2. DLO *has quantifier elimination.*

PROOF. Let A be a finite common substructure of the two models O_1 and O_2. We choose an ascending enumeration $A = \{a_1, \dots, a_n\}$. Let $\exists y\, \rho(y)$ be a simple existential $L(A)$-sentence, which is true in O_1 and assume $O_1 \models \rho(b_1)$. We want to extend the order preserving map $a_i \mapsto a_i$ to an order preserving map $A \cup \{b_1\} \to O_2$. For this we have to find an image b_2 of b_1. There are four cases:

i) $b_1 \in A$. We set $b_2 = b_1$.
ii) b_1 lies between a_i and a_{i+1}. We choose b_2 in O_2 with the same property.
iii) b_1 is smaller than all elements of A. We choose a $b_2 \in O_2$ of the same kind.
iv) b_1 is bigger that all a_i. Choose b_2 in the same manner.

This defines an isomorphism $A \cup \{b_1\} \to A \cup \{b_2\}$, which shows that $O_2 \models \rho(b_2)$. ⊣

Since $\mathrm{L_{Order}}$ has no constants, we have another proof that DLO is complete.

3.3.3. Modules. Let R be a (possibly non-commutative) ring with 1. An R-module

$$\mathfrak{M} = (M, 0, +, -, r)_{r \in R}$$

is an abelian group $(M, 0, +, -)$ together with operations $r\colon M \to M$ for every ring element $r \in R$ which satisfies certain axioms. We formulate the axioms in the language $\mathrm{L_{Mod}}(R) = \mathrm{L_{AbG}} \cup \{r \mid r \in R\}$. The theory $\mathsf{Mod}(R)$ of R-modules consists of

- AbG
- $\forall x, y\ r(x+y) \doteq rx + ry$
- $\forall x\ (r+s)x \doteq rx + sx$
- $\forall x\ (rs)x \doteq r(sx)$
- $\forall x\ 1x \doteq x$

for all $r, s \in R$. Then $\mathsf{Infset} \cup \mathsf{Mod}(R)$ is the theory of all infinite R-modules.

We start with the case where the ring is a field K. Of course, a K-module is just a vector space over K.

THEOREM 3.3.3. *Let K be a field. Then the theory of all infinite K-vector spaces has quantifier elimination and is complete.*

PROOF. Let A be a common finitely generated substructure (i.e., a subspace) of the two infinite K-vector spaces V_1 and V_2. Let $\exists y\, \rho(y)$ be a simple existential $L(A)$-sentence which holds in V_1. Choose a b_1 from V_1 which satisfies $\rho(y)$. If b_1 belongs to A, we are finished since then $V_2 \models \rho(b_1)$. If not, we choose a $b_2 \in V_2 \setminus A$. Possibly we have to replace V_2 by an elementary extension. The vector spaces $A + Kb_1$ and $A + Kb_2$ are isomorphic by an isomorphism which maps b_1 to b_2 and fixes A elementwise. Hence $V_2 \models \rho(b_2)$.

The theory is complete since a quantifier-free sentence is true in a vector space if and only if it is true in the zero-vector space. ⊣

For arbitrary rings R, we only get a relative elimination result down to *positive primitive formulas.*

DEFINITION 3.3.4. An *equation* is an $\mathrm{L}_{\mathrm{Mod}}(R)$-formula $\gamma(\overline{x})$ of the form

$$r_1x_1 + r_2x_2 + \cdots + r_mx_m = 0.$$

A *positive primitive* formula (*pp*-formula) is of the form

$$\exists\overline{y}(\gamma_1 \wedge \cdots \wedge \gamma_n)$$

where the $\gamma_i(\overline{x}\overline{y})$ are equations.

THEOREM 3.3.5. *For every ring R and any R-module M, every $\mathrm{L}_{\mathrm{Mod}}(R)$-formula is equivalent (modulo the theory of M) to a Boolean combination of positive primitive formulas.*

REMARK 3.3.6. 1. We assume the class of positive primitive formulas to be closed under $\wedge$.

2. A pp-formula $\varphi(x_1, \ldots, x_n)$ defines a subgroup $\varphi(M^n)$ of M^n:

$$M \models \varphi(0) \text{ and } M \models \varphi(x) \wedge \varphi(y) \rightarrow \varphi(x - y).$$

LEMMA 3.3.7. *Let $\varphi(x, y)$ be a pp-formula and $a \in M$. Then $\varphi(M, a)$ is empty or a coset of $\varphi(M, 0)$.*

PROOF. $M \models \varphi(x, a) \rightarrow (\varphi(y, 0) \leftrightarrow \varphi(x + y, a))$. ⊣

COROLLARY 3.3.8. *Let $a, b \in M$, $\varphi(x, y)$ a pp-formula. Then (in M) $\varphi(x, a)$ and $\varphi(x, b)$ are equivalent or contradictory.*

For the proof of Theorem 3.3.5 we need two further lemmas.

LEMMA 3.3.9 (B. H. Neumann). *Let H_i denote subgroups of some abelian group. If $H_0 + a_0 \subseteq \bigcup_{i=1}^{n} H_i + a_i$ and $H_0/(H_0 \cap H_i)$ is infinite for $i > k$, then $H_0 + a_0 \subseteq \bigcup_{i=1}^{k} H_i + a_i$.*

For a proof see Exercise 6.1.16. The following is an easy calculation.

LEMMA 3.3.10. *Let $A_i, i \leq k$, be any sets. If A_0 is finite, then $A_0 \subseteq \bigcup_{i=1}^{k} A_i$ if and only if*

$$\sum_{\Delta \subseteq \{1,\ldots,k\}} (-1)^{|\Delta|} \Big| A_0 \cap \bigcap_{i \in \Delta} A_i \Big| = 0.$$

PROOF OF THEOREM 3.3.5. Fix M. It is enough to show that if $\psi(x, y)$ is in M equivalent to a Boolean combination of pp-formulas, then so is $\forall x \psi$. Since pp-formulas are closed under conjunction, ψ is M-equivalent to a conjunction of formulas $\varphi_0(x, y) \rightarrow \varphi_1(x, y) \vee \cdots \vee \varphi_n(x, y)$ where the $\varphi_i(x, y)$ are pp-formulas.

We may assume that ψ itself is of this form. Let $H_i = \varphi_i(M, 0)$, so the $\varphi_i(M, y)$ are empty or cosets of H_i. (Think of y as being fixed in M.) Let

$H_0/(H_0 \cap H_i)$ be finite for $i = 1, \ldots, k$ and infinite for $i = k+1, \ldots, n, k \geq 0$. By Neumann's Lemma we have

$$M \models \forall x \psi \leftrightarrow \forall x \left(\varphi_0(x, y) \rightarrow \varphi_1(x, y) \vee \cdots \vee \varphi_k(x, y)\right).$$

We apply Lemma 3.3.10 to the sets $A_i = \varphi_i(M, y)/(H_0 \cap \cdots \cap H_k)$: so $\varphi(M, y) \cap \bigcap_{i \in \Delta} \varphi_i(M, y)$ is empty or consists of N_Δ cosets of $H_0 \cap \cdots \cap H_k$ where

$$N_\Delta = \left| H_0 \cap \bigcap_{i \in \Delta} H_i/(H_0 \cap \cdots \cap H_k) \right|.$$

Whence

$$M \models \forall x \psi \leftrightarrow \sum_{\Delta \in \mathcal{N}} (-1)^{|\Delta|} N_\Delta = 0.$$

where

$$\mathcal{N} = \left\{ \Delta \subseteq \{1, \ldots, k\} \,\middle|\, \exists x \left(\varphi_0(x, y) \wedge \bigwedge_{i \in \Delta} \varphi_i(x, y)\right) \right\}.$$

⊣

3.3.4. Algebraically closed fields. As the next group of examples we consider fields.

THEOREM 3.3.11 (Tarski). *The theory* ACF *of algebraically closed fields has quantifier elimination.*

PROOF. Let K_1 and K_2 be two algebraically closed fields and R a common subring. Let $\exists y\ \rho(y)$ be a simple existential sentence with parameters in R which holds in K_1. We have to show that $\exists y\ \rho(y)$ is also true in K_2.

Let F_1 and F_2 be the quotient fields of R in K_1 and K_2, respectively, and let $f : F_1 \to F_2$ be an isomorphism which is the identity on R (see e.g., [35], Ch. II.4). Then f extends to an isomorphism $g : G_1 \to G_2$ between the relative algebraic closures G_i of F_i in K_i, $(i = 1, 2)$ (see e.g., [35], Ch. V.2). Choose an element b_1 of K_1 which satisfies $\rho(y)$.

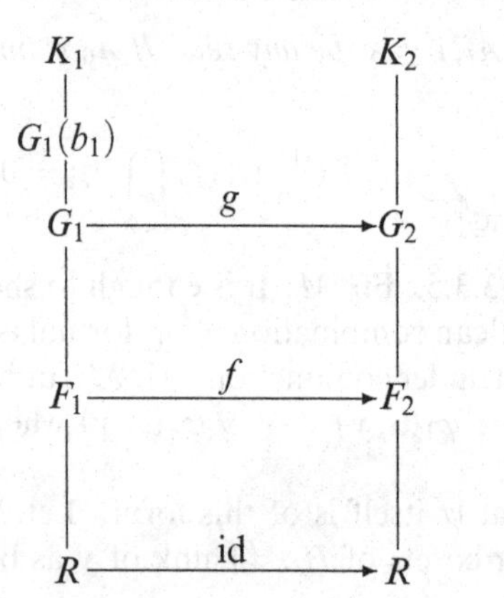

There are two cases:

Case 1: $b_1 \in G_1$. Then $b_2 = g(b_1)$ satisfies the formula $\rho(y)$ in K_2.

Case 2: $b_1 \notin G_1$. Then b_1 is transcendental over G_1 and the field extension $G_1(b_1)$ is isomorphic to the rational function field $G_1(X)$. If K_2 is a proper extension of G_2, we choose any element from $K_2 \setminus G_2$ for b_2. Then g extends to an isomorphism between $G_1(b_1)$ and $G_2(b_2)$ which maps b_1 to b_2. Hence b_2 satisfies $\rho(y)$ in K_2. In case that $K_2 = G_2$ we take a proper elementary extension K_2' of K_2. (Such a K_2' exists by 2.3.1(2) since K_2 is infinite.) Then (for the same reason) $\exists y\, \rho(y)$ holds in K_2' and therefore in K_2. ⊣

COROLLARY 3.3.12. ACF *is model complete.* ⊣

Obviously, ACF is not complete: for prime numbers p let

$$\mathsf{ACF}_p = \mathsf{ACF} \cup \{p \cdot 1 \doteq 0\}$$

be the theory of algebraically closed fields of characteristic p and

$$\mathsf{ACF}_0 = \mathsf{ACF} \cup \{\neg n \cdot 1 \doteq 0 \mid n = 1, 2, \dots\}$$

the theory of algebraically closed fields of characteristic 0. We use here the notation $n \cdot 1 = \underbrace{1 + \cdots + 1}_{n-\text{ times}}$.

COROLLARY 3.3.13. *The theories* ACF_p *and* ACF_0 *are complete.*

PROOF. This follows from Theorem 3.2.2 since the prime fields are prime structures for these theories. ⊣

COROLLARY 3.3.14 (Hilbert's Nullstellensatz). *Let K be a field. Then any proper ideal I in $K[X_1, \dots, X_n]$ has a zero in the algebraic closure* $\mathrm{acl}(K)$.

PROOF. As a proper ideal, I is contained in a maximal ideal P. Then $L = K[X_1, \dots, X_n]/P$ is an extension field of K in which the cosets of the X_i are a zero of I. If I is generated by $f_0, \dots, f_{k-1}$ and

$$\varphi = \exists x_1 \dots x_n \bigwedge_{i<k} f_i(x_1, \dots, x_n) \doteq 0,$$

then φ holds in L and therefore in $\mathrm{acl}(L)$. We can assume that $\mathrm{acl}(K)$ lies in $\mathrm{acl}(L)$. Since $\mathrm{acl}(K) \prec \mathrm{acl}(L)$, we have that φ holds in $\mathrm{acl}(K)$. ⊣

3.3.5. Real closed fields. The theory of real closed fields, RCF, will be discussed in Section B.1. It is axiomatised in the language $\mathrm{L}_{\mathrm{ORing}}$ of ordered rings.

THEOREM 3.3.15 (Tarski–Seidenberg). RCF *has quantifier elimination and is complete.*

PROOF. Let $(K_1, <)$ and $(K_2, <)$ be two real closed field with a common subring R. Consider an $\mathrm{L}_{\mathrm{ORing}}(R)$-sentence $\exists y\, \rho(y)$ (for a quantifier-free ρ) which holds in $(K_1, <)$. We have to show $\exists y\, \rho(y)$ also holds in $(K_2, <)$.

We build first the quotient fields F_1 and F_2 of R in K_1 and K_2. By B.1.1 there is an isomorphism $f\colon (F_1, <) \to (F_2, <)$ which fixes R. The relative algebraic closure G_i of F_i in K_i is a real closure of $(F_i, <)$, $(i = 1, 2)$. By B.1.5 f extends to an isomorphism $g\colon (G_1, <) \to (G_2, <)$.

Let b_1 be an element of K_1 which satisfies $\rho(y)$. There are two cases:

Case 1: $b_1 \in G_1$: Then $b_2 = g(b_1)$ satisfies $\rho(y)$ in K_2.

Case 2: $b_1 \notin G_1$: Then b_1 is transcendental over G_1 and the field extension $G_1(b_1)$ is isomorphic to the rational function field $G_1(X)$. Let G_1^ℓ be the set of all elements of G_1 which are smaller than b_1, and G_1^r the set of all elements of G_1 which are larger than b_1. Then all elements of $G_2^\ell = g(G_1^\ell)$ are smaller than all elements of $G_2^r = g(G_1^r)$. Since fields are densely ordered, we find in an elementary extension $(K_2', <)$ of $(K_2, <)$ an element b_2 which lies between the elements of G_2^ℓ and the elements of G_2^r. Since b_2 is not in G_2, it is transcendental over G_2. Hence g extends to an isomorphism $h\colon G_1(b_1) \to G_2(b_2)$ which maps b_1 to b_2.

In order to show that h is order preserving it suffices to show that h is order preserving on $G_1[b_1]$ (Lemma B.1.1). Let $p(b_1)$ be an element of $G_1[b_1]$. Corollary B.1.8 gives us a decomposition

$$p(X) = \varepsilon \prod_{i<m}(X - a_i) \prod_{j<n}((X - c_j)^2 + d_j)$$

with positive d_j. The sign of $p(b_1)$ depends only on the signs of the factors $\varepsilon, b_1 - a_0, \ldots, b_1 - a_{m-1}$. The sign of $h(p(b_1))$ depends in the same way on the signs of $g(\varepsilon), b_2 - g(a_0), \ldots, b_2 - g(a_{m-1})$. But b_2 was chosen in such a way that

$$b_1 < a_i \Leftrightarrow b_2 < g(a_i).$$

Hence $p(b_1)$ is positive if and only if $h(p(b_1))$ is positive.

Finally we have

$$(K_1, <) \models \rho(b_1) \Rightarrow (G_1(b_1), <) \models \rho(b_1) \Rightarrow (G_2(b_2), <) \models \rho(b_2) \Rightarrow$$
$$\Rightarrow (K_2', <) \models \exists y\, \rho(y) \Rightarrow (K_2, <) \models \exists y\, \rho(y),$$

which proves quantifier elimination.

RCF is complete since the ordered field of the rationals is a prime structure. ⊣

COROLLARY 3.3.16 (Hilbert's 17th Problem). *Let $(K, <)$ be a real closed field. A polynomial $f \in K[X_1, \ldots, X_n]$ is a sum of squares*

$$f = g_1^2 + \cdots + g_k^2$$

of rational functions $g_i \in K(X_1, \ldots, X_n)$ if and only if

$$f(a_1, \ldots, a_n) \geq 0$$

for all $a_1, \ldots, a_n \in K$.

PROOF. Clearly a sum of squares cannot have negative values. For the converse assume that f is not a sum of squares. Then, by Corollary B.1.3, $K(X_1, \ldots, X_n)$ has an ordering in which f is negative. Since in K the positive elements are squares, this ordering, which we denote also by $<$, extends the ordering of K. Let $(L, <)$ be the real closure of $\big(K(X_1, \ldots, X_n), <\big)$. In $(L, <)$ the sentence

$$\exists x_1, \ldots, x_n \; f(x_1, \ldots, x_n) < 0$$

is true. Hence it is also true in $(K, <)$. ⊣

3.3.6. Separably closed fields. A field is *separably closed* if every non-constant separable polynomial has a zero, or equivalently, if it has no proper separable algebraic extension. Clearly, separably closed fields which are also perfect are algebraically closed.

For any field K of characteristic $p > 0$, $K^p = \{a^p \mid a \in K\}$ is a subfield of K. If the degree $[K : K^p]$ is finite, it has the form p^e and e is called the *degree of imperfection* of K. If the degree $[K : K^p]$ is infinite, then we say that K has infinite degree of imperfection. See also page 207.

For any natural number e we denote by $\mathsf{SCF}_{p,e}$ the theory of separably closed fields with degree of imperfection e. By $\mathsf{SCF}_{p,\infty}$ we denote the theory of separably closed closed field of characteristic p with infinite degree of imperfection. We will prove below that $\mathsf{SCF}_{p,e}$ is complete. A proof of the completeness of $\mathsf{SCF}_{p,\infty}$ can be found in [19].

To study $\mathsf{SCF}_{p,e}$ we consider an expansion of it: $\mathsf{SCF}_p(c_1, \cdots, c_e)$, the theory of separably closed fields of characteristic p in the language $L(c_1, \ldots, c_e)$ of rings with constants $c_1, \ldots, c_e$ for a distinguished finite p-basis. We show

PROPOSITION 3.3.17. *$\mathsf{SCF}_p(c_1, \cdots, c_e)$ is model complete.*

For the proof we need the following lemma.

LEMMA 3.3.18. *Let K and L be extensions of F. Assume that K/F is separable and that L is separably closed. Then K embeds over F in an elementary extension of L.*

PROOF. Since L is infinite, it has arbitrarily large elementary extensions. So we may assume that the transcendence degree $\operatorname{tr.deg}(L/F)$ of L over F is infinite. Let K' be a finitely generated subfield of K over F. By compactness it suffices to show that all such K'/F can be embedded into L/F. By Lemma B.3.12 K'/F has a transcendence basis $x_1, \ldots, x_n$ so that $K'/F(x_1, \ldots, x_n)$ is separably algebraic. $F(x_1, \ldots, x_n)/F$ can be embedded into L/F. Since L is separably closed this embedding extend to K'. ⊣

PROOF OF PROPOSITION 3.3.17. Let $(F, b_1, \ldots, b_e) \subseteq (K, b_1, \ldots, b_e)$ be an extension of models of $\mathsf{SCF}_p(c_1, \cdots, c_e)$. Since F and K have the same p-basis, K is separable over F by Remark B.3.9. By Lemma 3.3.18, K embeds over F into an elementary extension of F, showing that F is existentially closed in K. Now the claim follows by Robinson's Test (Lemma 3.2.7). ⊣

COROLLARY 3.3.19 (Ershov). *$\mathsf{SCF}_{p,e}$ is complete.*

PROOF. Consider the polynomial ring $R = \mathbb{F}_p[x_1, \dots, x_e]$. It is easy to see that $x_1, \dots, x_e$ is a p-basis of R in the sense of Lemma B.3.11. The same lemma implies that $x_1, \dots, x_e$ is a p-basis of $F = \mathbb{F}_p(x_1, \dots, x_e)^{\text{sep}}$ (where L^{sep} denotes the separable algebraic closure of the field L). So $(F, x_1, \dots, x_e)$ is a model of $\mathsf{SCF}_p(c_1, \cdots, c_e)$. If $(L, b_1, \dots, b_e)$ is another model of $\mathsf{SCF}_p(c_1, \cdots, c_e)$, then by Lemma B.3.10 the b_i are algebraically independent over $\mathbb{F}_p$, so we can embed $(F, x_1, \dots, x_e)$ into $(L, b_1, \dots, b_e)$. This embedding is elementary by Proposition 3.3.17, so L is elementarily equivalent to F. ⊣

Let K be a field with p-basis $b_1, \dots, b_e$. The λ-functions $\lambda_\nu \colon K \to K$ are defined by

$$x = \sum \lambda_\nu(x)^p b^\nu,$$

where the ν are multi-indices $(\nu_1, \dots, \nu_e)$ with $0 \leq \nu_i < p$ and b^ν denotes $b_1^{\nu_1} \cdots b_e^{\nu_e}$.

THEOREM 3.3.20 (Delon). *$\mathsf{SCF}_p(c_1, \cdots, c_e)$ has quantifier elimination in the language $L(c_1, \dots, c_e, \lambda_\nu)_{\nu \in p^e}$.*

It can be shown that $\mathsf{SCF}_p(c_1, \cdots, c_e)$ has quantifier elimination already in the language $L(\lambda_\nu)_{\nu \in p^e}$ without naming a p-basis.

PROOF. Let $\mathcal{K} = (K, b_1, \dots, b_e)$ and $\mathcal{L} = (L, b_1, \dots, b_e)$ each be models of $\mathsf{SCF}_p(c_1, \cdots, c_e)$ and let R be a common subring which contains the b_i and is closed under the λ-functions of $\mathcal{K}$ and $\mathcal{L}$.[2] Since R is closed under the λ-functions, the b_i form a p-basis of R in the sense of Lemma B.3.11. Let F be the separable closure of the quotient field of R. By Lemma B.3.11 the b_i also form a p-basis of F. So F is an elementary subfield of K and of L by Proposition 3.3.17, and hence K_R and L_R are elementarily equivalent. Now the claim follows from Theorem 3.2.5. ⊣

3.3.7. Differentially closed fields. We next consider fields with a derivation in the language of fields expanded by a function symbol d. Differential fields are introduced and discussed in Section B.2.

DEFINITION 3.3.21. The theory of *differentially closed fields*, DCF_0, is the theory of differential fields (K, d) in characteristic 0 satisfying the following property:

For $f \in K[x_0, \dots, x_n] \setminus K[x_0, \dots, x_{n-1}]$ and $g \in K[x_0, \dots, x_{n-1}]$, $g \neq 0$, there is some $a \in K$ such that $f(a, da, \dots, d^n a) = 0$ and $g(a, da, \dots, d^{n-1} a) \neq 0$.

Clearly, models of DCF_0 are algebraically closed.

THEOREM 3.3.22.

1. *Any differential field can be extended to a model of* DCF_0.

[2] Note that the λ-functions of $\mathcal{K}$ and $\mathcal{L}$ agree automatically on R.

2. DCF_0 *is complete and has quantifier elimination.*

PROOF. Let (K, d) be a differential field and f and g as in the definition. We may assume that f is irreducible. Then f determines a field extension $F = K(t_0, \ldots, t_{n-1}, b)$ where the t_i are algebraically independent over K and

$$f(t_0, \ldots, t_{n-1}, b) = 0$$

(see Remark B.3.7). By Lemma B.2.2 and B.2.3 there is an extension of the derivation to F with $dt_i = t_{i+1}$ and $dt_{n-1} = b$. For $a = t_0$ we have $f(a, da, \ldots, d^n a) = 0$ and $g(a, da, \ldots, d^{n-1} a) \neq 0$. Let K_0 denote the differential field that we obtain from K by doing this for all pairs f, g as above with coefficients from K. Inductively, we define differential fields K_{i+1} satisfying the required condition for polynomials with coefficients in K_i. Their union $\bigcup_{i<\omega} K_i$ is a model of DCF_0.

Since the rational numbers with trivial derivation are a prime structure for DCF_0, by Lemma 3.2.2 it suffices for part 2 to prove quantifier elimination. For this, let K be a differential field with two extensions F_1 and F_2 which are models of DCF_0. Let a be an element of F_1 and let $K\{a\} = K(a, da, d^2a, \ldots)$ be the differential field generated by K and a. We have to show that $K\{a\}$ can be embedded over K into an elementary extension of F_2. We distinguish two cases.

1. The derivatives $a, da, d^2a, \ldots$ are algebraically independent over K: since F_2 is a model of DCF_0, there is an element b in some elementary extension such that $g(a, da, \ldots, d^{n-1}a) \neq 0$ for all n and all $g \in K[x_0, \ldots, x_{n-1}] \setminus 0$. The isomorphism from $K\{a\}$ to $K\{b\}$ defined by $d^i a \mapsto d^i b$ is the required embedding.

2. Let $d^n a$ be algebraic over $K(a, da, \ldots, d^{n-1}a)$ and n minimal: choose an irreducible $f \in K[x_0, \ldots, x_n]$ such that $f(a, da, \ldots, d^n a) = 0$ (see Remark B.3.7). We may find some b with $f(b, db, \ldots, d^n b) = 0$ and $g(b, ba, \ldots, d^{n-1}b) \neq 0$ for all $g \in K[x_0, \ldots, x_{n-1}] \setminus 0$ in an elementary extension of F_2. The field isomorphism from $K_1 = K(a, \ldots, d^n a)$ to $K_2 = K(b, \ldots, d^n b)$ fixing K and taking $d^i a$ to $d^i b$ takes the derivation of F_1 restricted to $K(a, \ldots, d^{n-1}a)$ to the derivation of F_2 restricted to $K(b, \ldots, d^{n-1}b)$. The uniqueness part of Lemma B.2.3 implies that K_1 and K_2 are closed under the respective derivations, and that K_1 and K_2 are isomorphic over K as differential fields. $\dashv$

EXERCISE 3.3.1. Let Graph be the theory of graphs. The theory RG of the *random graph* is the extension of Graph by the following axiom scheme:

$$\forall x_0 \ldots x_{m-1}\, y_1 \ldots y_{n-1} \Big(\bigwedge_{i \neq j} \neg x_i \doteq y_j \rightarrow$$

$$\exists z \big(\bigwedge_{i<m} zRx_i\big) \wedge \big(\bigwedge_{j<n} \neg zRy_j \wedge \neg z \doteq y_j\big)\Big)$$

Show that RG has quantifier elimination and is complete. Show also that RG is the model companion of Graph.

EXERCISE 3.3.2. In models of ACF, RCF and DCF_0, the model-theoretic algebraic closure of a set A coincides with the algebraic closure of the (differential) field generated by A.

EXERCISE 3.3.3. Show that the following is true in any algebraically closed field K: every injective polynomial map of a definable subset of K^n in itself is surjective.

In fact, more is true: use Exercise 6.1.14 to show that the previous statement holds for every injective *definable* map.

Chapter 4

COUNTABLE MODELS

4.1. The omitting types theorem

As we have seen in Corollary 2.2.9, it is not hard to realise a given type or in fact any number of them. But as Sacks [50] pointed out, it needs a model theorist to avoid realizing a given type.

DEFINITION 4.1.1. Let T be an L-theory and $\Sigma(x)$ a set of L-formulas. A model $\mathfrak{A}$ of T not realizing $\Sigma(x)$ is said to *omit* $\Sigma(x)$. A formula $\varphi(x)$ *isolates* $\Sigma(x)$ if

a) $\varphi(x)$ is consistent with T.
b) $T \vdash \forall x\ (\varphi(x) \rightarrow \sigma(x))$ for all $\sigma(x)$ in $\Sigma(x)$.

A set of formulas is often called a *partial type*. This explains the name of the following theorem.

THEOREM 4.1.2 (Omitting Types). *If T is countable*[1] *and consistent and if $\Sigma(x)$ is not isolated in T, then T has a model which omits $\Sigma(x)$.*

If $\Sigma(x)$ is isolated by $\varphi(x)$ and $\mathfrak{A}$ is a model of T, then $\Sigma(x)$ is realised in $\mathfrak{A}$ by all realisations of $\varphi(x)$. Therefore the converse of the theorem is true for complete theories T: if $\Sigma(x)$ is isolated in T, then it is realised in every model of T.

PROOF. We choose a countable set C of new constants and extend T to a theory T^* with the following properties:

a) T^* is a Henkin theory: for all $L(C)$-formulas $\psi(x)$ there exists a constant $c \in C$ with $\exists x\ \psi(x) \rightarrow \psi(c)\ \in\ T^*$.
b) For all $c \in C$ there is a $\sigma(x) \in \Sigma(x)$ with $\neg\sigma(c) \in T^*$.

We construct T^* inductively as the union of an ascending chain

$$T = T_0 \subseteq T_1 \subseteq \cdots$$

of consistent extensions of T by finitely many axioms from $L(C)$, in each step making an instance of a) or b) true.

[1] An L-theory is *countable* if L is at most countable.

Enumerate $C = \{c_i \mid i < \omega\}$ and let $\{\psi_i(x) \mid i < \omega\}$ be an enumeration of the $L(C)$-formulas.

Assume that T_{2i} is already constructed. Choose some $c \in C$ which does not occur in $T_{2i} \cup \{\psi_i(x)\}$ and set $T_{2i+1} = T_{2i} \cup \{\exists x\, \psi_i(x) \to \psi_i(c)\}$. Clearly T_{2i+1} is consistent.

Up to equivalence T_{2i+1} has the form $T \cup \{\delta(c_i, \overline{c})\}$ for an L-formula $\delta(x, \overline{y})$ and a tuple $\overline{c} \in C$ which does not contain c_i. Since $\exists \overline{y}\, \delta(x, \overline{y})$ does not isolate $\Sigma(x)$, for some $\sigma \in \Sigma$ the formula $\exists \overline{y}\, \delta(x, \bar{y}) \wedge \neg\sigma(x)$ is consistent with T. Thus, $T_{2i+2} = T_{2i+1} \cup \{\neg\sigma(c_i)\}$ consistent.

Take a model $(\mathfrak{A}', a_c)_{c \in C}$ of T^*. Since T^* is a Henkin theory, Tarski's Test 2.1.2 shows that $A = \{a_c \mid c \in C\}$ is the universe of an elementary substructure $\mathfrak{A}$ (see Lemma 2.2.3). By property b), $\Sigma(x)$ is omitted in $\mathfrak{A}$. ⊣

COROLLARY 4.1.3. *Let T be countable and consistent and let*

$$\Sigma_0(x_1, \ldots, x_{n_0}),\ \Sigma_1(x_1, \ldots, x_{n_1}), \ldots$$

be a sequence of partial types. If all Σ_i are not isolated, then T has a model which omits all Σ_i.

PROOF. Generalise the proof of the Omitting Types Theorem. ⊣

EXERCISE 4.1.1. Prove Corollary 4.1.3.

4.2. The space of types

We now endow the set of types of a given theory with a topology. The Compactness Theorem 2.2.1 then translates into the statement that this topology is compact, whence its name.

Fix a theory T. An *n-type* is a maximal set of formulas $p(x_1, \ldots, x_n)$ consistent with T. We denote by $\mathrm{S}_n(T)$ the set of all n-types of T. We also write $\mathrm{S}(T)$ for $\mathrm{S}_1(T)$.[2]

If B is a subset of an L-structure $\mathfrak{A}$, we recover $\mathrm{S}_n^{\mathfrak{A}}(B)$ (see p. 22) as $\mathrm{S}_n(\mathrm{Th}(\mathfrak{A}_B))$. In particular, if T is complete and $\mathfrak{A}$ is any model of T, we have $\mathrm{S}^{\mathfrak{A}}(\emptyset) = \mathrm{S}(T)$.

For any L-formula $\varphi(x_1, \ldots, x_n)$, let $[\varphi]$ denote the set of all types containing φ.

LEMMA 4.2.1.

1. $[\varphi] = [\psi]$ *if and only if φ and ψ are equivalent modulo T.*
2. *The sets $[\varphi]$ are closed under Boolean operations. In fact* $[\varphi] \cap [\psi] = [\varphi \wedge \psi]$, $[\varphi] \cup [\psi] = [\varphi \vee \psi]$, $\mathrm{S}_n(T) \setminus [\varphi] = [\neg\varphi]$, $\mathrm{S}_n(T) = [\top]$ *and* $\emptyset = [\bot]$. ⊣

[2] $\mathrm{S}_0(T)$ can be considered as the set of all complete extensions of T, up to equivalence.

PROOF. For the first part just notice that if φ and ψ are not equivalent modulo T, then $\varphi \wedge \neg\psi$ or $\neg\varphi \wedge \psi$ is consistent with T and hence $[\varphi] \neq [\psi]$. The rest is clear. ⊣

It follows that the collection of sets of the form $[\varphi]$ is closed under finite intersections and includes $\mathrm{S}_n(T)$. So these sets form a basis of a topology on $\mathrm{S}_n(T)$.

LEMMA 4.2.2. *The space* $\mathrm{S}_n(T)$ *is 0-dimensional and compact.*

PROOF. Being 0-dimensional means having a basis of clopen sets. Our basic open sets are clopen since their complements are also basic open.

If p and q are two different types, there is a formula φ contained in p but not in q. It follows that $[\varphi]$ and $[\neg\varphi]$ are open sets which separate p and q. This shows that $\mathrm{S}_n(T)$ is Hausdorff.

To show compactness consider a family $[\varphi_i]$, $(i \in I)$, with the finite intersection property. This means that all $\varphi_{i_1} \wedge \cdots \wedge \varphi_{i_k}$ are consistent with T. So, by Corollary 2.2.5, $\{\varphi_i \mid i \in I\}$ is consistent with T and can be extended to a type p, which then belongs to all $[\varphi_i]$. ⊣

LEMMA 4.2.3. *All clopen subsets of* $\mathrm{S}_n(T)$ *have the form* $[\varphi]$.

PROOF. It follows from Exercise 3.1.1 that we can separate any two disjoint closed subsets of $\mathrm{S}_n(T)$ by a basic open set. ⊣

REMARK. The Stone duality theorem asserts that the map

$$X \mapsto \{C \mid C \text{ clopen subset of } X\}$$

yields an equivalence between the category of 0-dimensional compact spaces and the category of Boolean algebras. The inverse map assigns to every Boolean algebra $\mathcal{B}$ its *Stone space* $\mathrm{S}(\mathcal{B})$, the set of all ultrafilters (see Exercise 1.2.4) of $\mathcal{B}$. For more on Boolean algebras see [21].

DEFINITION 4.2.4. A map f from a subset of a structure $\mathfrak{A}$ to a structure $\mathfrak{B}$ is *elementary* if it preserves the truth of formulas; i.e., $f : A_0 \to B$ is elementary if for every formula $\varphi(x_1, \ldots, x_n)$ and $\overline{a} \in A_0$ we have

$$\mathfrak{A} \models \varphi(\overline{a}) \Rightarrow \mathfrak{B} \models \varphi(f(\overline{a})).$$

Note that the empty map is elementary if and only if $\mathfrak{A}$ and $\mathfrak{B}$ are elementarily equivalent. An elementary embedding of $\mathfrak{A}$ is an elementary map which is defined on all of A.

LEMMA 4.2.5. *Let $\mathfrak{A}$ and $\mathfrak{B}$ be L-structures, A_0 and B_0 subsets of A and B, respectively. Any elementary map $A_0 \to B_0$ induces a continuous surjective map* $\mathrm{S}_n(B_0) \to \mathrm{S}_n(A_0)$.

PROOF. If $q(x) \in \mathrm{S}_n(B_0)$, we define

$$\mathrm{S}(f)(q) = \{\varphi(x_1, \ldots, x_n, \overline{a}) \mid \overline{a} \in A_0,\ \varphi(x_1, \ldots, x_n, f(\overline{a})) \in q\}.$$

It is easy to see that $S(f)$ defines a map from $S_n(B_0)$ to $S_n(A_0)$. Moreover it is surjective since $\{\varphi(x_1,\ldots,x_n,f(\bar{a})) \mid \varphi(x_1,\ldots,x_n,a) \in p\}$ is finitely satisfiable for all $p \in S_n(A_0)$. And $S(f)$ is continuous since $[\varphi(x_1,\ldots,x_n,f(\overline{a}))]$ is the preimage of $[\varphi(x_1,\ldots,x_n,\overline{a})]$ under $S(f)$. ⊣

There are two main cases:

- An elementary bijection $f\colon A_0 \to B_0$ defines a homeomorphism $S_n(A_0) \to S_n(B_0)$. We write $f(p)$ for the image of p.
- If $\mathfrak{A} = \mathfrak{B}$ and $A_0 \subseteq B_0$, the inclusion map induces the *restriction*[3] $S_n(B_0) \to S_n(A_0)$. We write $q \restriction A_0$ for the restriction of q to A_0. We call q an *extension* of $q \restriction A_0$.

We leave the following lemma as an exercise (see Exercise 4.2.1).

LEMMA 4.2.6. *A type p is isolated in T if and only if p is an isolated point in $S_n(T)$. In fact, φ isolates p if and only if $[\varphi] = \{p\}$. That is, $[\varphi]$ is an* atom *in the Boolean algebra of clopen subsets of $S_n(T)$.*

We call a formula $\varphi(x)$ *complete* if

$$\{\psi(\overline{x}) \mid T \vdash \forall\overline{x}\,(\varphi(\overline{x}) \to \psi(\overline{x}))\}$$

is a type. We have shown:

COROLLARY 4.2.7. *A formula isolates a type if and only if it is complete.*

EXERCISE 4.2.1. Show that a type p is isolated if and only if it is isolated as an element in the Stone space.

EXERCISE 4.2.2. a) Closed subsets of $S_n(T)$ have the form $\{p \in S_n(T) \mid \Sigma \subseteq p\}$, where Σ is any set of formulas.

b) Let T be countable and consistent. Then any meagre[4] subset X of $S_n(T)$ can be omitted, i.e., there is model which omits all $p \in X$.

EXERCISE 4.2.3. Consider the space $S_\omega(T)$ of all complete types in variables $v_0, v_1, \ldots$. Note that $S_\omega(T)$ is again a compact space and therefore not meagre by Baire's theorem.

1. Show that $\{\mathrm{tp}(a_0, a_2, \ldots) \mid \text{the } a_i \text{ enumerate a model of } T\}$ is comeagre in $S_\omega(T)$.
2. Use this to give a purely topological proof the Omitting Types Theorem (4.1.3).

EXERCISE 4.2.4. Let $L \subseteq L'$, T an L-theory, T' an L'-theory and $T \subseteq T'$. Show that there is a natural continuous map $S_n(T') \to S_n(T)$. This map is surjective if and only if T'/T is a conservative extension, i.e., if T' and T prove the same L-sentences.

---

[3]"restriction of parameters".

[4]A subset of a topological space is *nowhere dense* if its closure has no interior. A countable union of nowhere dense sets is *meagre*.

EXERCISE 4.2.5. Let B be a subset of $\mathfrak{A}$. Show that the *restriction*[5] map $\mathrm{S}_{m+n}(B) \to \mathrm{S}_n(B)$ is open, continuous and surjective. Let a be an n-tuple in A. Show that the fibre over $\mathrm{tp}(a/B)$ is canonically homeomorphic to $\mathrm{S}_m(aB)$.

EXERCISE 4.2.6. A theory T has quantifier elimination if and only if every type is implied by its quantifier-free part.

EXERCISE 4.2.7. Consider the structure $\mathfrak{M} = (\mathbb{Q}, <)$. Determine all types in $\mathrm{S}_1(\mathbb{Q})$. Which of these types are realised in $\mathbb{R}$? Which extensions does a type over $\mathbb{Q}$ have to a type over $\mathbb{R}$?

4.3. $\aleph_0$-categorical theories

In this section, we consider theories with a unique countable model (up to isomorphism, of course). These theories can be characterised by the fact that they have only finite many n-types for each n, see Exercise 4.3.3. We show the following equivalent statement.

THEOREM 4.3.1 (Ryll–Nardzewski). *Let T be a countable complete theory. Then T is $\aleph_0$-categorical if and only if for every n there are only finitely many formulas $\varphi(x_1, \ldots, x_n)$ up to equivalence relative to T.*

The proof will make use of the following notion.

DEFINITION 4.3.2. An L-structure $\mathfrak{A}$ is *ω-saturated* if all types over finite subsets of A are realised in $\mathfrak{A}$.

The types in the definition are meant to be 1-types. On the other hand, it is not hard to see that an ω-saturated structure realises all n-types over finite sets (see Exercise 4.3.9), for all $n \geq 1$. The following lemma is a generalisation of the $\aleph_0$-categoricity of DLO. The proof is essentially the same, see p. 25.

LEMMA 4.3.3. *Two elementarily equivalent, countable and ω-saturated structures are isomorphic.*

PROOF. Suppose $\mathfrak{A}$ and $\mathfrak{B}$ are as in the lemma. We choose enumerations $A = \{a_0, a_1, \ldots\}$ and $B = \{b_0, b_1, \ldots\}$. Then we construct an ascending sequence $f_0 \subseteq f_1 \subseteq \cdots$ of finite elementary maps

$$f_i : A_i \to B_i$$

between finite subsets of $\mathfrak{A}$ and $\mathfrak{B}$. We will choose the f_i in such a way that A is the union of the A_i and B the union of the B_i. The union of the f_i is then the desired isomorphism between $\mathfrak{A}$ and $\mathfrak{B}$.

The empty map $f_0 = \emptyset$ is elementary since $\mathfrak{A}$ and $\mathfrak{B}$ are elementarily equivalent. Assume that f_i is already constructed. There are two cases:

[5] "restriction of variables".

$i = 2n$: We will extend f_i to $A_{i+1} = A_i \cup \{a_n\}$.Consider the type

$$p(x) = \mathrm{tp}(a_n/A_i).$$

Since f_i is elementary, $f_i(p)(x)$ is in $\mathfrak{B}$ a type over B_i. Since $\mathfrak{B}$ is ω-saturated, there is a realisation b' of this type. So for $\overline{a} \in A_i$

$$\mathfrak{A} \models \varphi(a_n, \overline{a}) \Rightarrow \mathfrak{B} \models \varphi(b', f_i(\overline{a})).$$

This shows that $f_{i+1}(a_n) = b'$ defines an elementary extension of f_i.

$i = 2n + 1$: We exchange $\mathfrak{A}$ and $\mathfrak{B}$: since $\mathfrak{A}$ is ω-saturated, we find an elementary map f_{i+1} with image $B_{i+1} = B_i \cup \{b_n\}$. ⊣

PROOF OF THEOREM 4.3.1. Assume that there are only finitely many $\varphi(x_1, \ldots, x_n)$ relative to T for every n. By Lemma 4.3.3 it suffices to show that all models of T are ω-saturated. Let $\mathfrak{M}$ be a model of T and A an n-element subset. If there are only N many formulas, up to equivalence, in the variables $x_1, \ldots, x_{n+1}$, there are, up to equivalence in $\mathfrak{M}$, at most N many $L(A)$-formulas $\varphi(x)$. Thus, each type $p(x) \in \mathrm{S}(A)$ is isolated (with respect to $\mathrm{Th}(\mathfrak{M}_A)$) by a "smallest" formula $\varphi_p(x)$. Each element of M which realises $\varphi_p(x)$ also realises $p(x)$, so $\mathfrak{M}$ is ω-saturated.

Conversely, if there are infinitely many $\varphi(x_1, \ldots, x_n)$ modulo T for some n, then – as the type space $S_n(T)$ is compact – there must be some non-isolated type p. By the Omitting Types Theorem (4.1.2) there is a countable model of T in which this type is not realised. On the other hand, there also exists a countable model of T realizing this type. So T is not $\aleph_0$-categorical. ⊣

REMARK 4.3.4. The proof shows that a countable complete theory with infinite models is $\aleph_0$-categorical if and only if all countable models are ω-saturated.

In Theorem 5.2.11 this characterisation will be extended to theories categorical in uncountable cardinalities.

REMARK 4.3.5. The proof of Lemma 4.3.3 also shows that ω-saturated models are ω-homogeneous in the following sense.

DEFINITION 4.3.6. An L-structure $\mathfrak{M}$ is *ω-homogeneous* if for every elementary map f_0 defined on a finite subset A of M and for any $a \in M$ there is some $b \in M$ such that

$$f = f_0 \cup \{\langle a, b \rangle\}$$

is elementary.

Note that $f = f_0 \cup \{\langle a, b \rangle\}$ is elementary if and only if b realises $f_0(\mathrm{tp}(a/A))$.

COROLLARY 4.3.7. *Let $\mathfrak{A}$ be a structure and $a_1, \ldots, a_n$ elements of $\mathfrak{A}$. Then* $\mathrm{Th}(\mathfrak{A})$ *is $\aleph_0$-categorical if and only if* $\mathrm{Th}(\mathfrak{A}, a_1, \ldots, a_n)$ *is $\aleph_0$-categorical.* ⊣

EXAMPLES. The following theories are $\aleph_0$-categorical:

- Infset, the theory of infinite sets.
- For every finite field $\mathbb{F}_q$, the theory of infinite $\mathbb{F}_q$-vector spaces. (Indeed, this theory is categorical in all infinite cardinals. This follows directly from the fact that vector spaces over the same field and of the same dimension are isomorphic.)
- The theory RG of the random graph (see Exercise 3.3.1): this follows from Theorem 4.3.1 since RG has quantifier elimination and for any n there are only finitely many graphs on n elements.
- The theory DLO of dense linear orders without endpoints. This follows from Theorem 4.3.1 since DLO has quantifier elimination: for every n there are only finitely many (say N_n) ways to order n (not necessarily distinct) elements. For $n = 2$ for example there are the three possibilities $a_1 < a_2$, $a_1 = a_2$ and $a_2 < a_1$. Each of these possibilities corresponds to a complete formula $\psi(x_1, \ldots, x_n)$. Hence there are, up to equivalence, exactly 2^{N_n} many formulas $\varphi(x_1, \ldots, x_n)$.

Next we study the existence of countable ω-saturated structures.

DEFINITION 4.3.8. A theory T is *small* if $\mathrm{S}_n(T)$ are at most countable for all $n < \omega$.

A countable theory with at most countably many non-isomorphic at most countable models is always small. The converse is not true.

LEMMA 4.3.9. *A countable[6] complete theory is small if and only if it has a countable ω-saturated model.*

PROOF. If T has a finite model $\mathfrak{A}$, T is small and $\mathfrak{A}$ is ω-saturated. So we may assume that T has infinite models.

If all types can be realised in a single countable model, there can be at most countably many types.

If conversely all $\mathrm{S}_{n+1}(T)$ are at most countable, then over any n-element subset of a model of T there are at most countably many types. We construct an elementary chain

$$\mathfrak{A}_0 \prec \mathfrak{A}_1 \prec \cdots$$

of models of T. For $\mathfrak{A}_0$ we take any countable model. If $\mathfrak{A}_i$ is already constructed, we use Corollary 2.2.9 and Theorem 2.3.1.1 to construct a countable model $\mathfrak{A}_{i+1}$ in such a way that all types over finite subsets of A_i are realised in $\mathfrak{A}_{i+1}$. This can be done since there are only countable many such types. The union $\mathfrak{A} = \bigcup_{i\in\omega} \mathfrak{A}_i$ is countable and ω-saturated since every type over a finite subset B of $\mathfrak{A}$ is realised in $\mathfrak{A}_{i+1}$ if $B \subseteq A_i$. $\dashv$

[6] The statement is true even for uncountable L.

THEOREM 4.3.10 (Vaught). *A countable complete theory cannot have exactly two countable models.*

PROOF. We can assume that T is small and not $\aleph_0$-categorical. We will show that T has at least three non-isomorphic countable models. First, T has an ω-saturated countable model $\mathfrak{A}$ and there is a non-isolated type $p(\overline{x})$, which can be omitted in a countable model $\mathfrak{B}$. Let $p(\overline{x})$ be realised in $\mathfrak{A}$ by $\overline{a}$. Since $\mathrm{Th}(\mathfrak{A}, \overline{a})$ is not $\aleph_0$-categorical, $\mathrm{Th}(\mathfrak{A}, \overline{a})$ has a countable model $(\mathfrak{C}, \overline{c})$ which is not ω-saturated. Then $\mathfrak{C}$ is not ω-saturated and therefore not isomorphic to $\mathfrak{A}$. But $\mathfrak{C}$ realises $p(\overline{x})$ and is therefore not isomorphic to $\mathfrak{B}$. ⊣

Exercise 4.3.5 shows that for any $n \neq 2, n \leq \omega$, there is a countable complete theory with exactly n countable models. *Vaught's Conjecture* states that if a complete countable theory has fewer than continuum many countable non-isomorphic models, the number of countable models is at most countable (see [49] for a survey on what is currently known).

EXERCISE 4.3.1. 1. If T is $\aleph_0$-categorical, then in any model $\mathfrak{M}$ the algebraic closure of a finite set is finite (see Definition p. 79). In particular, $\mathfrak{M}$ is *locally finite*, i.e., any substructure generated by a finite subset is finite. (In many-sorted structures we mean that in each sort the trace of the algebraic closure is finite.)
2. There is no $\aleph_0$-categorical theory of fields, i.e., if T is a complete L_{Ring}-theory containing Field, then T is not $\aleph_0$-categorical.

EXERCISE 4.3.2. A theory T is small exactly if T has at most countably many completions, each of which is small.

EXERCISE 4.3.3. Show that T is $\aleph_0$-categorical if and only if $S_n(T)$ is finite for all n.

EXERCISE 4.3.4. Write down a theory with exactly two countable models.

EXERCISE 4.3.5. Show that for every $n > 2$ there is a countable complete theory with exactly n countable models. (Consider $(\mathbb{Q}, <, P_0, \dots, P_{n-2}, c_0, c_1, \dots)$, where the P_i form a partition of $\mathbb{Q}$ into dense subsets and the c_i are an increasing sequence of elements of P_0.)

EXERCISE 4.3.6. Give an example of an uncountable complete theory with exactly one countable model which does not satisfy the conclusion of Theorem 4.3.1.

EXERCISE 4.3.7. Suppose $\mathfrak{M}$ is countable and $\aleph_0$-categorical. Show that if $X \subseteq M^n$ is invariant under all automorphisms of $\mathfrak{M}$, then X is definable.

EXERCISE 4.3.8. Let $\mathfrak{M}$ be a structure and assume that for some n only finitely many n-types are realised in $\mathfrak{M}$. Then any structure elementarily equivalent to $\mathfrak{M}$ satisfies exactly the same n-types.

EXERCISE 4.3.9. If $\mathfrak{A}$ is ω-saturated, all n-types over finite sets are realised. More generally prove the following: If $\mathfrak{A}$ is *κ-saturated* i.e., if all 1-types over sets of cardinality less than κ are realised in $\mathfrak{A}$, then the same is true for all n-types. See also Exercise 6.1.6.

EXERCISE 4.3.10. Show:

1. The theory of $(\mathbb{R}, 0, +)$ has exactly two 1-types but $\aleph_0$ many 2-types.
2. The theory of $(\mathbb{R}, 0, +, <)$ has exactly three 1-types but $2^{\aleph_0}$ many 2-types.

EXERCISE 4.3.11. Show that all models of an $\aleph_0$-categorical theory are partially isomorphic.

EXERCISE 4.3.12. Show that two countable partially isomorphic structures are isomorphic.

EXERCISE 4.3.13. Let $\mathfrak{A}$ be ω-saturated. Show that $\mathfrak{B}$ is partially isomorphic to $\mathfrak{A}$ if and only if $\mathfrak{B}$ is ω-saturated and elementarily equivalent to $\mathfrak{A}$.

4.4. The amalgamation method

In this section we will present one of the main methods for constructing new and interesting examples of first order structures. It goes back to Fraïssé, but has more recently been modified by Hrushovski [28]. We here focus mainly on the $\aleph_0$-categorical examples and return to the fancier version in Section 10.4.

DEFINITION 4.4.1. For any language L, the *skeleton*[7] $\mathcal{K}$ of an L-structure $\mathfrak{M}$ is the class of all finitely-generated L-structures which are isomorphic to a substructure of $\mathfrak{M}$. We say that an L-structure $\mathfrak{M}$ is *$\mathcal{K}$-saturated* if its skeleton is $\mathcal{K}$ and if for all $\mathfrak{A}$, $\mathfrak{B}$ in $\mathcal{K}$ and all embeddings $f_0 : \mathfrak{A} \to \mathfrak{M}$ and $f_1 : \mathfrak{A} \to \mathfrak{B}$ there is an embedding $g_1 : \mathfrak{B} \to \mathfrak{M}$ with $f_0 = g_1 \circ f_1$.

THEOREM 4.4.2. *Let L be a countable language. Any two countable $\mathcal{K}$-saturated structures are isomorphic.*

PROOF. Let $\mathfrak{M}$ and $\mathfrak{N}$ be countable L-structures with the same skeleton $\mathcal{K}$, and assume that $\mathfrak{M}$ and $\mathfrak{N}$ are $\mathcal{K}$-saturated. As in the proof of Lemma 4.3.3 we construct an isomorphism between $\mathfrak{M}$ and $\mathfrak{N}$ as the union of an ascending sequence of isomorphisms between finitely-generated substructures of M and N. This can be done because if $f_1 : \mathfrak{A} \to \mathfrak{N}$ is an embedding of a finitely-generated substructure $\mathfrak{A}$ of $\mathfrak{M}$ into $\mathfrak{N}$, and a is an element of $\mathfrak{M}$, then by $\mathcal{K}$-saturation f_1 can be extended to an embedding $g_1 : \mathfrak{A}' \to \mathfrak{N}$ where $\mathfrak{A}' = \langle Aa \rangle^{\mathfrak{M}}$. Now interchange the roles of $\mathfrak{M}$ and $\mathfrak{N}$. ⊣

REMARK 4.4.3. The proof shows that any countable $\mathcal{K}$-saturated structure $\mathfrak{M}$ is *ultrahomogeneous* i.e., any isomorphism between finitely generated substructures extends to an automorphism of $\mathfrak{M}$.

---

[7] This is also called the *age* of $\mathfrak{M}$.

THEOREM 4.4.4. *Let L be a countable language and $\mathcal{K}$ a countable class of finitely-generated L-structures. There is a countable $\mathcal{K}$-saturated L-structure $\mathfrak{M}$ if and only if*

a) (Heredity) *If $\mathfrak{A}_0$ belongs to $\mathcal{K}$, then all elements of the skeleton of $\mathfrak{A}_0$ also belong to $\mathcal{K}$.*
b) (Joint Embedding) *For $\mathfrak{B}_0, \mathfrak{B}_1 \in \mathcal{K}$ there are some $\mathfrak{D} \in \mathcal{K}$ and embeddings $g_i : \mathfrak{B}_i \to \mathfrak{D}$.*
c) (Amalgamation) *If $\mathfrak{A}, \mathfrak{B}_0, \mathfrak{B}_1 \in \mathcal{K}$ and $f_i : \mathfrak{A} \to \mathfrak{B}_i$, $(i = 0, 1)$ are embeddings, there is some $\mathfrak{D} \in \mathcal{K}$ and two embeddings $g_i : \mathfrak{B}_i \to \mathfrak{D}$ such that $g_0 \circ f_0 = g_1 \circ f_1$.*

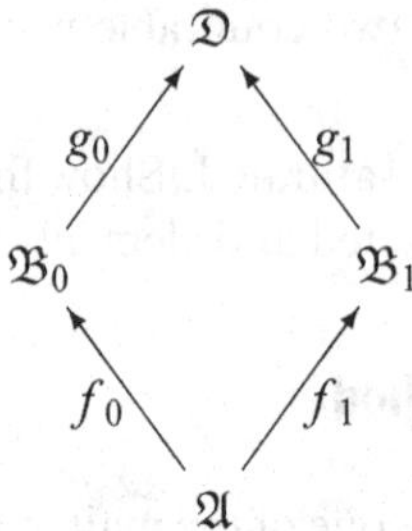

In this case, $\mathfrak{M}$ is unique up to isomorphism and is called the Fraïssé limit of $\mathcal{K}$.

PROOF. Let $\mathcal{K}$ be the skeleton of a countable $\mathcal{K}$-saturated structure $\mathfrak{M}$. Clearly, $\mathcal{K}$ has the Hereditary Property. To see that $\mathcal{K}$ has the Amalgamation Property let $\mathfrak{A}$, $\mathfrak{B}_0$, $\mathfrak{B}_1$, f_0 and f_1 be as in c). We may assume that $\mathfrak{B}_0 \subseteq \mathfrak{M}$ and f_0 is the inclusion map. Furthermore we can assume $\mathfrak{A} \subseteq \mathfrak{B}_1$ and that f_1 is the inclusion map. Now the embedding $g_1 : \mathfrak{B}_1 \to \mathfrak{M}$ is the extension of the isomorphism $f_0 : \mathfrak{A} \to f_0(\mathfrak{A})$ to $\mathfrak{B}_1$ and satisfies $f_0 = g_1 \circ f_1$. For $\mathfrak{D}$ we choose a finitely-generated substructure of $\mathfrak{M}$ which contains $\mathfrak{B}_0$ and the image of g_1. For $g_0 : \mathfrak{B}_0 \to \mathfrak{D}$ take the inclusion map. The Joint Embedding Property is proved similarly.

For the converse assume that $\mathcal{K}$ has properties a), b), and c). Choose an enumeration $(\mathfrak{B}_i)_{i \in \omega}$ of all isomorphism types in $\mathcal{K}$. We construct $\mathfrak{M}$ as the union of an ascending chain

$$\mathfrak{M}_0 \subseteq \mathfrak{M}_1 \subseteq \cdots \subseteq \mathfrak{M}$$

of elements of $\mathcal{K}$. Suppose that $\mathfrak{M}_i$ is already constructed. If $i = 2n$ is even, we choose $\mathfrak{M}_{i+1}$ as the top of a diagram

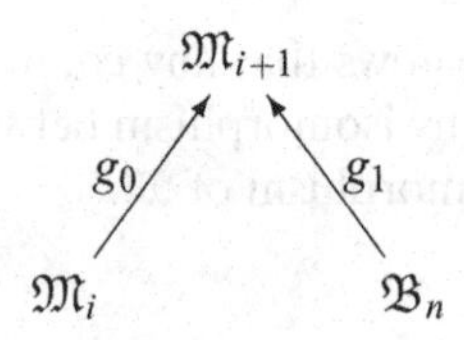

where we can assume that g_0 is the inclusion map. If $i = 2n + 1$ is odd, let $\mathfrak{A}$ and $\mathfrak{B}$ from $\mathcal{K}$ and two embeddings $f_0 : \mathfrak{A} \to \mathfrak{M}_i$ and $f_1 : \mathfrak{A} \to \mathfrak{B}$ be given. We construct $\mathfrak{M}_{i+1}$ using the diagram

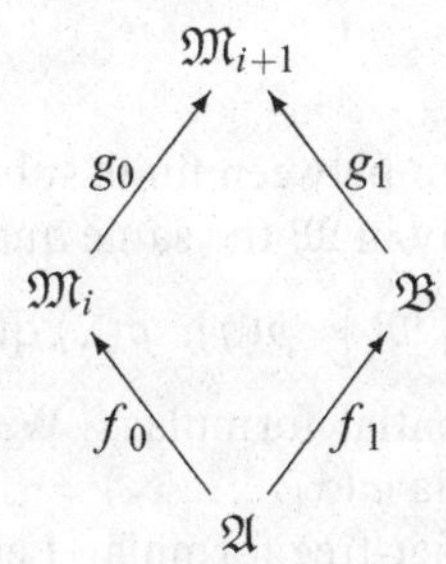

To ensure that $\mathfrak{M}$ is $\mathcal{K}$-saturated we have in the odd steps to make the right choice of $\mathfrak{A}$, $\mathfrak{B}$, f_0 and f_1. Assume that we have $\mathfrak{A}, \mathfrak{B} \in \mathcal{K}$ and embeddings $f_0 : \mathfrak{A} \to \mathfrak{M}$ and $f_1 : \mathfrak{A} \to \mathfrak{B}$. For large j the image of f_0 will be contained in $\mathfrak{M}_j$. During the construction of the the $\mathfrak{M}_i$, in order to guarantee the $\mathcal{K}$-saturation of $\mathfrak{M}$, we have to ensure that eventually, for some odd $i \geq j$, the embeddings $f_0 : \mathfrak{A} \to \mathfrak{M}_i$ and $f_1 : \mathfrak{A} \to \mathfrak{B}$ were used in the construction of $\mathfrak{M}_{i+1}$. This can be done since for each j there are – up to isomorphism – at most countably many possibilities. Thus there exists an embedding $g_1 : \mathfrak{B} \to \mathfrak{M}_{i+1}$ with $f_0 = g_1 \circ f_1$.

Clearly, $\mathcal{K}$ is the skeleton of $\mathfrak{M}$: the finitely-generated substructures of $\mathfrak{M}$ are the substructures of the $\mathfrak{M}_i$. Since the $\mathfrak{M}_i$ belong to $\mathcal{K}$, their finitely-generated substructures also belong to $\mathcal{K}$. On the other hand each $\mathfrak{B}_n$ is isomorphic to a substructure of $\mathfrak{M}_{2n+1}$.

Uniqueness follows from Theorem 4.4.2 $\dashv$

For finite relational languages L, any non-empty finite subset is itself a (finitely-generated) substructure. For such languages, the construction yields $\aleph_0$-categorical structures. We now take a closer look at $\aleph_0$-categorical theories with quantifier elimination in a *finite relational language*.

REMARK 4.4.5. A complete theory T in a finite relational language with quantifier elimination is $\aleph_0$-categorical. So all its models are ω-homogeneous by Remarks 4.3.4 and 4.3.5.

PROOF. For every n, there is only a finite number of non-equivalent quantifier-free formulas $\rho(x_1, \ldots, x_n)$. If T has quantifier elimination, this number is also the number of all formulas $\varphi(x_1, \ldots, x_n)$ modulo T and so T is $\aleph_0$-categorical by Theorem 4.3.1. $\dashv$

Clearly, if a theory has quantifier elimination, any isomorphism between substructures is elementary. For relational languages we can say more.

LEMMA 4.4.6. *Let T be a complete theory in a finite relational language and $\mathfrak{M}$ an infinite model of T. The following are equivalent:*

a) *T has quantifier elimination.*
b) *Any isomorphism between finite substructures is elementary.*
c) *The domain of any isomorphism between finite substructures can be extended to any further element.*

PROOF. a) $\Rightarrow$ b) is clear.

b) $\Rightarrow$ a): If any isomorphism between finite substructures of $\mathfrak{M}$ is elementary, all n-tuples $\overline{a}$ which satisfy in $\mathfrak{M}$ the same quantifier-free type

$$\mathrm{tp}_{\mathrm{qf}}(\overline{a}) = \{\rho(\overline{x}) \mid \mathfrak{M} \models \rho(\overline{a}),\ \rho(\overline{x}) \text{ quantifier-free}\}$$

satisfy the same simple existential formulas. We will show from this that every simple existential formula $\varphi(x_1, \ldots, x_n) = \exists y\, \rho(x_1, \ldots, x_n, y)$ is, modulo T, equivalent to a quantifier-free formula. Let $r_1(\overline{x}), \ldots, r_{k-1}(\overline{x})$ be the quantifier-free types of all n-tuples in $\mathfrak{M}$ which satisfy $\varphi(\overline{x})$. Let $\rho_i(\overline{x})$ be equivalent to the conjunction of all formulas from $r_i(\overline{x})$. Then

$$T \vdash \forall \overline{x}\ (\varphi(\overline{x}) \leftrightarrow \bigvee_{i<k} \rho_i(\overline{x})).$$

a) $\Rightarrow$ c): The theory T is $\aleph_0$-categorical and hence all models are ω-homogeneous. Since any isomorphism between finite substructures is elementary by the equivalence of a) and b) the claim follows.

c) $\Rightarrow$ b): If the domain of any finite isomorphism can be extended to any further element, it is easy to see that every finite isomorphism is elementary. $\dashv$

We have thus established the following.

THEOREM 4.4.7. *Let L be a finite relational language and $\mathcal{K}$ a class of finite L-structures. If the Fraïssé limit of $\mathcal{K}$ exists, its theory is $\aleph_0$-categorical and has quantifier elimination.*

EXAMPLE. The class of finite linear orders obviously has the Amalgamation Property. Their Fraïssé limit is the dense linear order without endpoints.

EXERCISE 4.4.1. Show that two $\mathcal{K}$-saturated structures are partially isomorphic.

EXERCISE 4.4.2. Prove Remark 4.4.3.

EXERCISE 4.4.3. Let $\mathcal{K}$ be the class of finite graphs. Show that its Fraïssé limit is the countable random graph. This yields another proof that the theory of the random graph has quantifier elimination.

4.5. Prime models

Some, but not all, theories have models which are smallest in the sense that they elementarily embed into any other model of the theory. For countable complete theories these are the models realizing only the 'necessary' types. If they exist, they are unique and ω-homogeneous.

In this section – unless explicitly stated otherwise – we let T be a countable complete theory with infinite models.

DEFINITION 4.5.1. Let T be a countable theory with infinite models, not necessarily complete.

1. We call $\mathfrak{A}_0$ a *prime model* of T if $\mathfrak{A}_0$ can be elementarily embedded into all models of T.
2. A structure $\mathfrak{A}$ is called *atomic* if all n-tuples $\overline{a}$ of elements of $\mathfrak{A}$ are atomic. This means that the types $\operatorname{tp}(\overline{a})$ are isolated in $\mathrm{S}_n^{\mathfrak{A}}(\emptyset)$.

Prime models need not exist, see the example on p. 60. By Corollary 4.2.7, a tuple $\overline{a}$ is atomic if and only if it satisfies a complete formula. For the terminology see Lemma 4.2.6.

Since T has countable models, prime models must be countable and since non-isolated types can be omitted in suitable models by Theorem 4.1.2, only isolated types can be realised in prime models. Thus, one direction of the following theorem is clear.

THEOREM 4.5.2. *A model of T is prime if and only if it is countable and atomic.*

PROOF. As just noted, a prime model has to be countable and atomic. For the converse let $\mathfrak{M}_0$ be a countable and atomic model of T and $\mathfrak{M}$ any model of T. We construct an elementary embedding of $\mathfrak{M}_0$ to $\mathfrak{M}$ as a union of an ascending sequence of elementary maps

$$f : A \to B$$

between finite subsets A of M_0 and B of M. We start with the empty map, which is elementary since $\mathfrak{M}_0$ and $\mathfrak{M}$ are elementarily equivalent.

It is enough to show that every f can be extended to any given $A \cup \{a\}$. Let $p(x)$ be the type of a over A and $f(p)$ the image of p under f (see Lemma 4.2.5). We will show that $f(p)$ has a realisation $b \in M$. Then $f \cup \{\langle a, b\rangle\}$ is an elementary extension of f.

Let $\overline{a}$ be a tuple which enumerates the elements of A and $\varphi(x, \overline{x})$ an L-formula which isolates the type of $a\overline{a}$. Then p is isolated by $\varphi(x, \overline{a})$: clearly $\varphi(x, \overline{a}) \in \operatorname{tp}(a/\overline{a})$ and if $\rho(x, \overline{a}) \in \operatorname{tp}(a/\overline{a})$, we have $\rho(x, y) \in \operatorname{tp}(a, \overline{a})$. This implies $\mathfrak{M}_0 \models \forall x, y\, (\varphi(x, y) \to \rho(x, y))$ and $\mathfrak{M} \models \forall x\, (\varphi(x, \overline{a}) \to \rho(x, \overline{a}))$. Thus $f(p)$ is isolated by $\varphi(x, f(\overline{a}))$ and, since $\varphi(x, f(\overline{a}))$ can be realised in $\mathfrak{M}$, so can be $f(p)$. $\dashv$

THEOREM 4.5.3. *All prime models of T are isomorphic.*

PROOF. Let $\mathfrak{M}_0$ and $\mathfrak{M}_0'$ be two prime models. Since prime models are atomic, elementary maps between finite subsets of $\mathfrak{M}_0$ and $\mathfrak{M}_0'$ can be extended to all finite extensions. Since $\mathfrak{M}_0$ and $\mathfrak{M}_0'$ are countable, it follows exactly as in the proof of Lemma 4.3.3 that $\mathfrak{M}_0$ and $\mathfrak{M}_0'$ are isomorphic. ⊣

The previous proof also shows the following.

COROLLARY 4.5.4. *Prime models are ω-homogeneous.*

PROOF. Let $\mathfrak{M}_0$ be prime and $\overline{a}$ any tuple of elements from M_0. By Theorem 4.5.2, $(\mathfrak{M}_0, \overline{a})$ is a prime model of its theory. The claim follows now from Theorem 4.5.3. ⊣

DEFINITION 4.5.5. The isolated types are *dense* in T if every consistent L-formula $\psi(x_1, \ldots, x_n)$ belongs to an isolated type $p(x_1, \ldots, x_n) \in S_n(T)$.

REMARK 4.5.6. By Corollary 4.2.7 this definition is equivalent to asking that every consistent L-formula $\psi(x_1, \ldots, x_n)$ contains a complete formula $\varphi(x_1, \ldots, x_n)$:

$$T \vdash \forall \overline{x} \, (\varphi(\overline{x}) \to \psi(\overline{x})).$$

THEOREM 4.5.7. *T has a prime model if and only if the isolated types are dense.*

PROOF. Suppose T has a prime model $\mathfrak{M}$ (so $\mathfrak{M}$ is atomic by Theorem 4.5.2). Since consistent formulas $\psi(\overline{x})$ are realised in all models of T, $\psi(\overline{x})$ is realised by an atomic tuple $\overline{a}$ and $\psi(\overline{x})$ belongs to the isolated type $\text{tp}(\overline{a})$.

For the other direction notice that a structure $\mathfrak{M}_0$ is atomic if and only if for all n the set

$$\Sigma_n(x_1, \ldots, x_n) = \{\neg\varphi(x_1, \ldots, x_n) \mid \varphi(x_1, \ldots, x_n) \text{ complete}\}$$

is not realised in $\mathfrak{M}_0$. Hence, by Corollary 4.1.3, it is enough to show that the Σ_n are not isolated in T. This is the case if and only if for every consistent $\psi(x_1, \ldots, x_n)$ there is a complete formula $\varphi(x_1, \ldots, x_n)$ with $T \not\vdash \forall \overline{x} \, (\psi(\overline{x}) \to \neg\varphi(\overline{x}))$. Since $\varphi(\overline{x})$ is complete, this is equivalent to $T \vdash \forall \overline{x} \, (\varphi(\overline{x}) \to \psi(\overline{x}))$. We conclude that Σ_n is not isolated if and only if the isolated n-types are dense. ⊣

Notice that the last part shows in fact the equivalence directly. (Because if Σ_n is isolated for some n, then it is realised in every model and no atomic model can exist.)

EXAMPLE. Let L be the language having a unary predicate P_s for every finite 0–1-sequence $s \in {}^{<\omega}2$. The axioms of Tree say that the P_s, $s \in {}^{<\omega}2$, form a binary decomposition of the universe:

- $\forall x \, P_\emptyset(x)$
- $\exists x \, P_s(x)$
- $\forall x \;\; ((P_{s0}(x) \vee P_{s1}(x)) \leftrightarrow P_s(x))$
- $\forall x \;\; \neg(P_{s0}(x) \wedge P_{s1}(x))$.

Tree is complete and has quantifier elimination. There are no complete formulas and no prime model.

DEFINITION 4.5.8. A family of formulas $\varphi_s(\overline{x})$, $s \in {}^{<\omega}2$, is a *binary tree* if for all $s \in {}^{<\omega}2$ the following holds:

a) $T \vdash \forall \overline{x} \; ((\varphi_{s0}(\overline{x}) \vee \varphi_{s1}(\overline{x})) \rightarrow \varphi_s(\overline{x}))$
b) $T \vdash \forall \overline{x} \; \neg(\varphi_{s0}(\overline{x}) \wedge \varphi_{s1}(\overline{x}))$.

THEOREM 4.5.9. *Let T be a complete theory.*

1. *If T is small, it has no binary tree of consistent L-formulas. If T is countable, the converse holds as well.*
2. *If T has no binary tree of consistent L-formulas, the isolated types are dense.*

PROOF. 1. Let $(\varphi_s(x_1, \ldots, x_n))$ be a binary tree of consistent formulas. Then, for all $\eta \in {}^{\omega}2$, the set

$$\{\varphi_s(\overline{x}) \mid s \subseteq \eta\}$$

is consistent and therefore is contained in some type $p_\eta(\overline{x}) \in \mathrm{S}_n(T)$. The $p_\eta(\overline{x})$ are all different showing that T is not small. We leave the converse as Exercise 4.5.1.

2. If the isolated types are not dense, there is a consistent $\varphi(x_1, \ldots, x_n)$ which does not contain a complete formula. Call such a formula *perfect*. Since perfect formulas are not complete, they can be decomposed into disjoint[8] consistent formulas, which again have to be perfect. This allows us to construct a binary tree of perfect formulas. ⊣

EXERCISE 4.5.1. Countable theories without a binary tree of consistent formulas are small.

EXERCISE 4.5.2. Show that isolated types being dense is equivalent to isolated types being (topologically) dense in the Stone space $S_n(T)$.

EXERCISE 4.5.3. Let T be the theory of $(\mathbb{R}, <, Q)$ where Q is a predicate for the rational numbers. Does T have a prime model?

[8] We call two formulas *disjoint* if their conjunction is not consistent with T.

Definition 4.5.3 A family of formulas [illegible] is a binary tree [illegible] of formulas if the following holds:

[illegible]

Theorem 4.5.4 Let T be a countable theory.

[illegible] a binary tree of consistent [illegible] formulas [illegible] containing the constants [illegible]

[illegible] are [illegible] isolated types are [illegible]

Proof [illegible] be a binary tree of consistent formulas. [illegible] Then [illegible] the set

[illegible]

[illegible] therefore is contained in some type [illegible] are all different showing that [illegible] is not small. We leave the [illegible] Exercise 4.5.[illegible]

[illegible] isolated types [illegible] there is a [illegible] complete formula [illegible] such a formula [illegible] be described [illegible] which [illegible]

[illegible] formulas [illegible]

[illegible] Exercise 4.5.[illegible] Construct examples [illegible]

[illegible]

Exercise 4.5.[illegible] Show that isolated types [illegible] the whole space [illegible]

Exercise 4.5.[illegible] Let T be the theory of [illegible] the language [illegible]. Does T have a prime model?

[illegible] We call [illegible] formulas [illegible] if [illegible] is not consistent with [illegible]

Chapter 5

$\aleph_1$-CATEGORICAL THEORIES

We have already seen examples of $\aleph_0$-categorical theories (e.g., the theory of dense linear orderings without endpoints) and of theories categorical in all infinite κ (e.g., the theory of infinite dimensional vector spaces over finite fields) and all uncountably infinite κ (e.g., the theory of algebraically closed fields of fixed characteristic).

The aim of this chapter is to understand the structure of $\aleph_1$-categorical theories and to prove, in Corollary 5.8.2, Morley's theorem that a countable theory categorical in some uncountable cardinality is categorical in all uncountable cardinalities (but not necessarily countably categorical).

As in the case of $\aleph_0$-categorical theories, we will see that the number of complete types in an $\aleph_1$-categorical theory is rather small (the theory is ω-stable) albeit not always finite. We will define a geometry associated to a strongly minimal set whose dimension determines the isomorphism type of a model of such a theory. This then implies Morley's theorem.

5.1. Indiscernibles

In this section we begin with a few facts about 'indiscernible' elements. We will see that structures generated by them realise only few types.

Definition 5.1.1. Let I be a linear order and $\mathfrak{A}$ an L-structure. A family $(a_i)_{i \in I}$ of elements[1] of A is called a *sequence of indiscernibles* if for all L-formulas $\varphi(x_1, \ldots, x_n)$ and all $i_1 < \cdots < i_n$ and $j_1 < \cdots < j_n$ from I

$$\mathfrak{A} \models \varphi(a_{i_1}, \ldots, a_{i_n}) \leftrightarrow \varphi(a_{j_1}, \ldots, a_{j_n}).$$

If two of the a_i are equal, all a_i are the same. Therefore it is often assumed that the a_i are distinct.

Sometimes sequences of indiscernibles are also called *order indiscernible* to distinguish them from *totally indiscernible* sequences in which the ordering of the index set does not matter. However, in stable theories (see Section 5.2 and

[1] or, more generally, of tuples of elements, all of the same length.

Chapter 8), these notions coincide. So if nothing else is said, indiscernible elements will always be order indiscernible in the sense just defined.

DEFINITION 5.1.2. Let I be an infinite linear order and $\mathcal{I} = (a_i)_{i \in I}$ a sequence of k-tuples in $\mathfrak{M}$, $A \subseteq M$. The *Ehrenfeucht–Mostowski type* $\mathrm{EM}(\mathcal{I}/A)$ of $\mathcal{I}$ over A is the set of $L(A)$-formulas $\varphi(x_1, \dots, x_n)$ with $\mathfrak{M} \models \varphi(a_{i_1}, \dots, a_{i_n})$ for all $i_1 < \dots < i_n \in I$, $n < \omega$.

LEMMA 5.1.3 (The Standard Lemma). *Let I and J be two infinite linear orders and $\mathcal{I} = (a_i)_{i \in I}$ a sequence of elements of a structure $\mathfrak{M}$. Then there is a structure $\mathfrak{N} \equiv \mathfrak{M}$ with an indiscernible sequence $(b_j)_{j \in J}$ realizing the Ehrenfeucht–Mostowski type* $\mathrm{EM}(\mathcal{I})$ *of* $\mathcal{I}$.

COROLLARY 5.1.4. *Assume that T has an infinite model. Then, for any linear order I, T has a model with a sequence $(a_i)_{i \in I}$ of distinct indiscernibles.* ⊣

For the proof of the Standard Lemma we need Ramsey's Theorem. Let $[A]^n$ denote the set of all n-element subsets of A.

THEOREM 5.1.5 (Ramsey). *Let A be infinite and $n \in \omega$. Partition the set of n-element subsets $[A]^n$ into subsets $C_1, \dots, C_k$. Then there is an infinite subset of A whose n-element subsets all belong to the same subset C_i.*

PROOF. Thinking of the partition as a colouring on $[A]^n$, we are looking for an infinite subset B of A such that $[B]^n$ is monochromatic. We prove the theorem by induction on n. For $n = 1$, the statement is evident from the pigeonhole principle. Assuming the theorem is true for n, we now prove it for $n + 1$. Let $a_0 \in A$. Then any colouring of $[A]^{n+1}$ induces a colouring of the n-element subsets of $A' = A \setminus \{a_0\}$: just colour $x \in [A']^n$ by the colour of $\{a_0\} \cup x \in [A]^{n+1}$. By the induction hypothesis, there exists an infinite monochromatic subset B_1 of A' in the induced colouring. Thus, all the $(n + 1)$-element subsets of A consisting of a_0 and n elements of B_1 have the same colour. Now pick any $a_1 \in B_1$. By the same argument we obtain an infinite subset B_2 of B_1 with the same properties. Inductively, we thus construct an infinite sequence $A = B_0 \supset B_1 \supset B_2 \supset \cdots$, and elements $a_i \in B_i \setminus B_{i+1}$ such that the colour of each $(n+1)$-element subset $\{a_{i(0)}, a_{i(1)}, \dots, a_{i(n)}\}$ with $i(0) < i(1) < \cdots < i(n)$ depends only on the value of $i(0)$. Again by the pigeonhole principle there are infinitely many values of $i(0)$ for which this colour will be the same. These $a_{i(0)}$ then yield the desired monochromatic set. ⊣

PROOF OF LEMMA 5.1.3. Choose a set C of new constants with an ordering isomorphic to J. Consider the theories

$$T' = \{\varphi(\overline{c}) \mid \varphi(\overline{x}) \in \mathrm{EM}(\mathcal{I})\} \text{ and}$$
$$T'' = \{\varphi(\overline{c}) \leftrightarrow \varphi(\overline{d}) \mid \overline{c}, \overline{d} \in C\}.$$

Here the $\varphi(\overline{x})$ are L-formulas and $\overline{c}, \overline{d}$ tuples in increasing order. We have to show that $T \cup T' \cup T''$ is consistent. It is enough to show that

$$T_{C_0,\Delta} = T \cup \{\varphi(\overline{c}) \in T' \mid \overline{c} \in C_0\} \cup \{\varphi(\overline{c}) \leftrightarrow \varphi(\overline{d}) \mid \varphi(\overline{x}) \in \Delta,\ \overline{c}, \overline{d} \in C_0\}$$

is consistent for finite sets C_0 and Δ. We can assume that the elements of Δ are formulas with free variables $x_1, \ldots, x_n$ and that all tuples $\overline{c}$ and $\overline{d}$ have the same length n.

For notational simplicity we assume that all a_i are different. So we may consider $A = \{a_i \mid i \in I\}$ as an ordered set. We define an equivalence relation on $[A]^n$ by

$$\overline{a} \sim \overline{b} \ \Leftrightarrow\ \mathfrak{M} \models \varphi(\overline{a}) \leftrightarrow \varphi(\overline{b}) \text{ for all } \varphi(x_1, \ldots, x_n) \in \Delta$$

where $\overline{a}, \overline{b}$ are tuples in increasing order. Since this equivalence relation has at most $2^{|\Delta|}$ many classes, by Ramsey's Theorem there is an infinite subset B of A with all n-element subsets in the same equivalence class. We interpret the constants $c \in C_0$ by elements b_c in B ordered in the same way as the c. Then $(\mathfrak{M}, b_c)_{c \in C_0}$ is a model of $T_{C_0,\Delta}$. ⊣

LEMMA 5.1.6. *Assume L is countable. If the L-structure $\mathfrak{M}$ is generated by a well-ordered sequence (a_i) of indiscernibles, then $\mathfrak{M}$ realises only countably many types over every countable subset of M.*

PROOF. If $A = \{a_i \mid i \in I\}$, then every element b of M has the form $b = t(\overline{a})$, where t is an L-term and $\overline{a}$ is a tuple from A.

Consider a countable subset S of M. Write

$$S = \{t_n^{\mathfrak{M}}(\overline{a}^n) \mid n \in \omega\}.$$

Let $A_0 = \{a_i \mid i \in I_0\}$ be the (countable) set of elements of A which occur in the $\overline{a}^n$. Then every type $\mathrm{tp}(b/S)$ is determined by $\mathrm{tp}(b/A_0)$ since every $L(S)$-formula

$$\varphi(x, t_{n_1}^{\mathfrak{M}}(a^{n_1}), \ldots)$$

can be replaced by the $L(A_0)$-formula $\varphi(x, t_{n_1}(a^{n_1}), \ldots)$.

Now the type of $b = t(\overline{a})$ over A_0 depends only on $t(\overline{x})$ (countably many possibilities) and the type $\mathrm{tp}(\overline{a}/A_0)$. Write $\overline{a} = a_{\overline{i}}$ for a tuple $\overline{i}$ from I. Since the a_i are indiscernible, the type depends only on the quantifier-free type $\mathrm{tp}_{\mathrm{qf}}(\overline{i}/I_0)$ in the structure $(I, <)$. This type again depends on $\mathrm{tp}_{\mathrm{qf}}(\overline{i})$ (finitely many possibilities) and on the types $p(x) = \mathrm{tp}_{\mathrm{qf}}(i/I_0)$ of the elements i of $\overline{i}$. There are three kinds of such types:

1. i is bigger than all elements of I_0.
2. i is an element i_0 of I_0.
3. For some $i_0 \in I_0$, i is smaller than i_0 but bigger than all elements of $\{j \in I_0 \mid j < i_0\}$.

There is only one type in the first case, in the other cases the type is determined by i_0. This results in countably many possibilities for each component of $\overline{i}$. ⊣

DEFINITION 5.1.7. Let L be a language. A *Skolem theory* $\mathsf{Skolem}(L)$ is a theory in a bigger language L_{Skolem} with the following properties:

a) $\mathsf{Skolem}(L)$ has quantifier elimination.
b) $\mathsf{Skolem}(L)$ is universal.
c) Every L-structure can be expanded to a model of $\mathsf{Skolem}(L)$.
d) $|L_{\mathsf{Skolem}}| \leq \max(|L|, \aleph_0)$.

THEOREM 5.1.8. *Every language L has a Skolem theory.*

PROOF. We define an ascending sequence of languages

$$L = L_0 \subseteq L_1 \subseteq L_2 \subseteq \cdots,$$

by introducing for every quantifier-free L_i-formula $\varphi(x_1, \ldots, x_n, y)$ a new n-place *Skolem function*[2] f_φ and defining L_{i+1} as the union of L_i and the set of these function symbols. The language L_{Skolem} is the union of all L_i. We now define the Skolem theory as

$$\mathsf{Skolem} = \left\{ \forall \bar{x}\ (\exists y\, \varphi(\bar{x}, y) \to \varphi(\bar{x}, f_\varphi(\bar{x}))) \;\middle|\; \varphi(\bar{x}, y) \text{ q.f. } L_{\mathsf{Skolem}}\text{-formula} \right\}.$$

$\dashv$

COROLLARY 5.1.9. *Let T be a countable theory with an infinite model and let κ be an infinite cardinal. Then T has a model of cardinality κ which realises only countably many types over every countable subset.*

PROOF. Consider the theory $T^* = T \cup \mathsf{Skolem}(L)$. Then T^* is countable, has an infinite model and quantifier elimination.

CLAIM. *T^* is equivalent to a universal theory.*

PROOF OF CLAIM. Modulo $\mathsf{Skolem}(L)$ every axiom φ of T is equivalent to a quantifier-free L_{Skolem}-sentence φ^*. Therefore T^* is equivalent to the universal theory $\{\varphi^* \mid \varphi \in T\} \cup \mathsf{Skolem}(L)$. $\dashv$

Let I be a well-ordering of cardinality κ and $\mathfrak{N}^*$ a model of T^* with indiscernibles $(a_i)_{i \in I}$. The claim implies that the substructure $\mathfrak{M}^*$ generated by the a_i is a model of T^* and $\mathfrak{M}^*$ has cardinality κ. Since T^* has quantifier elimination, $\mathfrak{M}^*$ is an elementary substructure of $\mathfrak{N}^*$ and (a_i) is indiscernible in $\mathfrak{M}^*$. By Lemma 5.1.6, there are only countably many types over every countable set realised in $\mathfrak{M}^*$. The same is then true for the reduct $\mathfrak{M} = \mathfrak{M}^* \upharpoonright L$. $\dashv$

EXERCISE 5.1.1. A sequence of elements in $(\mathbb{Q}, <)$ is indiscernible if and only if it is either constant, strictly increasing or strictly decreasing.

EXERCISE 5.1.2. Prove Ramsey's Theorem 5.1.5 by induction on n similarly to the proof of C.3.2 using a non-principal ultrafilter on A. (For ultrafilters see Exercise 1.2.4. An ultrafilter is non-principal if it contains no finite sets.)

[2] If $n = 0$, f_φ is a constant.

5.2. ω-stable theories

In this section we fix a complete theory T with infinite models.

In the previous section we saw that we may add indiscernible elements to a model without changing the number of realised types. We will now use this to show that $\aleph_1$-categorical theories have a small number of types, i.e., they are ω-stable. Conversely, with few types it is easier to be saturated and since saturated structures are unique we find the connection to categorical theories.

DEFINITION 5.2.1. Let κ be an infinite cardinal. We say T is *κ-stable* if in each model of T, over every set of parameters of size at most κ, and for each n, there are at most κ many n-types, i.e.,

$$|A| \leq \kappa \Rightarrow |\mathrm{S}_n(A)| \leq \kappa.$$

Note that if T is κ-stable, then – up to logical equivalence – we have $|T| \leq \kappa$, see Exercise 5.2.6.

LEMMA 5.2.2. *T is κ-stable if and only if T is κ-stable for* 1*-types, i.e.,*

$$|A| \leq \kappa \Rightarrow |\mathrm{S}(A)| \leq \kappa.$$

PROOF. Assume that T is κ-stable for 1-types. We show that T is κ-stable for n-types by induction on n. Let A be a subset of the model $\mathfrak{M}$ and $|A| \leq \kappa$. We may assume that all types over A are realised in $\mathfrak{M}$. Consider the restriction map $\pi : \mathrm{S}_n(A) \to \mathrm{S}_1(A)$. By assumption the image $\mathrm{S}_1(A)$ has cardinality at most κ. Every $p \in \mathrm{S}_1(A)$ has the form $\mathrm{tp}(a/A)$ for some $a \in M$. By Exercise 4.2.5 the fibre $\pi^{-1}(p)$ is in bijection with $\mathrm{S}_{n-1}(aA)$ and so has cardinality at most κ by induction. This shows $|\mathrm{S}_n(A)| \leq \kappa$. $\dashv$

EXAMPLE 5.2.3 (Algebraically closed fields). The theories ACF_p for p a prime or 0 are κ-stable for all κ.

Note that by Theorem 5.2.6 below it would suffice to prove that the theories ACF_p are ω-stable. The converse holds as well: any infinite ω-stable field is in fact algebraically closed (see [38]).

PROOF. Let K be a subfield of an algebraically closed field. By quantifier elimination the type of an element a over K is determined by the isomorphism type of the extension $K[a]/K$. If a is transcendental over K, $K[a]$ is isomorphic to the polynomial ring $K[X]$. If a is algebraic with minimal polynomial $f \in K[X]$, then $K[a]$ is isomorphic to $K[X]/(f)$. So there is one more 1-type over K than there are irreducible polynomials. $\dashv$

That ACF_p is κ-stable for n-types has a direct algebraic proof: the isomorphism type of $K[a_1, \ldots, a_n]/K$ is determined by the vanishing ideal P of $a_1, \ldots, a_n$ (see Lemma B.3.6). By Hilbert's Basis Theorem, P is finitely generated. So, if K has cardinality κ, the polynomial ring $K[X_1, \ldots, X_n]$ has only κ many ideals.

THEOREM 5.2.4. *A countable theory T which is categorical in an uncountable cardinal κ is ω-stable* [3].

PROOF. Let $\mathfrak{N}$ be a model and $A \subseteq N$ countable with $\mathrm{S}(A)$ uncountable. Let $(b_i)_{i \in I}$ be a sequence of $\aleph_1$ many elements with pairwise distinct types over A. (Note that we can assume that all types over A are realised in $\mathfrak{N}$.) We choose first an elementary substructure $\mathfrak{M}_0$ of cardinality $\aleph_1$ which contains A and all b_i. Then we choose an elementary extension $\mathfrak{M}$ of $\mathfrak{M}_0$ of cardinality κ. The model $\mathfrak{M}$ is of cardinality κ and realises uncountably many types over the countable set A. By Corollary 5.1.9, T has another model in which this is not the case. So T cannot be κ-categorical. ⊣

DEFINITION 5.2.5. A countable theory T is *totally transcendental* if it has no model $\mathfrak{M}$ with a binary tree of consistent $L(M)$-formulas.

THEOREM 5.2.6. 1. *ω-stable theories are totally transcendental.*
2. *Totally transcendental theories are κ-stable for all $\kappa \geq |T|$.*

It follows that a countable theory T is ω-stable if and only if it is totally transcendental.

PROOF. 1. Let $\mathfrak{M}$ be a model with a binary tree of consistent $L(M)$-formulas with free variables among $x_1, \ldots, x_n$. The set A of parameters which occur in the tree's formulas is countable but $\mathrm{S}_n(A)$ has cardinality $2^{\aleph_0}$. (see Theorem 4.5.9).

2. Assume that there are more than κ many n-types over some set A of cardinality κ. Let us call an $L(A)$-formula $\varphi(\overline{x})$ *big* if it belongs to more than κ many types over A and *thin* otherwise. By assumption the true formula is big. If we can show that each big formula decomposes into two big formulas, we can construct a binary tree of big formulas, which finishes the proof.

So assume that φ is big. Since each thin formula belongs to at most κ types and since there are at most κ formulas, there are at most κ types which contain thin formulas. Therefore φ belongs to two distinct types p and q which contain only big formulas. If we separate p and q by $\psi \in p$ and $\neg\psi \in q$, we decompose φ into the big formulas $\varphi \wedge \psi$ and $\varphi \wedge \neg\psi$. ⊣

The proof and Lemma 5.2.2 show that T is totally transcendental if and only if there is no binary tree of consistent formulas in one free variable. This is clear for countable T; the general case follows from Exercise 5.2.5.

The following definition generalises the notion of ω-saturation.

DEFINITION 5.2.7. Let κ be an infinite cardinal. An L-structure $\mathfrak{A}$ is *κ-saturated* if in $\mathfrak{A}$ all types over sets of cardinality less than κ are realised. An infinite structure $\mathfrak{A}$ is *saturated* if it is $|\mathfrak{A}|$-saturated.

[3] ω-stable and $\aleph_0$-stable are synonymous.

Even though saturation requires only that 1-types are realised, as in the ω-saturated case this easily implies that all n-types are realised as well (see Exercise 4.3.9).

Lemma 4.3.3 generalises to sets.

LEMMA 5.2.8. *Elementarily equivalent saturated structures of the same cardinality are isomorphic.*

PROOF. Let $\mathfrak{A}$ and $\mathfrak{B}$ be elementarily equivalent saturated structures each of cardinality κ. We choose enumerations $(a_\alpha)_{\alpha<\kappa}$ and $(b_\alpha)_{\alpha<\kappa}$ of A and B and construct an increasing sequence of elementary maps $f_\alpha\colon A_\alpha \to B_\alpha$. Assume that the f_β are constructed for all $\beta < \alpha$. The union of the f_β is an elementary map $f^*_\alpha\colon A^*_\alpha \to B^*_\alpha$. The construction will imply that A^*_α and B^*_α have cardinality at most $|\alpha|$, which is smaller than κ.

We write $\alpha = \lambda + n$ (as in p. 187) and distinguish two cases:

$n = 2i$: In this case we consider $p(x) = \text{tp}(a_{\lambda+i}/A^*_\alpha)$. Realise $f^*_\alpha(p)$ by $b \in B$ and define

$$f_\alpha = f^*_\alpha \cup \{\langle a_{\lambda+i}, b\rangle\}.$$

$n = 2i + 1$: Similarly. We find an extension

$$f_\alpha = f^*_\alpha \cup \{\langle a, b_{\lambda+i}\rangle\}.$$

Then $\bigcup_{\alpha<\kappa} f_\alpha$ is the desired isomorphism between $\mathfrak{A}$ and $\mathfrak{B}$. ⊣

LEMMA 5.2.9. *If T is κ-stable, then for all regular $\lambda \leq \kappa$ there is a model of cardinality κ which is λ-saturated.*

PROOF. By Exercise 5.2.6 we may assume that $|T| \leq \kappa$. Consider a model $\mathfrak{M}$ of cardinality κ. Since $\text{S}(M_\alpha)$ has cardinality κ, Corollary 2.2.9 and the Löwenheim–Skolem Theorem give an elementary extension of cardinality κ in which all types over $\mathfrak{M}$ are realised. So we can construct a continuous elementary chain

$$\mathfrak{M}_0 \prec \mathfrak{M}_1 \cdots \prec \mathfrak{M}_\alpha \prec \cdots \quad (\alpha < \lambda),$$

of models of T with cardinality κ such that all $p \in \text{S}(M_\alpha)$ are realised in $\mathfrak{M}_{\alpha+1}$. Let $\mathfrak{M}$ be the union of this chain. Then $\mathfrak{M}$ is λ-saturated. In fact, if $|A| < \lambda$ and if $a \in A$ is contained in $M_{\alpha(a)}$ then $\Lambda = \bigcup_{a\in A} \alpha(a)$ is an initial segment of λ of smaller cardinality than λ. So Λ has an upper bound $\mu < \lambda$. It follows that $A \subseteq M_\mu$ and all types over A are realised in $\mathfrak{M}_{\mu+1}$. ⊣

REMARK 5.2.10. If T is κ-stable for a regular cardinal κ, the previous lemma yields a saturated model of cardinality κ. Harnik [22] showed that this holds in fact for arbitrary κ. See also Corollary 6.1.3 for more general constructions.

THEOREM 5.2.11. *A countable theory T is κ-categorical if and only if all models of cardinality κ are saturated.*

PROOF. If all models of cardinality κ are saturated, it follows from Lemma 5.2.8 that T is κ-categorical.

Assume, for the converse, that T is κ-categorical. For $\kappa = \aleph_0$ the theorem follows from (the proof of) Theorem 4.3.1. So we may assume that κ is uncountable. Then T is totally transcendental by Theorems 5.2.4 and 5.2.6 and therefore κ-stable by Theorem 5.2.6.

By Lemma 5.2.9, all models of T of cardinality κ are μ^+-saturated for all $\mu < \kappa$. i.e., κ-saturated. ⊣

EXERCISE 5.2.1. Use Exercise 8.2.8 to show that a theory with an infinite definable linear ordering (like DLO and RCF) cannot be κ-stable for any κ.

EXERCISE 5.2.2. Show that the theory of an equivalence relation with two infinite classes has quantifier elimination and is ω-stable. Is it $\aleph_1$-categorical?

EXERCISE 5.2.3. Let L be at most countable, $\mathfrak{A}_0, \mathfrak{A}_1, \ldots$ a sequence of L-structures and $\mathcal{F}$ a non-principal ultrafilter on ω. Show that $\prod_{i<\omega} \mathfrak{A}_i/\mathcal{F}$ is $\aleph_1$-saturated. If we assume the Continuum Hypothesis, this implies that if $\mathfrak{A}$ and $\mathfrak{B}$ are two countable and elementarily equivalent L-structures, the two ultrapowers $\mathfrak{A}^\omega/\mathcal{F}$ and $\mathfrak{B}^\omega/\mathcal{F}$ are isomorphic.

Shelah has shown in [52] that for any two elementarily equivalent structures there is a set I and an ultrafilter $\mathcal{F}$ on I such that $\mathfrak{A}^I/\mathcal{F}$ and $\mathfrak{B}^I/\mathcal{F}$ are isomorphic.

EXERCISE 5.2.4. If $\mathfrak{A}$ is κ-saturated, then all definable subsets are either finite or have cardinality at least κ.

EXERCISE 5.2.5. If T is an L-theory and K is a sublanguage of L, the *reduct* $T \upharpoonright K$ is the set of all K–sentences which follow from T. Show that T is totally transcendental if and only if $T \upharpoonright K$ is ω-stable for all at most countable $K \subseteq L$.

EXERCISE 5.2.6. If T is κ-stable, then *essentially* (i.e., up to logical equivalence) $|T| \leq \kappa$.

5.3. Prime extensions

As with prime models, prime extensions are the smallest ones in the sense of elementary embeddings. We will see here (and in Sections 9.2 and 9.3) that prime extensions, if they exist, share a number of important properties with prime models.

DEFINITION 5.3.1. Let $\mathfrak{M}$ be a model of T and $A \subseteq M$.

1. $\mathfrak{M}$ is a *prime extension* of A (or *prime over* A) if every elementary map $A \to \mathfrak{N}$ extends to an elementary map $\mathfrak{M} \to \mathfrak{N}$.

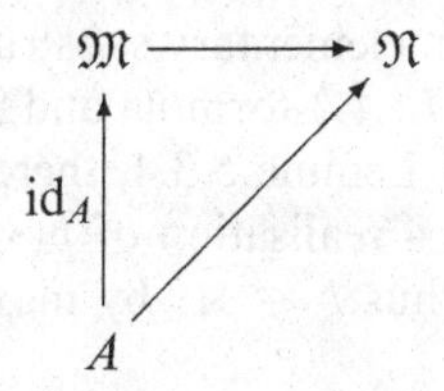

2. $B \subseteq M$ is *constructible* over A if B has an enumeration

$$B = \{b_\alpha \mid \alpha < \lambda\},$$

where each b_α is atomic over $A \cup B_\alpha$, with $B_\alpha = \{b_\mu \mid \mu < \alpha\}$.

So $\mathfrak{M}$ is a prime extension of A if and only if $\mathfrak{M}_A$ is a prime model of $\mathrm{Th}(\mathfrak{M}_A)$.

Notice the following.

LEMMA 5.3.2. *If a model M is constructible over A, then $\mathfrak{M}$ is prime over A.*

PROOF. Let $(m_\alpha)_{\alpha<\lambda}$ an enumeration of M, such that each m_α is atomic over $A \cup M_\alpha$. Let $f : A \to \mathfrak{N}$ be an elementary map. We define inductively an increasing sequence of elementary maps $f_\alpha : A \cup M_{\alpha+1} \to \mathfrak{N}$ with $f_0 = f$. Assume that f_β is defined for all $\beta < \alpha$. The union of these f_β is an elementary map $f'_\alpha : A \cup M_\alpha \to \mathfrak{N}$. Since $p(x) = \mathrm{tp}(a_\alpha/A \cup M_\alpha)$ is isolated, $f'_\alpha(p) \in \mathrm{S}(f'_\alpha(A \cup M_\alpha))$ is also isolated and has a realisation b in $\mathfrak{N}$. We set $f_\alpha = f'_\alpha \cup \{\langle a_\alpha, b\rangle\}$.

Finally, the union of all f_α $(\alpha < \lambda)$ is an elementary embedding $\mathfrak{M} \to \mathfrak{N}$. ⊣

We will see below that in totally transcendental theories prime extensions are atomic.

THEOREM 5.3.3. *If T is totally transcendental, every subset of a model of T has a constructible prime extension.*

We will see in Section 9.2 that in totally transcendental theories, prime extensions are unique up to isomorphism (see Theorem 4.5.3).

For the proof we need the following lemma which generalises Theorem 4.5.7.

LEMMA 5.3.4. *If T is totally transcendental, the isolated types are dense over every subset of any model.*

PROOF. Consider a subset A of a model $\mathfrak{M}$. Then $\mathrm{Th}(\mathfrak{M}_A)$ has no binary tree of consistent formulas. By Theorem 4.5.9, the isolated types in $\mathrm{Th}(\mathfrak{M}_A)$ are dense. ⊣

We can now prove Theorem 5.3.3.

PROOF. By Lemma 5.3.2 it suffices to construct an elementary substructure $\mathfrak{M}_0 \prec \mathfrak{M}$ which contains A and is constructible over A. An application of Zorn's Lemma gives us a maximal construction $(a_\alpha)_{\alpha<\lambda}$, which cannot be prolonged by an element $a_\lambda \in M \setminus A_\lambda$. Clearly A is contained in A_λ. We show that A_λ is the universe of an elementary substructure $\mathfrak{M}_0$ using Tarski's Test. So assume that $\varphi(x)$ is an $L(A_\lambda)$-formula and $\mathfrak{M} \models \exists x\, \varphi(x)$. Since isolated types over A_λ are dense by Lemma 5.3.4, there is an isolated $p(x) \in \mathrm{S}(A_\lambda)$ containing $\varphi(x)$. Let b be a realisation of $p(x)$ in $\mathfrak{M}$. We can prolong our construction by $a_\lambda = b$; thus $b \in A_\lambda$ by maximality and $\varphi(x)$ is realised in A_λ. ⊣

To prove that in totally transcendental theories prime extensions are atomic, we need the following.

LEMMA 5.3.5. *Let a and b be two finite tuples of elements of a structure $\mathfrak{M}$. Then* $\mathrm{tp}(ab)$ *is atomic if and only if* $\mathrm{tp}(a/b)$ *and* $\mathrm{tp}(b)$ *are atomic.*

PROOF. First assume that $\varphi(x, y)$ isolates $\mathrm{tp}(a, b)$. As in the proof of Theorem 4.5.2, $\varphi(x, b)$ isolates $\mathrm{tp}(a/b)$ and we claim that $\exists x\, \varphi(x, y)$ isolates $p(y) = \mathrm{tp}(b)$: we have $\exists x\, \varphi(x, y) \in p(y)$ and if $\sigma(y) \in p(y)$, then

$$\mathfrak{M} \models \forall x, y\, (\varphi(x, y) \rightarrow \sigma(y)).$$

Hence $\mathfrak{M} \models \forall y\, (\exists x \varphi(x, y) \rightarrow \sigma(y))$.

Now, conversely, assume that $\rho(x, b)$ isolates $\mathrm{tp}(a/b)$ and that $\sigma(y)$ isolates $p(y) = \mathrm{tp}(b)$. Then $\rho(x, y) \wedge \sigma(y)$ isolates $\mathrm{tp}(a, b)$. For, clearly, we have $\rho(x, y) \wedge \sigma(y) \in \mathrm{tp}(a, b)$. If, on the other hand, $\varphi(x, y) \in \mathrm{tp}(a, b)$, then $\varphi(x, b)$ belongs to $\mathrm{tp}(a/b)$ and

$$\mathfrak{M} \models \forall x\, (\rho(x, b) \rightarrow \varphi(x, b)).$$

Hence

$$\forall x\, (\rho(x, y) \rightarrow \varphi(x, y)) \in p(y)$$

and it follows that

$$\mathfrak{M} \models \forall y\, (\sigma(y) \rightarrow \forall x\, (\rho(x, y) \rightarrow \varphi(x, y))).$$

Thus $\mathfrak{M} \models \forall x, y\, (\rho(x, y) \wedge \sigma(y) \rightarrow \varphi(x, y))$. ⊣

COROLLARY 5.3.6. *Constructible extensions are atomic.*

PROOF. Let $\mathfrak{M}_0$ be a constructible extension of A and let $\overline{a}$ be a tuple from M_0. We have to show that $\overline{a}$ is atomic over A. We can clearly assume that the elements of $\overline{a}$ are pairwise distinct and do not belong to A. We can also permute the elements of $\overline{a}$ so that

$$\overline{a} = a_\alpha \overline{b}$$

for some tuple $\overline{b} \in A_\alpha$. Let $\varphi(x, \overline{c})$ be an $L(A_\alpha)$-formula which is complete over A_α and satisfied by a_α. Then a_α is also atomic over $A \cup \{\overline{b}\overline{c}\}$. Using induction, we know that $\overline{b}\overline{c}$ is atomic over A. By Lemma 5.3.5 applied to $(\mathfrak{M}_0)_A$, $a_\alpha \overline{b}\overline{c}$ is atomic over A, which implies that $\overline{a} = a_\alpha \overline{b}$ is atomic over A. ⊣

COROLLARY 5.3.7. *If T is totally transcendental, prime extensions are atomic.*

PROOF. Let $\mathfrak{M}$ be a model of T and $A \subseteq M$. Since A has at least one constructible extension $\mathfrak{M}_0$ and since all prime extensions of A are contained in $\mathfrak{M}_0$[4], all prime extensions are atomic. ⊣

A structure $\mathfrak{M}$ is called a *minimal* extension of the subset A if M has no proper elementary substructure which contains A.

LEMMA 5.3.8. *Let $\mathfrak{M}$ be a model of T and $A \subseteq M$. If A has a prime extension and a minimal extension, they are isomorphic over A, i.e., there is an isomorphism fixing A elementwise.*

PROOF. A prime extension embeds elementarily in the minimal extension. This embedding must be surjective by minimality. ⊣

EXERCISE 5.3.1. For Theorem 5.3.3 we used only that isolated types are dense in all $S_1(A)$. Prove for arbitrary T that this implies that the isolated types are dense in all $S_n(A)$.

EXERCISE 5.3.2. For every countable T the following are equivalent (see Theorem 4.5.7):

a) Every parameter set has a prime extension. (We say that T *has prime extensions.*)
b) Over every countable parameter set the isolated types are dense.
c) Over every parameter set the isolated types are dense.

EXERCISE 5.3.3. Lemma 5.3.5 follows from Exercise 4.2.5 and the following observation: let $\pi\colon X \to Y$ be a continuous open map between topological spaces. Then a point $x \in X$ is isolated if and only if $\pi(x)$ is isolated in Y and x is isolated in $\pi^{-1}(\pi(x))$.

5.4. Lachlan's Theorem

Using the fact (established in Section 5.2) that uncountably categorical theories are totally transcendental, we will prove the downward direction of Morley's theorem. We use Lachlan's result that, in totally transcendental theories, models have arbitrary large elementary extensions realizing few types.

THEOREM 5.4.1 (Lachlan). ([2] Lemma 10) *Let T be totally transcendental and $\mathfrak{M}$ an uncountable model of T. Then $\mathfrak{M}$ has arbitrarily large elementary extensions which omit every countable set of $L(M)$-formulas that is omitted in $\mathfrak{M}$.*

PROOF. For the proof, we call an $L(M)$-formula $\varphi(x)$ *large* if its realisation set $\varphi(\mathfrak{M})$ is uncountable. Since there is no infinite binary tree of large formulas,

[4]More precisely, they are isomorphic over A to elementary substructures of $\mathfrak{M}_0$.

there exists a *minimal* large formula $\varphi_0(x)$. This means that for every $L(M)$-formula $\psi(x)$ either $\varphi_0(x) \wedge \psi(x)$ or $\varphi_0(x) \wedge \neg\psi(x)$ is at most countable. Now it is easy to see that

$$p(x) = \{\psi(x) \mid \varphi_0(x) \wedge \psi(x) \text{ large}\}$$

is a type in $\mathrm{S}(M)$.

Clearly $p(x)$ contains no formula of the form $x \doteq a$ for $a \in M$, so $p(x)$ is not realised in M. On the other hand, every countable subset $\Pi(x)$ of $p(x)$ is realised in $\mathfrak{M}$: since $\varphi_0(\mathfrak{M}) \setminus \psi(\mathfrak{M})$ is at most countable for every $\psi(x) \in \Pi(x)$, the elements of $\varphi_0(\mathfrak{M})$ which do not belong to the union of these sets realise $\Pi(x)$.

Let a be a realisation of $p(x)$ in a (proper) elementary extension $\mathfrak{N}$. By Theorem 5.3.3 we can assume that $\mathfrak{N}$ is atomic over $M \cup \{a\}$.

Fix $b \in N$. We have to show that every countable subset $\Sigma(y)$ of $\mathrm{tp}(b/M)$ is realised in M.

Let $\chi(x, y)$ be an $L(M)$-formula such that $\chi(a, y)$ isolates $q(y) = \mathrm{tp}(b/M \cup \{a\})$. If b realises an $L(M)$-formula $\sigma(y)$, we have $\mathfrak{N} \models \forall y\, (\chi(a, y) \to \sigma(y))$. Hence the formula

$$\sigma^*(x) = \forall y\, (\chi(x, y) \to \sigma(y))$$

belongs to $p(x)$. Note that $\exists y\, \chi(x, y)$ belongs also to $p(x)$.

Choose an element $a' \in M$ which satisfies

$$\{\sigma^*(x) \mid \sigma \in \Sigma\} \cup \{\exists y\, \chi(x, y)\}$$

and choose $b' \in M$ with $\mathfrak{M} \models \chi(a', b')$. Since $\sigma^*(a')$ is true in $\mathfrak{M}$, $\sigma(b')$ is true in $\mathfrak{M}$. So b' realises $\Sigma(y)$.

We have shown that $\mathfrak{M}$ has a proper elementary extension which realises no new countable set of $L(M)$-formulas. By iteration we obtain arbitrarily long chains of elementary extensions with the same property. $\dashv$

The corollary is the *downwards* part of Morley's Theorem, p. 63.

COROLLARY 5.4.2. *A countable theory which is κ-categorical for some uncountable κ, is $\aleph_1$-categorical.*

PROOF. Let T be κ-categorical and assume that T is not $\aleph_1$-categorical. Then T has a model $\mathfrak{M}$ of cardinality $\aleph_1$ which is not saturated. So there is a type p over a countable subset of M which is not realised in $\mathfrak{M}$. By Theorems 5.2.4 and 5.2.6 T is totally transcendental. Theorem 5.4.1 gives an elementary extension $\mathfrak{N}$ of $\mathfrak{M}$ of cardinality κ which omits all countable sets of formulas which are omitted in $\mathfrak{M}$. Thus also p is omitted. Since $\mathfrak{N}$ is not saturated, T is not κ-categorical, a contradiction. $\dashv$

EXERCISE 5.4.1. Prove in a similar way: if a countable theory T is κ-categorical for some uncountable κ, it is λ-categorical for every uncountable $\lambda \leq \kappa$.

5.5. Vaughtian pairs

A crucial fact about uncountably categorical theories is the absence of definable sets whose size is independent of the size of the model in which they live (captured in the notion of a Vaughtian pair). In fact, in an uncountably categorical theory each model is prime over any infinite definable subset. This will allow us in Section 5.7 to attach a dimension to the models of an uncountably categorical theory. *In this section, T is a countable complete theory with infinite models.*

DEFINITION 5.5.1. We say that T has a *Vaughtian pair* if there are two models $\mathfrak{M} \prec \mathfrak{N}$ and an $L(M)$-formula $\varphi(x)$ such that

a) $\mathfrak{M} \neq \mathfrak{N}$,
b) $\varphi(\mathfrak{M})$ is infinite,
c) $\varphi(\mathfrak{M}) = \varphi(\mathfrak{N})$.

If $\varphi(x)$ does not contain parameters, we say that T has a Vaughtian pair for $\varphi(x)$.

REMARK. Notice that T does not have a Vaughtian pair if and only if every model $\mathfrak{M}$ is a minimal extension of $\varphi(\mathfrak{M}) \cup A$ for any formula $\varphi(x)$ with parameters in $A \subseteq M$ which defines an infinite set in $\mathfrak{M}$.

Let $\mathfrak{N}$ be a model of T where $\varphi(\mathfrak{N})$ is infinite but has smaller cardinality than $\mathfrak{N}$. The Löwenheim–Skolem Theorem yields an elementary substructure $\mathfrak{M}$ of $\mathfrak{N}$ which contains $\varphi(\mathfrak{N})$ and has the same cardinality as $\varphi(\mathfrak{N})$. Then $\mathfrak{M} \prec \mathfrak{N}$ is a Vaughtian pair for $\varphi(x)$. The next theorem shows that a converse of this observation is also true.

THEOREM 5.5.2 (Vaught's Two-cardinal Theorem). *If T has a Vaughtian pair, it has a model $\overline{\mathfrak{M}}$ of cardinality $\aleph_1$ with $\varphi(\overline{\mathfrak{M}})$ countable for some formula $\varphi(x) \in L(\bar{M})$.*

For the proof of Theorem 5.5.2 we need the following.

LEMMA 5.5.3. *Let T be complete, countable, and with infinite models.*

1. *Every countable model of T has a countable ω-homogeneous elementary extension.*
2. *The union of an elementary chain of ω-homogeneous models is ω-homogeneous.*
3. *Two ω-homogeneous countable models of T realizing the same n-types for all $n < \omega$ are isomorphic.*

PROOF. 1. Let $\mathfrak{M}_0$ be a countable model of T. We realise the countably many types

$$\{f(\mathrm{tp}(a/A)) \mid a, A \subseteq M_0, A \text{ finite}, f : A \to M_0 \text{ elementary}\}$$

in a countable elementary extension $\mathfrak{M}_1$. By iterating this process we obtain an elementary chain

$$\mathfrak{M}_0 \prec \mathfrak{M}_1 \prec \cdots,$$

whose union is ω-homogeneous.

2. Clear.

3. Suppose $\mathfrak{A}$ and $\mathfrak{B}$ are ω-homogeneous, countable and realise the same n-types. We show that we can extend any finite elementary map $f\colon \{a_1, \ldots, a_i\} \to \{b_1, \ldots, b_i\}$; $a_j \mapsto b_j$ to any $a \in A \setminus A_i$. Realise the type $\mathrm{tp}(a_1, \ldots, a_i, a)$ by some tuple $\overline{b'} = b'_1, \ldots, b'_{i+1}$ in B. Using the ω-homogeneity of B we may extend the finite partial isomorphism $g = \{(b'_j, b_j) \mid 1 \leq j \leq i\}$ by (b'_{i+1}, b) for some $b \in B$. Then $f_{i+1} = f_i \cup \{(a, b)\}$ is the required extension. Reversing the roles of B and A we construct the desired isomorphism. ⊣

Proof. (of Theorem 5.5.2) Suppose that the Vaughtian pair is witnessed (in certain models) by some formula $\varphi(x)$. For simplicity we assume that $\varphi(x)$ does not contain parameters (see Exercise 5.5.4). Let P be a new unary predicate. It is easy to find an $L(P)$-theory T_{VP} whose models $(\mathfrak{N}, M)$ consist of a model $\mathfrak{N}$ of T and a subset M defined by the new predicate P which is the universe of an elementary substructure $\mathfrak{M}$ which together with $\mathfrak{N}$ forms a Vaughtian pair for $\varphi(x)$. The Löwenheim–Skolem Theorem applied to T_{VP} yields a Vaughtian pair $\mathfrak{M}_0 \prec \mathfrak{N}_0$ for $\varphi(x)$ with $\mathfrak{M}_0, \mathfrak{N}_0$ countable.

We first construct an elementary chain

$$(\mathfrak{N}_0, M_0) \prec (\mathfrak{N}_1, M_1) \prec \cdots$$

of countable Vaughtian pairs, with the aim that both components of the union pair

$$(\mathfrak{N}, M)$$

are ω-homogeneous and realise the same n-types. If $(\mathfrak{N}_i, M_i)$ is given, we first choose a countable elementary extension $(\mathfrak{N}', M')$ such that $\mathfrak{M}'$ realises all n-types which are realised in $\mathfrak{N}_i$. Then we choose as in the proof of Lemma 5.5.3(1) a countable elementary extension $(\mathfrak{N}_{i+1}, M_{i+1})$ of $(\mathfrak{N}', M')$ for which $\mathfrak{N}_{i+1}$ and $\mathfrak{M}_{i+1}$ are ω-homogeneous.

It follows from Lemma 5.5.3(3) that $\mathfrak{M}$ and $\mathfrak{N}$ are isomorphic.

Next we construct a continuous elementary chain

$$\mathfrak{M}^0 \prec \mathfrak{M}^1 \prec \cdots \prec \mathfrak{M}^\alpha \prec \cdots \quad (\alpha < \omega_1)$$

with $(\mathfrak{M}^{\alpha+1}, M^\alpha) \cong (\mathfrak{N}, M)$ for all α. We start with $\mathfrak{M}^0 = \mathfrak{M}$. If $\mathfrak{M}^\alpha$ is constructed, we choose an isomorphism $\mathfrak{M} \to \mathfrak{M}^\alpha$ and extend it to an isomorphism $\mathfrak{N} \to \mathfrak{M}^{\alpha+1}$ (see Lemma 1.1.8). For a countable limit ordinal λ, $\mathfrak{M}^\lambda$ is the union of the $\mathfrak{M}^\alpha$ $(\alpha < \lambda)$. So $\mathfrak{M}^\lambda$ is isomorphic to $\mathfrak{M}$ by Lemma 5.5.3(2) and 5.5.3(3).

Finally we set

$$\overline{\mathfrak{M}} = \bigcup_{\alpha<\omega_1} \mathfrak{M}^\alpha.$$

Since the $\mathfrak{M}^\alpha$ are growing, $\overline{\mathfrak{M}}$ has cardinality $\aleph_1$ while $\varphi(\overline{\mathfrak{M}}) = \varphi(\mathfrak{M}^\alpha) = \varphi(\mathfrak{M}^0)$ is countable. ⊣

COROLLARY 5.5.4. *If T is categorical in an uncountable cardinality, it does not have a Vaughtian pair.*

PROOF. If T has a Vaughtian pair, then by Theorem 5.5.2 it has a model $\mathfrak{M}$ of cardinality $\aleph_1$ such that for some $\varphi(x) \in L(M)$ the set $\varphi(\mathfrak{M})$ is countable. On the other hand, if T is categorical in an uncountable cardinal, it is $\aleph_1$-categorical by Corollary 5.4.2 and by Theorem 5.2.11, all models of T of cardinality $\aleph_1$ are saturated. In particular, each formula is either satisfied by a finite number or by $\aleph_1$ many elements, a contradiction. ⊣

COROLLARY 5.5.5. *Let T be categorical in an uncountable cardinal, $\mathfrak{M}$ a model, and $\varphi(\mathfrak{M})$ infinite and definable over $A \subseteq M$. Then $\mathfrak{M}$ is – the unique – prime extension of $A \cup \varphi(\mathfrak{M})$.*

PROOF. By Corollary 5.5.4, T does not have a Vaughtian pair, so $\mathfrak{M}$ is minimal over $A \cup \varphi(\mathfrak{M})$. If $\mathfrak{N}$ is a prime extension over $A \cup \varphi(\mathfrak{M})$, which exists by Theorem 5.3.3, $\mathfrak{N}$ is isomorphic to $\mathfrak{M}$ over $A \cup \varphi(\mathfrak{M})$ by Lemma 5.3.8. ⊣

DEFINITION 5.5.6. We say that T *eliminates the quantifier* $\exists^\infty x$, *there are infinitely many* x, if for every L-formula $\varphi(x, \overline{y})$ there is a finite bound n_φ such that in all models $\mathfrak{M}$ of T and for all parameters $\overline{a} \in M$,

$$\varphi(\mathfrak{M}, \overline{a})$$

is either infinite or has or at most n_φ elements.

REMARK. This means that for all $\varphi(x, \overline{y})$ there is a $\psi(\overline{y})$ such that in all models $\mathfrak{M}$ of T and for all $\overline{a} \in M$

$$\mathfrak{M} \models \exists^\infty x \, \varphi(x, \overline{a}) \iff \mathfrak{M} \models \psi(\overline{a}).$$

We denote this by

$$T \vdash \forall \overline{y} (\exists^\infty x \, \varphi(x, \overline{y}) \leftrightarrow \psi(\overline{y})).$$

PROOF. If n_φ exists, we can use $\psi(\overline{y}) = \exists^{>n_\varphi} x \, \varphi(x, \overline{y})$ (*there are more than n_φ many x such that $\varphi(x, y)$*). If, conversely, $\psi(\overline{y})$ is a formula which is implied by $\exists^\infty x \, \varphi(x, \overline{y})$, a compactness argument shows that there must be a bound n_φ such that

$$T \vdash \exists^{>n_\varphi} x \, \varphi(x, \overline{y}) \to \psi(\overline{y}).$$

⊣

LEMMA 5.5.7. *A theory T without Vaughtian pair eliminates the quantifier $\exists^\infty x$.*

PROOF. Let P be a new unary predicate and $c_1, \dots, c_n$ new constants. Let T^* be the theory of all $L \cup \{P, c_1, \dots, c_n\}$-structures

$$(\mathfrak{M}, N, a_1, \dots, a_n),$$

where $\mathfrak{M}$ is a model of T, N the universe of a proper elementary substructure, $a_1, \dots, a_n$ elements of N and $\varphi(\mathfrak{M}, \overline{a}) \subseteq N$. Suppose that the bound n_φ does not exist. Then, for any n, there is a model $\mathfrak{N}$ of T and $\overline{a} \in N$ such that $\varphi(\mathfrak{N}, \overline{a})$ is finite, but has more than n elements. Let $\mathfrak{M}$ be a proper elementary extension of $\mathfrak{N}$. Then $\varphi(\mathfrak{M}, \overline{a}) = \varphi(\mathfrak{N}, \overline{a})$ and the pair $(\mathfrak{M}, N, \overline{a})$ is a model of T^*. This shows that the theory

$$T^* \cup \{\exists^{>n} x\, \varphi(x, \overline{c}) \mid n = 1, 2, \dots\}$$

is finitely satisfiable. A model of this theory gives a Vaughtian pair for T. ⊣

EXERCISE 5.5.1. If T is totally transcendental and has a Vaughtian pair for $\varphi(x)$, then it has, for all uncountable κ, a model of cardinality κ with countable $\varphi(\mathfrak{M})$. Prove Corollary 5.5.4 from this. (Use Theorem 5.4.1.)

EXERCISE 5.5.2. Show directly (without using Lemma 5.2.9) that a theory T which is categorical in some uncountable cardinality, has a model $\mathfrak{M}$ of cardinality $\aleph_1$ in which each $L(M)$-formula is either satisfied by a finite number or by $\aleph_1$ many elements.

EXERCISE 5.5.3. Show that the theory RG of the random graph has a Vaughtian pair.

EXERCISE 5.5.4. Let T be a theory, $\mathfrak{M}$ a model of T and $\overline{a} \subseteq M$ a finite tuple of parameters. Let $q(\overline{x})$ be the type of $\overline{a}$ in $\mathfrak{M}$. Then for new constants $\overline{c}$, the $L(\overline{c})$-theory

$$T(q) = \mathrm{Th}(\mathfrak{M}, \overline{a}) = T \cup \{\varphi(\overline{c}) |\, \varphi(\overline{x}) \in q(\overline{x})\}$$

is complete. Show that T is λ-stable (or without Vaughtian pair etc.) if and only if $T(q)$ is. For countable languages, this implies that T is categorical in some uncountable cardinal if and only if $T(q)$ is.

EXERCISE 5.5.5. If T eliminates $\exists^\infty$, then T eliminates for every n the quantifier "there are infinitely many n-tuples $x_1, \dots, x_n$".

EXERCISE 5.5.6. Assume that T eliminates the quantifier $\exists^\infty$. Then for every formula $\varphi(x_1, \dots, x_n, \overline{y})$ there is a formula $\theta(\overline{y})$ such that in all models $\mathfrak{M}$ of T a tuple $\overline{b}$ satisfies θ if and only if $\mathfrak{M}$ has an elementary extension $\mathfrak{M}'$ with elements $a_1, \dots, a_n \in M' \setminus M$ such that $\mathfrak{M}' \models \varphi(a_1, \dots, a_n, \overline{b})$.

EXERCISE 5.5.7. Let T_1 and T_2 be two model complete theories in disjoint languages L_1 and L_2. Assume that both theories eliminate $\exists^\infty$. Then $T_1 \cup T_2$ has a model companion.

5.6. Algebraic formulas

Formulas defining a finite set are called algebraic. In this section we collect a few facts and a bit of terminology around this concept which will be crucial in the following sections.

DEFINITION 5.6.1. Let $\mathfrak{M}$ be a structure and A a subset of M. A formula $\varphi(x) \in L(A)$ is called *algebraic* if $\varphi(\mathfrak{M})$ is finite. An element $a \in M$ is algebraic over A if it realises an algebraic $L(A)$-formula. We call an element algebraic if it is algebraic over the empty set. The *algebraic closure* of A, $\mathrm{acl}(A)$, is the set of all elements of $\mathfrak{M}$ algebraic over A, and A is called *algebraically closed* if it equals its algebraic closure.

REMARK. Note that the algebraic closure of A does not grow in elementary extensions of $\mathfrak{M}$ because an $L(A)$-formula which defines a finite set in $\mathfrak{M}$ defines the same set in every elementary extension. As a special case we have that elementary substructures are algebraically closed.

It is easy to see that

$$|\mathrm{acl}(A)| \leq \max(|T|, |A|) \tag{5.1}$$

(see Theorem 2.3.1).

In algebraically closed fields, an element a is algebraic over A precisely if a is algebraic (in the field-theoretical sense) over the field generated by A. This follows easily from quantifier elimination in ACF.

We call a type $p(x) \in \mathrm{S}(A)$ algebraic if (and only if) p contains an algebraic formula. Any algebraic type p is isolated by an algebraic formula $\varphi(x) \in L(A)$, namely by any $\varphi \in p$ having the minimal number of solutions in $\mathfrak{M}$. This number is called the *degree* $\deg(p)$ of p. As isolated types are realised in every model, the algebraic types over A are exactly those of the form $\mathrm{tp}(a/A)$ where a is algebraic over A. The *degree* of a over A $\deg(a/A)$ is the degree of $\mathrm{tp}(a/A)$.

LEMMA 5.6.2. *Let $p \in \mathrm{S}(A)$ be non-algebraic and $A \subseteq B$. Then p has a non-algebraic extension $q \in \mathrm{S}(B)$.*

PROOF. The extension $q_0(x) = p(x) \cup \{\neg\psi(x) \mid \psi(x)$ algebraic $L(B)$-formula$\}$ is finitely satisfiable. For otherwise there are $\varphi(x) \in p(x)$ and algebraic $L(B)$-formulas $\psi_1(x), \ldots, \psi_n(x)$ with

$$\mathfrak{M} \models \forall x\, (\varphi(x) \rightarrow \psi_1(x) \vee \cdots \vee \psi_n(x)).$$

But then $\varphi(x)$ and hence $p(x)$ is algebraic. So we can take for q any type containing q_0. $\dashv$

REMARK 5.6.3. Since algebraic types are isolated by algebraic formulas, an easy compactness argument shows that a type $p \in \mathrm{S}(A)$ is algebraic if and only if p has only finitely many realisations (namely $\deg(p)$ many) in all elementary extensions of $\mathfrak{M}$.

LEMMA 5.6.4. *Let $\mathfrak{M}$ and $\mathfrak{N}$ be two structures and $f: A \to B$ an elementary bijection between two subsets. Then f extends to an elementary bijection between* $\mathrm{acl}(A)$ *and* $\mathrm{acl}(B)$.

PROOF. Let $g: A' \to B'$ a maximal extension of f to two subsets of $\mathrm{acl}(A)$ and $\mathrm{acl}(B)$. Let a be an element of $\mathrm{acl}(A)$. Since a is algebraic over A', a is atomic over A'. We can therefore realise the type $g(\mathrm{tp}(a/A'))$ in $\mathfrak{N}$ – by an element $b \in \mathrm{acl}(B)$ – and obtain an extension $g \cup \{\langle a, b\rangle\}$ of g. It follows that $a \in A'$. So g is defined on the whole of $\mathrm{acl}(A)$. Interchanging A and B shows that g is surjective. (See Lemma 6.1.9 for an alternative proof.) ⊣

The algebraic closure operation will be used to study models of $\aleph_1$-categorical theories in further detail.

DEFINITION 5.6.5. A *pregeometry*[5] (or *matroid*) (X, cl) is a set X with a closure operator $\mathrm{cl}: \mathfrak{P}(X) \to \mathfrak{P}(X)$, where $\mathfrak{P}$ denotes the power set, such that for all $A \subseteq X$ and $a, b \in X$:

a) (REFLEXIVITY) $A \subseteq \mathrm{cl}(A)$.
b) (FINITE CHARACTER) $\mathrm{cl}(A)$ is the union of all $\mathrm{cl}(A')$, where the A' range over all finite subsets of A.
c) (TRANSITIVITY) $\mathrm{cl}(\mathrm{cl}(A)) = \mathrm{cl}(A)$.
d) (EXCHANGE) $a \in \mathrm{cl}(Ab) \setminus \mathrm{cl}(A) \Rightarrow b \in \mathrm{cl}(Aa)$.

A set A is called *closed* (or cl-closed) if $A = \mathrm{cl}(A)$. Note that the closure $\mathrm{cl}(A)$ of A is the smallest cl-closed set containing A. So a pregeometry is given by the system of cl-closed subsets.

The operator $\mathrm{cl}(A) = A$ for all $A \subseteq X$ is a trivial example of a pregeometry. The three standard examples from algebra are vector spaces with the linear closure operator, for a field K with prime field F, the relative algebraic closure $\mathrm{cl}(A) = F(A)^{\mathrm{alg}} \cap K$,[6] and for a field K of characteristic p, the p-closure given by $\mathrm{cl}(A) = K^p(A)$ (see Remark C.1.1).

LEMMA 5.6.6. *If X is the universe of a structure,* acl *satisfies* REFLEXIVITY, FINITE CHARACTER *and* TRANSITIVITY.

PROOF. REFLEXIVITY and FINITE CHARACTER are clear. For TRANSITIVITY assume that c is algebraic over $b_1, \dots, b_n$ and the b_i are algebraic over A. We have to show that c is algebraic over A. Choose an algebraic formula $\varphi(x, b_1, \dots, b_n)$ satisfied by c and algebraic $L(A)$-formulas $\varphi_i(y)$ satisfied by the b_i. Let $\varphi(x, b_1, \dots, b_n)$ be satisfied by exactly k elements. Then the $L(A)$-formula

$$\exists y_1 \dots y_n (\varphi_1(y_1) \wedge \dots \wedge \varphi_n(y_n) \wedge \exists^{\leq k} z \varphi(z, y_1, \dots, y_n) \wedge \varphi(x, y_1, \dots, y_n))$$

is algebraic and realised by c. ⊣

[5] Pregeometries were introduced by van der Waerden (1930) and Whitney (1934).
[6] L^{alg} denotes the algebraic closure of the field L.

EXERCISE 5.6.1. If $A \subseteq M$ and $\mathfrak{M}$ is $|A|^+$-saturated, then $p \in \mathrm{S}(A)$ is algebraic if and only if $p(M)$ is finite.

EXERCISE 5.6.2 (P. M. Neumann). Let A, B be subsets of $\mathfrak{M}$ and $(c_0, \dots, c_n)$ a sequence of elements which are not algebraic over A. If $\mathfrak{M}$ is $|A \cup B|^+$-saturated, the type $\mathrm{tp}(c_0, \dots, c_n/A)$ has a realisation which is disjoint from B. (Hint: Use induction on n. Distinguish between whether or not c_i is algebraic over Ac_n for some $i < n$.)

5.7. Strongly minimal sets

Strongly minimal theories defined below turn out to be uncountably categorical: the isomorphism type of the model is determined by the dimension of an associated geometry. While this appears to be a very special case of such theories, we will see in the next section that we can always essentially reduce to this situation.

Throughout this section we fix a complete theory T with infinite models. For Corollary 5.7.9 we have to assume T countable.

DEFINITION 5.7.1. Let $\mathfrak{M}$ be a model of T and $\varphi(\overline{x})$ a non-algebraic $L(M)$-formula.

1. The set $\varphi(\mathfrak{M})$ is called *minimal in* $\mathfrak{M}$ if for all $L(M)$-formulas $\psi(\overline{x})$ the intersection $\varphi(\mathfrak{M}) \wedge \psi(\mathfrak{M})$ is either finite or cofinite in $\varphi(\mathfrak{M})$.
2. The formula $\varphi(\overline{x})$ is *strongly minimal* if $\varphi(\overline{x})$ defines a minimal set in all elementary extensions of $\mathfrak{M}$. In this case, we also call the definable set $\varphi(\mathfrak{M})$ strongly minimal. A non-algebraic type containing a strongly minimal formula is called strongly minimal.
3. A theory T is strongly minimal if the formula $x \doteq x$ is strongly minimal.

Clearly, strong minimality is preserved under definable bijections; i.e., if A and B are definable subsets of $\mathfrak{M}^k$, $\mathfrak{M}^m$ defined by φ and ψ, respectively, such that there is a definable bijection between A and B, then if φ is strongly minimal so is ψ.

EXAMPLES. 1. The following theories are strongly minimal, which is easily seen in each case using quantifier elimination.

- Infset, the theory of infinite sets. The sets which are definable over a parameter set A in a model M are the finite subsets S of A and their complements $M \setminus S$.
- For a field K, the theory of infinite K-vector spaces. The sets definable over a set A are the finite subsets of the subspace spanned by A and their complements.
- The theories ACF_p of algebraically closed fields of fixed characteristic. The definable sets of any model K are Boolean combinations

of zero-sets

$$\{a \in K \mid f(a) = 0\}$$

of polynomials $f(X) \in K[X]$. Zero-sets are finite or, if $f = 0$, all of K.

2. If K is a model ACF_p, for any $a, b \in K$, the formula $ax_1 + b = x_2$ defining an affine line A in K^2 is strongly minimal as there is a definable bijection between A and K.
3. For any strongly minimal formula $\varphi(x_1, \ldots, x_n)$, the *induced theory* $T \restriction \varphi$ is strongly minimal. Here, for any model $\mathfrak{M}$ of T, the induced theory is the theory of $\varphi(\mathfrak{M})$ with the structure given by all intersections of 0-definable subsets of M^{nm} with $\varphi(\mathfrak{M})^m$ for all $m \in \omega$. This theory depends only on T and φ, not on $\mathfrak{M}$.

Whether $\varphi(\overline{x}, \overline{a})$ is strongly minimal depends only on the type of the parameter tuple $\overline{a}$ and not on the actual model: observe that $\varphi(\overline{x}, \overline{a})$ is strongly minimal if and only if for all L-formulas $\psi(x, \overline{z})$ the set

$$\Sigma_\psi(\overline{z}, \overline{a}) = \{\exists^{>k}\overline{x}\,(\varphi(\overline{x}, \overline{a}) \wedge \psi(\overline{x}, \overline{z})) \wedge \exists^{>k}\overline{x}\,(\varphi(\overline{x}, \overline{a}) \wedge \neg\psi(\overline{x}, \overline{z})) \mid k = 1, 2, \ldots\}$$

cannot be realised in any elementary extension. This means that for all $\psi(\overline{x}, \overline{z})$ there is a bound k_ψ such that

$$\mathfrak{M} \models \forall \overline{z}\ \left(\exists^{\leq k_\psi}\overline{x}\,(\varphi(\overline{x}, \overline{a}) \wedge \psi(\overline{x}, \overline{z})) \vee \exists^{\leq k_\psi}\overline{x}\,(\varphi(\overline{x}, \overline{a}) \wedge \neg\psi(\overline{x}, \overline{z}))\right).$$

This is an *elementary property* of $\overline{a}$, i.e., expressible by a first-order formula. So it makes sense to call $\varphi(\overline{x}, \overline{a})$ a strongly minimal formula without specifying a model.

LEMMA 5.7.2. *If $\mathfrak{M}$ is ω-saturated, or if T eliminates the quantifier $\exists^\infty$, any minimal formula is strongly minimal. If T is totally transcendental, every infinite definable subset of $\mathfrak{M}^n$ contains a minimal set $\varphi(\mathfrak{M})$.*

PROOF. If $\mathfrak{M}$ is ω-saturated and $\varphi(\overline{x}, \overline{a})$ not strongly minimal, then for some L-formula $\psi(\overline{x}, \overline{z})$ the set $\Sigma_\psi(\overline{z}, \overline{a})$ is realised in $\mathfrak{M}$, so φ is not minimal.

If on the other hand $\varphi(\overline{x}, \overline{a})$ is minimal and T eliminates the quantifier $\exists^\infty$, then for all L-formulas $\psi(\overline{x}, \overline{z})$

$$\neg(\exists^\infty\overline{x}(\varphi(\overline{x}, \overline{a}) \wedge \psi(\overline{x}, \overline{z})) \wedge \exists^\infty\overline{x}(\varphi(\overline{x}, \overline{a}) \wedge \neg\psi(\overline{x}, \overline{z})))$$

is an elementary property of $\overline{z}$.

If $\varphi_0(\mathfrak{M})$ does not contain a minimal set, one can construct from $\varphi_0(\overline{x})$ a binary tree of $L(M)$-formulas defining infinite subsets of $\mathfrak{M}$. This contradicts ω-stability. $\dashv$

From now on we will only consider strongly minimal formulas in one variable. It should be clear how to extend everything to the more general context.

LEMMA 5.7.3. *The formula $\varphi(\mathfrak{M})$ is minimal if and only if there is a unique non-algebraic type $p \in S(M)$ containing $\varphi(x)$.*

PROOF. If $\varphi(\mathfrak{M})$ is minimal, then clearly

$$p = \{\psi \mid \psi(x) \in L(M) \text{ such that } \varphi \wedge \neg\psi \text{ is algebraic}\}$$

is the unique non-algebraic type in $S(M)$ containing it. If $\varphi(\mathfrak{M})$ is not minimal, there is some L-formula ψ with both $\varphi \wedge \psi$ and $\varphi \wedge \neg\psi$ non-algebraic. By Lemma 5.6.2 there are at least two non-algebraic types in $S(M)$ containing φ. ⊣

COROLLARY 5.7.4. *A strongly minimal type $p \in S(A)$ has a unique non-algebraic extension to all supersets B of A in elementary extensions of $\mathfrak{M}$. Consequently, the type of m realisations $a_1, \ldots, a_m$ of p with $a_i \notin \operatorname{acl}(a_1 \ldots a_{i-1}A)$, $i = 1, \ldots, m$, is uniquely determined.*

PROOF. Existence of non-algebraic extensions follows from Lemma 5.6.2, which also allows us to assume that B is a model. Uniqueness now follows from Lemma 5.7.3 applied to any strongly minimal formula of p. The last sentence follows by induction. ⊣

THEOREM 5.7.5. *If $\varphi(x)$ is a strongly minimal formula in $\mathfrak{M}$ without parameters, the operation*

$$\mathrm{cl} : \mathfrak{P}(\varphi(\mathfrak{M})) \to \mathfrak{P}(\varphi(\mathfrak{M}))$$

defined by

$$\mathrm{cl}(A) = \operatorname{acl}^M(A) \cap \varphi(\mathfrak{M})$$

is a pregeometry $(\varphi(\mathfrak{M}), \mathrm{cl})$.

PROOF. We have to verify EXCHANGE. For notational simplicity we assume $A = \emptyset$. Let $a \in \varphi(\mathfrak{M})$ be not algebraic over $\emptyset$ and $b \in \varphi(\mathfrak{M})$ not algebraic over a. By Corollary 5.7.4, all such pairs a, b have the same type $p(x, y)$. Let A' be an infinite set of non-algebraic elements realising φ (which exists in an elementary extension of $\mathfrak{M}$) and b' non-algebraic over A'. Since all $a' \in A'$ have the same type $p(x, b')$ over b', no a' is algebraic over b'. Thus also a is not algebraic over b. ⊣

The same proof shows that algebraic closure defines a pregeometry on the set of realizations of a *minimal* type, i.e., a non-algebraic type $p \in S_1(A)$ having a unique non-algebraic extension to all supersets B of A in elementary extensions of $\mathfrak{M}$. Here is an example to show that a minimal type need not be strongly minimal.

Let T be the theory of $\mathfrak{M} = (M, P_i)_{i<\omega}$ in which the P_i form a proper descending sequence of subsets. The type $p = \{x \in P_i \mid i < \omega\} \in S_1(\emptyset)$ is minimal. If all P_{i+1} are coinfinite in P_i, then p does not contain a minimal formula and is not strongly minimal.

In pregeometries there is a natural notion of independence and dimension (see Definition C.1.2), so in light of Theorem 5.7.5 we may define the following.

If $\varphi(x)$ is strongly minimal without parameters, the *φ-dimension* of a model $\mathfrak{M}$ of T is the dimension of the pregeometry $(\varphi(\mathfrak{M}), \mathrm{cl})$

$$\dim_\varphi(\mathfrak{M}).$$

If $\mathfrak{M}$ is the model of a strongly minimal theory, we just write $\dim(\mathfrak{M})$.

If $\varphi(x)$ is defined over $A_0 \subseteq M$, the closure operator of the pregeometry $\varphi(\mathfrak{M}_{A_0})$ is given by

$$\mathrm{cl}(A) = \mathrm{acl}^M(A_0 \cup A) \cap \varphi(\mathfrak{M})$$

and

$$\dim_\varphi(\mathfrak{M}/A_0) := \dim_\varphi(\mathfrak{M}_{A_0})$$

is called the φ-dimension of $\mathfrak{M}$ over A_0.

LEMMA 5.7.6. *Let $\varphi(x)$ be defined over A_0 and strongly minimal, and let $\mathfrak{M}$ and $\mathfrak{N}$ be models containing A_0. Then there exists an A_0-elementary map between $\varphi(\mathfrak{M})$ and $\varphi(\mathfrak{N})$ if and only if $\mathfrak{M}$ and $\mathfrak{N}$ have the same φ-dimension over A_0.*

PROOF. An A_0-elementary map between $\varphi(\mathfrak{M})$ and $\varphi(\mathfrak{N})$ maps bases to bases, so one direction is clear.

For the other direction we use Corollary 5.7.4: if $\varphi(\mathfrak{M})$ and $\varphi(\mathfrak{N})$ have the same dimension over A_0, let U and V be bases of $\varphi(\mathfrak{M})$ and $\varphi(\mathfrak{N})$, respectively, and let $f: U \to V$ be a bijection. By Corollary 5.7.4, f is A_0-elementary and by Lemma 5.6.4 f extends to an elementary bijection $g\colon \mathrm{acl}(A_0U) \to \mathrm{acl}(A_0V)$. Thus, $g \restriction \varphi(\mathfrak{M})$ is an A_0-elementary map from $\varphi(\mathfrak{M})$ to $\varphi(\mathfrak{N})$. ⊣

We now turn to showing that strongly minimal theories are categorical in all uncountable cardinals. For reference we first note the following special cases of the preceding lemmas.

COROLLARY 5.7.7.

1. *A theory T is strongly minimal if and only if over every parameter set there is exactly one non-algebraic type.*
2. *In models of a strongly minimal theory the algebraic closure defines a pregeometry.*
3. *Bijections between independent subsets of models of a strongly minimal theory are elementary. In particular, the type of n independent elements is uniquely determined.* ⊣

If T is strongly minimal, by the preceding we have

$$|\mathrm{S}(A)| \le |\mathrm{acl}(A)| + 1.$$

Strongly minimal theories are therefore λ-stable for all $\lambda \ge |T|$. Also there can be no binary tree of finite or cofinite sets. So by the remark after the proof of Theorem 5.2.6 T is totally transcendental. If $\varphi(\mathfrak{M})$ is cofinite and $\mathfrak{N}$ a proper elementary extension of $\mathfrak{M}$, then $\varphi(\mathfrak{N})$ is a proper extension of $\varphi(\mathfrak{M})$. Thus strongly minimal theories have no Vaughtian pairs.

THEOREM 5.7.8. *Let T be strongly minimal. Models of T are uniquely determined by their dimensions. The set of possible dimensions is an end segment of the cardinals. A model $\mathfrak{M}$ is ω-saturated if and only if $\dim(\mathfrak{M}) \geq \aleph_0$. All models are ω-homogeneous.*

PROOF. Let $\mathfrak{M}_0, \mathfrak{M}_1$ be models of the same dimension, and let B_0, B_1 be bases for $\mathfrak{M}_0$ and $\mathfrak{M}_1$, respectively. Then any bijection $f: B_0 \to B_1$ is an elementary map by Corollary 5.7.7, which extends to an isomorphism of the algebraic closures $\mathfrak{M}_0$ and $\mathfrak{M}_1$ by Lemma 5.6.4.

The next claim implies that the possible dimensions form an end segment of the cardinals.

CLAIM. *Every infinite algebraically closed subset S of M is the universe of an elementary substructure.*

PROOF OF CLAIM. By Theorem 2.1.2 it suffices to show that every consistent $L(S)$-formula $\varphi(x)$ can be realised in S. If $\varphi(\mathfrak{M})$ is finite, all realisations are algebraic over S and belong therefore to S. If $\varphi(\mathfrak{M})$ is cofinite, $\varphi(\mathfrak{M})$ meets all infinite sets.

Let A be a finite subset of $\mathfrak{M}$ and p the non-algebraic type in $S(A)$. Thus, p is realised in $\mathfrak{M}$ exactly if $M \neq \operatorname{acl}(A)$, i.e., if and only if $\dim(\mathfrak{M}) > \dim(A)$. Since all algebraic types over A are always realised in $\mathfrak{M}$, this shows that $\mathfrak{M}$ is ω-saturated if and only if $\mathfrak{M}$ has infinite dimension. ⊣

It remains to show that all models are ω-homogeneous. Let $f: A \to B$ be an elementary bijection between two finite subsets of M. By Lemma 5.6.4, f extends to an elementary bijection between $\operatorname{acl}(A)$ and $\operatorname{acl}(B)$. If $a \in M \setminus \operatorname{acl}(A)$, then $p = \operatorname{tp}(a/A)$ is the unique non-algebraic type over A and $f(p)$ is the unique non-algebraic type over B. Since $\dim(A) = \dim(B)$, the argument in the previous paragraph shows that $f(p)$ is realised in $\mathfrak{M}$. ⊣

COROLLARY 5.7.9. *If T is countable and strongly minimal, it is categorical in all uncountable cardinalities.*

PROOF. Let $\mathfrak{M}_1$ and $\mathfrak{M}_2$ be two models of cardinality $\kappa > \aleph_0$. Choose two bases B_1 and B_2 of $\mathfrak{M}_1$ and $\mathfrak{M}_2$ respectively. By p. 79, equation (5.1), B_1 and B_2 both have cardinality κ. Then any bijection $f: B_1 \to B_2$ is an elementary map by Corollary 5.7.7, which extends to an isomorphism of the algebraic closures M_1 and M_2 by Lemma 5.6.4. ⊣

EXERCISE 5.7.1. If $\mathfrak{M}$ is minimal and ω-saturated, then $\operatorname{Th}(\mathfrak{M})$ is strongly minimal.

EXERCISE 5.7.2. Show that the theory of an infinite set equipped with a bijection without finite cycles is strongly minimal and that the associated geometry is trivial.

EXERCISE 5.7.3. Show directly that strongly minimal theories eliminate $\exists^\infty$ (without using Corollaries 5.5.4 and 5.7.9.)

EXERCISE 5.7.4. A type is minimal if and only if its set of realisations in any model is minimal (i.e., has no infinite and coinfinite relatively definable subsets).

EXERCISE 5.7.5. Show that acl defines a pregeometry on $\mu(\mathfrak{M})$ if $\mu(\mathfrak{M})$ is minimal. In fact the following is true: if $b \in \mu(\mathfrak{M})$, $a \in \operatorname{acl}(Ab)$, $b \notin \operatorname{acl}(Aa)$, then $a \in \operatorname{acl}(A)$. Furthermore we have $\deg(a/A) = \deg(a/Ab)$.

5.8. The Baldwin–Lachlan Theorem

In this section, we present the characterisation of uncountably categorical theories due to Baldwin and Lachlan [2]. Since this characterisation is independent of the uncountable cardinal, it implies Morley's Theorem. The crucial point is the existence of a strongly minimal formula φ in a totally transcendental theory. By Corollary 5.5.5, each model $\mathfrak{M}$ is prime over the set of realisations $\varphi(\mathfrak{M})$ whose dimension determines the isomorphism type of the model.

THEOREM 5.8.1 (Baldwin–Lachlan). *Let κ be an uncountable cardinal. A countable theory T is κ-categorical if and only if T is ω-stable and has no Vaughtian pairs.*

PROOF. If T is categorical in some uncountable cardinal, then T is ω-stable by Theorem 5.2.4 and has no Vaughtian pair by Corollary 5.5.4.

For the other direction we first obtain a strongly minimal formula: since T is totally transcendental, it has a prime model $\mathfrak{M}_0$. (This follows from Theorems 4.5.7 and 4.5.9 or from Theorem 5.3.3.) Let $\varphi(x)$ be a minimal formula in $L(M_0)$, which exists by Lemma 5.7.2. Since T has no Vaughtian pairs, $\exists^\infty$ can be eliminated by Lemma 5.5.7 and hence $\varphi(x)$ is strongly minimal by Lemma 5.7.2.

Let $\mathfrak{M}_1, \mathfrak{M}_2$ be models of cardinality κ. We may assume that $\mathfrak{M}_0$ is an elementary submodel of both $\mathfrak{M}_1$ and $\mathfrak{M}_2$. Since T has no Vaughtian pair, $\mathfrak{M}_i$ is a minimal extension of $M_0 \cup \varphi(\mathfrak{M}_i)$, $i = 1, 2$. Therefore, $\varphi(\mathfrak{M}_i)$ has cardinality κ and hence (since κ is uncountable) we conclude that $\dim_\varphi(\mathfrak{M}_1/M_0) = \kappa = \dim_\varphi(\mathfrak{M}_2/M_0)$. By Lemma 5.7.6 there exists an M_0-elementary map from $\varphi(\mathfrak{M}_0)$ to $\varphi(\mathfrak{M}_1)$, which by Lemma 5.3.8 extends to an isomorphism from $\mathfrak{M}_1$ to $\mathfrak{M}_2$. $\dashv$

COROLLARY 5.8.2 (Morley). *Let κ be an uncountable cardinal. Then T is $\aleph_1$-categorical if and only if T is κ-categorical.*

Notice that the proof of Theorem 5.8.1 shows in fact the following.

COROLLARY 5.8.3. *Suppose T is $\aleph_1$-categorical, $\mathfrak{M}_1, \mathfrak{M}_2$ are models of T, $a_i \in \mathfrak{M}_i$ and $\varphi(x, a_i)$ strongly minimal, $i = 1, 2$, with* $\text{tp}(a_1) = \text{tp}(a_2)$. *If $\mathfrak{M}_1$ and $\mathfrak{M}_2$ have the same respective φ-dimension, then they are isomorphic.*

For uncountable models, the φ-dimension equals the cardinality of the model, so clearly does not depend on the realisation of $\text{tp}(a_i)$. We will show in Section 6.3 that the converse to Corollary 5.8.3 holds also for countable models, i.e., if $\varphi(x, a_0)$ is strongly minimal, then the φ-dimension of M is the same for all a realizing $\text{tp}(a_0)$. Thus, also the countable models of an uncountably categorical theory are in one-to-one correspondence with the possible φ-dimensions.

Chapter 6

MORLEY RANK

In this chapter we collect a number of further results about totally transcendental theories, in particular we will introduce Morley rank. We then finish the analysis of the countable models of uncountably categorical theories.

For convenience, we first introduce the 'monster model' (for arbitrary theories), a very large, very saturated, very homogeneous model. From now on, all models we consider will be elementary submodels of this monster model.

Furthermore, we often simplify the notation by assuming that we are working in a many-sorted structure where for each $n \in \omega$ we have an extra sort for n-tuples of elements. While we are then working in a many-sorted language in which we have to specify the sorts for all variables involved in a formula, this allows us to treat n-tuples exactly like 1-tuples, i.e., elements. We emphasise that this is purely a notational convention. In Section 8.4 we will show how to systematically extend a structure by introducing new sorts in a big way without changing those properties of the theories we are interested in.

6.1. Saturated models and the monster

The importance of saturated structures was already visible in Section 5.2 where we showed that saturated structures of fixed cardinality are unique up to isomorphism. Saturated structures need not exist (think about why not), but by considering special models instead, we can preserve many of the important properties – and prove their existence.

DEFINITION 6.1.1. A structure $\mathfrak{M}$ of cardinality $\kappa \geq \omega$ is *special* if $\mathfrak{M}$ is the union of an elementary chain $\mathfrak{M}_\lambda$ where λ runs over all cardinals less than κ and each $\mathfrak{M}_\lambda$ is λ^+-saturated.

We call $(\mathfrak{M}_\lambda)$ a *specialising chain.*

REMARK. Saturated structures are special. If $|\mathfrak{M}|$ is regular, the converse is true.

LEMMA 6.1.2. *Let λ be an infinite cardinal $\geq |L|$. Then every L-structure $\mathfrak{M}$ of cardinality 2^λ has a λ^+-saturated elementary extension of cardinality 2^λ.*

PROOF. Every set of cardinality 2^λ has 2^λ many subsets of cardinality at most λ. This allows us to construct a continuous elementary chain

$$\mathfrak{M} = \mathfrak{M}_0 \prec \mathfrak{M}_1 \cdots \prec \mathfrak{M}_\alpha \prec \cdots \quad (\alpha < \lambda^+)$$

of structures of cardinality 2^λ such that all $p \in \mathrm{S}(A)$, for $A \subseteq M_\alpha$, $|A| \leq \lambda$, are realised in $\mathfrak{M}_{\alpha+1}$. The union of this chain has the desired properties. ⊣

COROLLARY 6.1.3. *Let* $\kappa > |L|$ *be an uncountable cardinal. Assume that*

$$\lambda < \kappa \Rightarrow 2^\lambda \leq \kappa \tag{6.1}$$

Then every infinite L-structure $\mathfrak{M}$ *of cardinality smaller than* κ *has a special extension of cardinality* κ.

Let α be a limit ordinal. Then for any cardinal μ, $\kappa = \beth_\alpha(\mu)$ satisfies (6.1) and we have $\mathrm{cf}(\kappa) = \mathrm{cf}(\alpha)$ (see p. 190).

The following is a generalisation of Lemma 5.2.8

THEOREM 6.1.4. *Two elementarily equivalent special structures of the same cardinality are isomorphic.*

PROOF. The proof is a refined version of the proof of Lemma 5.2.8. Let $\mathfrak{A}$ and $\mathfrak{B}$ be two elementarily equivalent special structures of cardinality κ with specialising chains $(\mathfrak{A}_\lambda)$ and $(\mathfrak{B}_\lambda)$, respectively. The well-ordering defined in the proof of Lemma A.3.7 can be used to find enumerations $(a_\alpha)_{\alpha<\kappa}$ and $(b_\alpha)_{\alpha<\kappa}$ of A and B such that $a_\alpha \in A_{|\alpha|}$ and $b_\alpha \in B_{|\alpha|}$. We construct an increasing sequence of elementary maps $f^\alpha \colon A^\alpha \to B^\alpha$ such that for all α which are zero or limit ordinals we have $a_{\alpha+i} \in A^{\alpha+2i}$, $b_{\alpha+i} \in B^{\alpha+2i+1}$, and also $|A^\alpha| \leq |\alpha|$, $A^\alpha \subseteq A_{|\alpha|}$, $|B^\alpha| \leq |\alpha|$, $B^\alpha \subseteq B_{|\alpha|}$. ⊣

DEFINITION 6.1.5. A structure $\mathfrak{M}$ is

- κ*-universal* if every structure of cardinality $< \kappa$ which is elementarily equivalent to $\mathfrak{M}$ can be elementarily embedded into $\mathfrak{M}$.
- κ*-homogeneous* if for every subset A of M of cardinality smaller than κ and for every $a \in M$, every elementary map $A \to M$ can be extended to an elementary map $A \cup \{a\} \to M$.
- *strongly* κ*-homogeneous* if for every subset A of M of cardinality less than κ, every elementary map $A \to M$ can be extended to an automorphism of $\mathfrak{M}$.

THEOREM 6.1.6. *Special structures of cardinality* κ *are* κ^+*-universal and strongly* $\mathrm{cf}(\kappa)$*-homogeneous.*

PROOF. Let $\mathfrak{M}$ be a special structure of cardinality κ. The κ^+-universality of $\mathfrak{M}$ can be proved in the same way as Theorem 6.1.4. Let A be a subset of M of cardinality less than $\mathrm{cf}(\kappa)$ and let $f : A \to M$ an elementary map. Fix a specialising sequence $(\mathfrak{M}_\lambda)$. For λ_0 sufficiently large, $\mathfrak{M}_{\lambda_0}$ contains A. The sequence

$$M_\lambda^* = \begin{cases} (\mathfrak{M}_\lambda, a)_{a \in A}, & \text{if } \lambda_0 \leq \lambda \\ (\mathfrak{M}_{\lambda_0}, a)_{a \in A}, & \text{if } \lambda < \lambda_0 \end{cases}$$

is then a specialising sequence of $(\mathfrak{M}, a)_{a \in A}$. For the same reason $(\mathfrak{M}, f(a))_{a \in A}$ is special. By Theorem 6.1.4 these two structures are isomorphic under an automorphism of $\mathfrak{M}$ which extends f. ⊣

The monster model. Let T be a complete theory with infinite models. For convenience, we would like to work in a very large saturated structure, large enough so that any model of T can be considered as an elementary substructure. If T is totally transcendental, by Remark 5.2.10 we can choose such a *monster model* as a saturated model of cardinality κ where κ is a regular cardinal greater than all the models we ever consider otherwise. Using Exercise 8.2.7, this also works for stable theories and regular κ with $\kappa^{|T|} = \kappa$. For any infinite λ, $\kappa = (\lambda^{|T|})^+$ has this property.

In order to construct the *monster model* $\mathfrak{C}$ for an arbitrary theory T we will work in BGC (Bernays–Gödel + Global Choice). This is a conservative extension of ZFC (see Appendix A) which adds classes to ZFC. Then *being a model of* T is interpreted as being the union of an elementary chain of (set-size) models of T. The universe of our monster model $\mathfrak{C}$ will be a proper class.

THEOREM 6.1.7 (BGC). *There is a class-size model $\mathfrak{C}$ of T such that all types over all subsets of C are realised in $\mathfrak{C}$. Moreover $\mathfrak{C}$ is uniquely determined up to isomorphism.*

PROOF. Global choice allows us to construct a long continuous elementary chain $(M_\alpha)_{\alpha \in \mathrm{On}}$ of models of T such that all types over M_α are realised in $M_{\alpha+1}$. Let $\mathfrak{C}$ be the union of this chain. The uniqueness is proved as in Lemma 5.2.8. ⊣

We call $\mathfrak{C}$ the *monster model* of T. Note that Global Choice implies that $\mathfrak{C}$ can be well-ordered.

COROLLARY 6.1.8.

- *$\mathfrak{C}$ is κ-saturated for all cardinals κ.*
- *Any model of T is elementarily embeddable in $\mathfrak{C}$*
- *Any elementary bijection between two subsets of $\mathfrak{C}$ can be extended to an automorphism of $\mathfrak{C}$.* ⊣

We say that two elements are *conjugate over* some parameter set A if there is an automorphism of $\mathfrak{C}$ fixing A elementwise and taking one to the other. Note that $a, b \in \mathfrak{C}$ are conjugate over A if and only if they have the same type over A. We call types $p \in \mathrm{S}(A), q \in \mathrm{S}(B)$ *conjugate over* D if there is an automorphism f of $\mathfrak{C}$ fixing D and taking A to B and such that $q = \{\varphi(x, f(a)) \mid \varphi(x, a) \in p\}$. Note that strictly speaking $\mathrm{Aut}(\mathfrak{C})$ is not an object in Bernays–Gödel Set theory but we will nevertheless use this term as a way of talking about automorphisms.

Readers who mistrust set theory can fix a regular cardinal γ bigger than the cardinality of all models and parameter sets they want to consider. For $\mathfrak{C}$ they may then use a special model of cardinality $\kappa = \beth_\gamma(\aleph_0)$. This is κ^+-universal and strongly γ-homogeneous.

We will use the following convention throughout the rest of this book:

- Any *model* of T is an elementary substructure of $\mathfrak{C}$. We identify models with their universes and denote them by $M, N, \ldots$.
- *Parameter sets* $A, B, \ldots$ *are* subsets of $\mathfrak{C}$.
- Formulas $\varphi(x)$ with parameters define a *subclass* $\mathbb{F} = \varphi(\mathfrak{C})$ of $\mathfrak{C}$. Two formulas are *equivalent* if they define the same class.
- We write $\models \varphi$ for $\mathfrak{C} \models \varphi$.
- A set of formulas with parameters from $\mathfrak{C}$ is *consistent* if it is realised in $\mathfrak{C}$.
- If $\pi(x)$ and $\sigma(x)$ are partial types we write $\pi \vdash \sigma$ for $\pi(\mathfrak{C}) \subseteq \sigma(\mathfrak{C})$.
- A *global* type is a type p over $\mathfrak{C}$; we denote this by $p \in S(\mathfrak{C})$.

This convention changes the flavour of quite a number of proofs. As an example look at the following

LEMMA 6.1.9. *An elementary bijection* $f: A \to B$ *extends to an elementary bijection between* $\mathrm{acl}(A)$ *and* $\mathrm{acl}(B)$.

PROOF. Extend f to an automorphism f' of $\mathfrak{C}$. Clearly f' maps $\mathrm{acl}(A)$ to $\mathrm{acl}(B)$. ⊣

This implies Lemma 5.6.4 and the second claim in the proof of Theorem 5.7.6. Note that by the remark on p. 79 for any model M and any $A \subseteq M$ the algebraic closure of A in the sense of M equals the algebraic closure in the sense of $\mathfrak{C}$.

LEMMA 6.1.10. *Let* $\mathbb{D}$ *be a definable class and* A *a set of parameters. Then the following are equivalent*:

a) $\mathbb{D}$ *is definable over* A.
b) $\mathbb{D}$ *is invariant under all automorphisms of* $\mathfrak{C}$ *which fix* A *pointwise.*

PROOF. Let $\mathbb{D}$ be defined by φ, defined over $B \supset A$. Consider the maps

$$\mathfrak{C} \xrightarrow{\tau} S(B) \xrightarrow{\pi} S(A),$$

where $\tau(c) = \mathrm{tp}(c/B)$ and π is the restriction map. Let Y be the image of $\mathbb{D}$ in $S(A)$. Since $Y = \pi[\varphi]$, Y is closed.

Assume that $\mathbb{D}$ is invariant under all automorphisms of $\mathfrak{C}$ which fix A pointwise. Since elements which have the same type over A are conjugate by an automorphism of $\mathfrak{C}$, this means that $\mathbb{D}$-membership depends only on the type over A i.e., $\mathbb{D} = (\pi\tau)^{-1}(Y)$.

This implies that $[\varphi] = \pi^{-1}(Y)$, or $S(A) \setminus Y = \pi[\neg\varphi]$, hence $S(A) \setminus Y$ is also closed and we conclude that Y is clopen. By Lemma 4.2.3 $Y = [\psi]$ for some $L(A)$-formula ψ. This ψ defines $\mathbb{D}$. ⊣

The same proof shows that the same is true for definable *relations* $R \subseteq \mathfrak{C}^n$: namely, R is A-definable if and only if it is invariant under all $\alpha \in \mathrm{Aut}(\mathfrak{C}/A)$.

DEFINITION 6.1.11. The *definable closure* $\mathrm{dcl}(A)$ of A is the set of elements c for which there is an $L(A)$-formula $\varphi(x)$ such that c is the unique element satisfying φ. Elements or tuples a and b are said to be *interdefinable* if $a \in \mathrm{dcl}(b)$ and $b \in \mathrm{dcl}(a)$.

COROLLARY 6.1.12. 1. $a \in \mathrm{dcl}(A)$ *if and only if a has only one conjugate over A.*

2. $a \in \mathrm{acl}(A)$ *if and only if a has finitely many conjugates over A.*

PROOF. 1. is clear, since $a \in \mathrm{dcl}(A)$ means that $\{a\}$ is A-definable. 2. follows from Remark 5.6.3, since the realisations of $\mathrm{tp}(a/A)$ are exactly the conjugates of a over A. ⊣

EXERCISE 6.1.1. Finite structures are saturated.

EXERCISE 6.1.2. $\mathrm{acl}(A)$ is the intersection of all models which contain A.

EXERCISE 6.1.3. Prove *Robinson's Joint Consistency Lemma*: extend the complete L-theory T to an L_1-theory T_1 and an L_2-theory T_2 such that $L = L_1 \cap L_2$. If T_1 and T_2 are both consistent, show that $T_1 \cup T_2$ is consistent.

EXERCISE 6.1.4. Prove *Beth's Interpolation Theorem*: if $\vdash \varphi_1 \to \varphi_2$ for L_i-sentences φ_i, there is an $L = L_1 \cap L_2$-sentence θ such that $\vdash \varphi_1 \to \theta$ and $\vdash \theta \to \varphi_2$.

EXERCISE 6.1.5. A class $\mathcal{C}$ of L-structures is a PC_Δ-*class* if there is an extension L' of L and an L'-theory T' such that $\mathcal{C}$ consists of all reducts to L of models of T'. Show that a PC_Δ-class is elementary if and only if it is closed under elementary substructures.

EXERCISE 6.1.6. If M is κ-saturated, then over every set of cardinality smaller than κ every type in κ many variables is realised in M.

EXERCISE 6.1.7. Let κ be an infinite cardinal, not smaller than the cardinality of L and M an L-structure. Show that the following are equivalent

a) M is κ-saturated

b) M is κ^+-universal and κ-homogeneous

If $\max(|L|, \aleph_0) < \kappa$ this is also equivalent to

c) M is κ-universal and κ-homogeneous.

EXERCISE 6.1.8. Let κ be a an uncountable regular cardinal $> |L|$. We use the notation $2^{<\kappa}$ for $\sup_{\lambda<\kappa} 2^\lambda$. Show that

1. Every L-structure $\mathfrak{M}$ of cardinality $2^{<\kappa}$ has a κ-saturated elementary extension of cardinality $2^{<\kappa}$.
2. Assume that $\lambda < \kappa$ implies $2^{\lambda} \leq \kappa$. Then every L-structure $\mathfrak{M}$ of cardinality κ has a saturated elementary extension of cardinality κ.

EXERCISE 6.1.9. Let $\mathfrak{P}_{\omega}(\mathbb{N})$ be the set of all finite subsets of $\mathbb{N}$. Show that the theory of $(\mathfrak{P}_{\omega}(\mathbb{N}), \subseteq)$ has a saturated model of cardinality κ if and only if κ is regular and $\lambda < \kappa$ implies $2^{\lambda} \leq \kappa$.

EXERCISE 6.1.10. A *type-definable class* is the class of all realisations of a set of formulas. Show that a type-definable class is invariant under all automorphisms of $\mathfrak{C}$ which fix A pointwise if and only if it can be defined by a set of $L(A)$-formulas. (Use Exercise 4.2.2(a) and the proof of Lemma 6.1.10.)

EXERCISE 6.1.11. Let A be contained in B. Show that the following are equivalent:

1. $B \subseteq \operatorname{dcl}(A)$.
2. Every type over A extends uniquely to B.

EXERCISE 6.1.12.

1. b is in the definable closure of a if and only if there is a 0-definable class $\mathbb{D}$ containing a and a 0-definable map $\mathbb{D} \to \mathfrak{C}$ which maps a to b.
2. a and b are interdefinable if and only if a and b are contained in 0-definable classes $\mathbb{D}$ and $\mathbb{E}$ and there is a 0-definable bijection between $\mathbb{D}$ and $\mathbb{E}$ which maps a to b.

EXERCISE 6.1.13. Let K be a model of ACF_0, ACF_p for $p > 0$, of RCF or of DCF_0 and let A be a (differential) subfield of K. Prove that the definable closure of A is

1. (ACF_0) A itself,
2. (ACF_p) the perfect hull (see Definition B.3.8) of A,
3. (RCF) the relative algebraic closure of A,
4. (DCF_0) A itself.

EXERCISE 6.1.14. Use Exercise 6.1.13 to show the following:

1. Let K be a model of ACF_0 and $f : K^n \to K$ a definable function. Then K^n can be decomposed into a finite number of definable subsets X_i such that, on each X_i, f is given by a rational function.
2. The same is true for models of ACF_p, $p > 0$. But on each X_i, f is of the form $h^{p^{-m}}$, for some rational function h.
3. In models of DCF_0, f is given on each X_i by a differential rational function.

EXERCISE 6.1.15 (P. Neumann). Let X be an infinite set, $G \leq \operatorname{Sym}(X)$, and $B \subseteq X$ finite. Suppose that the orbit of each of the elements $c_0, \ldots, c_n \in X$

under G is infinite. Then there is some $g \in G$ with $g(c_i) \notin B$ for $i = 0, \ldots, n$. (Hint: Proceed as in Exercise 5.6.2. In fact, Exercise 5.6.2 follows from this by compactness.)

EXERCISE 6.1.16 (B. Neumann). Let G be a group (not necessarily abelian), and $H_0, \ldots, H_n$, subgroups of infinite index. Show that G is not a finite union of cosets of the H_i (but see Exercise 6.1.17, which is easy). Deduce Lemma 3.3.9 from this.

EXERCISE 6.1.17 (B. & P. Neumann). Deduce Exercise 6.1.15 from 6.1.16 and conversely.

6.2. Morley rank

The Morley rank is a rather natural notion of dimension on the formulas of a theory T or, equivalently, on the definable subsets of the monster model, defined inductively very much like the dimension of algebraic sets. It is ordinal valued for all consistent formulas if and only if T is totally transcendental. *In this section, we let T be a complete (possibly uncountable) theory.*

We now define the Morley rank MR for formulas $\varphi(x)$ with parameters in the monster model. We remind the reader that by the conventions introduced at the beginning of this chapter, a variable or element in the many-sorted language described there may refer to an n-tuple of the original single sort. We begin by defining the relation $\mathrm{MR}\,\varphi \geq \alpha$ by induction on the ordinal α.[1]

$\mathrm{MR}\,\varphi \geq 0$ if φ is consistent;

$\mathrm{MR}\,\varphi \geq \beta + 1$ if there is an infinite family $(\varphi_i(x) \mid i < \omega)$ of formulas (in the same variable x) which imply φ, are pairwise inconsistent and such that $\mathrm{MR}\,\varphi_i \geq \beta$ for all i;

$\mathrm{MR}\,\varphi \geq \lambda$ (for a limit ordinal λ) if $\mathrm{MR}\,\varphi \geq \beta$ for all $\beta < \lambda$.

DEFINITION 6.2.1. To define $\mathrm{MR}\,\varphi$ we distinguish three cases

1. If there is no α with $\mathrm{MR}\,\varphi \geq \alpha$, we put $\mathrm{MR}\,\varphi = -\infty$.
2. $\mathrm{MR}\,\varphi \geq \alpha$ for all α, we put $\mathrm{MR}\,\varphi = \infty$.
3. Otherwise, by the definition of $\mathrm{MR}\,\varphi \geq \lambda$ for limit ordinals λ, there is a maximal α with $\mathrm{MR}\,\varphi \geq \alpha$, and we set $\mathrm{MR}\,\varphi = \max\{\alpha \mid \mathrm{MR}\,\varphi \geq \alpha\}$.

It is easy to see by induction on α that the relation $\mathrm{MR}\,\varphi \geq \alpha$ implies the relation $\mathrm{MR}\,\varphi \geq \beta$ for $\beta \leq \alpha$. It follows from this that indeed the Morley rank of φ is at least α if and only if the relation $\mathrm{MR}\,\varphi \geq \alpha$ holds.[2]

---

[1] See the set-theoretic caveat before Exercise 6.2.1.

[2] Here, of course $-\infty$ is considered as being smaller and ∞ as being bigger than all ordinals.

Note that

$$\begin{array}{lll} \mathrm{MR}\,\varphi = -\infty & \Leftrightarrow & \varphi \text{ is inconsistent} \\ \mathrm{MR}\,\varphi = 0 & \Leftrightarrow & \varphi \text{ is consistent and algebraic.} \end{array}$$

If a formula has ordinal-valued Morley rank, we also say that *this formula has* Morley rank.[3] The Morley rank $\mathrm{MR}(T)$ of T is the Morley rank of the formula $x \doteq x$. The Morley rank of a formula $\varphi(x, a)$ only depends on $\varphi(x, y)$ and the type of a. It follows that if a formula has Morley rank, then it is less than $(2^{|T|})^+$. We will see in Exercise 6.2.5 that in fact all ordinal ranks are smaller that $|T|^+$.

REMARK. Clearly, if φ implies ψ, then $\mathrm{MR}\,\varphi \le \mathrm{MR}\,\psi$. It is also clear from the definition that if φ has rank $\alpha < \infty$, then for every $\beta < \alpha$ there is a formula ψ which implies φ and has rank β.

EXAMPLE 6.2.2. In Infset the formula $x_1 \doteq a$ has Morley rank 0. If considered as a formula in two variables, $\varphi(x_1, x_2) = x_1 \doteq a$, it has Morley rank 1.

The next lemma expresses the fact that the formulas of rank less than α form an *ideal* in the Boolean algebra of equivalence classes of formulas.

LEMMA 6.2.3.

$$\mathrm{MR}(\varphi \vee \psi) = \max(\mathrm{MR}\,\varphi, \mathrm{MR}\,\psi).$$

PROOF. By the previous remark, we have $\mathrm{MR}(\varphi \vee \psi) \ge \max(\mathrm{MR}\,\varphi, \mathrm{MR}\,\psi)$. For the other inequality we show by induction on α that

$$\mathrm{MR}(\varphi \vee \psi) \ge \alpha + 1 \text{ implies } \max(\mathrm{MR}\,\varphi, \mathrm{MR}\,\psi) \ge \alpha + 1.$$

Let $\mathrm{MR}(\varphi \vee \psi) \ge \alpha + 1$. Then there is an infinite family of formulas (φ_i) that imply $\varphi \vee \psi$, are pairwise inconsistent and such that $\mathrm{MR}\,\varphi_i \ge \alpha$. By the induction hypothesis, for each i we have $\mathrm{MR}(\varphi_i \wedge \varphi) \ge \alpha$ or $\mathrm{MR}(\varphi_i \wedge \psi) \ge \alpha$. If the first case holds for infinitely many i, then $\mathrm{MR}\,\varphi \ge \alpha + 1$. Otherwise $\mathrm{MR}\,\psi \ge \alpha + 1$. ⊣

We call φ and ψ α-equivalent,

$$\varphi \sim_\alpha \psi,$$

if their symmetric difference $\varphi \triangle \psi$ has rank less than α. By our previous considerations it is clear that α-equivalence is in fact an equivalence relation.

We call a formula φ α-strongly minimal if it has rank α and for any formula ψ implying φ either ψ or $\varphi \wedge \neg\psi$, has rank less than α, (equivalently, if every $\psi \subseteq \varphi$ is α-equivalent to $\emptyset$ or to φ). In particular, φ is 0-strongly minimal if and only if φ is realised by a single element and φ is 1-strongly minimal if and only if φ is strongly minimal.

---

[3]Note that having Morley rank is a nontrivial property of a formula (see Theorem 6.2.7).

LEMMA 6.2.4. *Each formula φ of rank $\alpha < \infty$ is equivalent to a disjunction of finitely many pairwise disjoint α-strongly minimal formulas $\varphi_1, \ldots, \varphi_d$, the* α-strongly minimal components (*or just components*) *of φ. The components are uniquely determined up to α-equivalence.*

PROOF. Suppose φ is a formula of rank α without such a decomposition. Then φ can be written as the disjoint disjunction of a formula φ_1 of rank α and another formula ψ_1 of rank α not having such a decomposition. Inductively there are formulas $\varphi = \varphi_0, \varphi_1, \ldots$ of rank α and $\psi_1, \psi_2, \ldots$ so that φ_i is the disjoint union of φ_{i+1} and ψ_{i+1}. But then the rank of φ would be greater than α.

To see the uniqueness of this decomposition, let ψ be an α-strongly minimal formula implying φ and let $\varphi_1, \ldots, \varphi_d$ be the α-strongly minimal components. Then ψ can be decomposed into the formulas $\psi \wedge \varphi_i$, one of which must be α-equivalent to ψ. So up to α-equivalence the components of φ are exactly the α-strongly minimal formulas implying φ. ⊣

DEFINITION 6.2.5. For a formula φ of Morley rank $\alpha < \infty$, the *Morley degree* $\mathrm{MD}(\varphi)$ is the number of its α-strongly minimal components.

The Morley degree is not defined for inconsistent formulas or formulas not having Morley rank. The Morley degree of a consistent algebraic formula is the number of its realisations. Strongly minimal formulas are exactly the formulas of Morley rank and Morley degree 1. As with strongly minimal formulas it is easy to see that Morley rank and degree are preserved under definable bijections.

Defining $\mathrm{MD}_\alpha(\varphi)$ as the Morley degree for formulas φ of rank α, as 0 for formulas of smaller rank and as ∞ for formulas φ of higher rank, we obtain the following.

LEMMA 6.2.6. *If φ is the disjoint union of ψ_1 and ψ_2, then*

$$\mathrm{MD}_\alpha(\varphi) = \mathrm{MD}_\alpha(\psi_1) + \mathrm{MD}_\alpha(\psi_2). \quad ⊣$$

THEOREM 6.2.7. *The theory T is totally transcendental if and only if each formula has Morley rank.*

PROOF. Since there are not arbitrarily large ordinal Morley ranks, each formula $\varphi(x)$ without Morley rank can be decomposed into two disjoint formulas without Morley rank, yielding a binary tree of consistent formulas in the free variable x.

For the other direction let $(\varphi_s(x) \mid s \in {}^{<\omega}2)$ be a binary tree of consistent formulas. Then none of the φ_s has Morley rank. Otherwise there is a φ_s whose ordinal rank α is minimal and (among the formulas of rank α) of minimal degree. Then both φ_{s0} and φ_{s1} have rank α and therefore smaller degree than φ, a contradiction. ⊣

A group is said to have the the *descending chain condition* (dcc) on definable subgroups, if there is no infinite properly descending chain $H_0 \supset H_1 \supset H_2 \supset \cdots$ of definable subgroups.

REMARK 6.2.8. A totally transcendental group has the descending chain condition on definable subgroups.

PROOF. If H is a definable proper subgroup of a totally transcendental group G, then either the Morley rank or the Morley degree of H must be smaller than that of G since any coset of H has the same Morley rank and degree as H. Therefore the claim follows from the fact that the ordinals are well-ordered. ⊣

The previous remark is a crucial tool in the theory of totally transcendental groups. For example, it immediately implies that $(\mathbb{Z}, +)$ is not totally transcendental.

COROLLARY 6.2.9. *The theory of separably closed fields of degree of imperfection $e > 0$ is not totally transcendental.*

PROOF. The subfields $K \supset K^p \supset K^{p^2} \supset K^{p^3} \supset \cdots$ form an infinite definable chain of properly descending (additive) subgroups. In fact we will see later that the proof shows that that separably closed fields are not even *superstable* (see Exercise 8.6.10.) ⊣

DEFINITION 6.2.10. The *Morley rank* $\mathrm{MR}(p)$ of a type p is the minimal rank of any formula in p. If $\mathrm{MR}(p)$ is an ordinal α, then its *Morley degree* $\mathrm{MD}(p)$ is the minimal degree of a formula of p having rank α. If $p = \mathrm{tp}(a/A)$ we also write $\mathrm{MR}(a/A)$ and $\mathrm{MD}(a/A)$.

Algebraic types have Morley rank 0 and

$$\mathrm{MD}(p) = \deg(p).$$

Strongly minimal types are exactly the types of Morley rank and Morley degree 1.

Let $p \in \mathrm{S}(A)$ have Morley rank α and Morley degree d. Then by definition there is some $\varphi \in p$ of rank α and degree d. Clearly, φ is uniquely determined up to α-equivalence since for all ψ we have $\mathrm{MR}(\varphi \wedge \neg\psi) < \alpha$ if and only if $\psi \in p$. Thus, p is uniquely determined by φ:

$$p = \{\psi(x) \mid \psi\ L(A)\text{-formula},\ \mathrm{MR}(\varphi \wedge \neg\psi) < \alpha\}. \tag{6.2}$$

Obviously, α-equivalent formulas determine the same type (see Lemma 5.7.3).

Thus $\varphi \in L(A)$ belongs to a unique type of rank α if and only if φ is α-minimal over A; i.e., if φ has rank α and cannot be decomposed as the union of two $L(A)$-formulas of rank α.

LEMMA 6.2.11. *Let φ be a consistent $L(A)$-formula.*

1. $\mathrm{MR}\,\varphi = \max\{\mathrm{MR}(p) \mid \varphi \in p \in \mathrm{S}(A)\}$.

2. *Let* $\mathrm{MR}\,\varphi = \alpha$. *Then*

$$\mathrm{MD}\,\varphi = \sum(\mathrm{MD}(p) \mid \varphi \in p \in \mathrm{S}(A),\ \mathrm{MR}(p) = \alpha).$$

PROOF. 1: If $\mathrm{MR}\,\varphi = \infty$, then $\{\varphi\} \cup \{\neg\psi \mid \psi\ L(A)\text{-formula},\ \mathrm{MR}\,\psi < \infty\}$ is consistent. Any type over A containing this set of formulas has rank ∞.

If $\mathrm{MR}\,\varphi = \alpha$, there is a decomposition of φ in $L(A)$-formulas $\varphi_1, \dots, \varphi_k$, α-minimal over A. (Note that k is bounded by $\mathrm{MD}\,\varphi$.) By (6.2), the φ_i determine types p_i of rank α.

2: The p_i are exactly the types of rank α containing φ. Furthermore,

$$\mathrm{MD}\,\varphi_i = \mathrm{MD}(p_i). \qquad \dashv$$

COROLLARY 6.2.12. *If* $p \in \mathrm{S}(A)$ *has Morley rank and* $A \subseteq B$, *then*

$$\mathrm{MD}(p) = \sum(\mathrm{MD}(q) \mid p \subseteq q \in \mathrm{S}(B),\ \mathrm{MR}(p) = \mathrm{MR}(q)) \qquad \dashv$$

COROLLARY 6.2.13. *Let* $p \in \mathrm{S}(A)$ *have Morley rank and* $A \subseteq B$. *Then* $p \in \mathrm{S}(A)$ *has at least one and at most* $\mathrm{MD}(p)$ *many extensions to* B *of the same rank.* $\dashv$

We will later show that extensions of the same Morley rank are a special case of the non-forking extensions studied in Chapters 7 and 8. For types with Morley rank those of the same Morley rank are exactly the non-forking ones.

Caveat: Set-theoretically we defined the Morley rank as a function which maps each α to a class of formulas. In Bernays–Gödel set theory one cannot in general define functions from ordinals to classes by a recursion scheme. The more conscientious reader should therefore use a different definition: for each set A define the relation $\mathrm{MR}_A(\varphi) \geq \alpha$ using only formulas with parameters from A, and put $\mathrm{MR}\,\varphi \geq \alpha$ if $\mathrm{MR}_A(\varphi) \geq \alpha$ for some (sufficiently large) A. The following exercise shows that if φ is defined over an ω-saturated model M, we have $\mathrm{MR}\,\varphi = \mathrm{MR}_M\,\varphi$.

EXERCISE 6.2.1. Let φ be a formula with parameters in the ω-saturated model M. If $\mathrm{MR}\,\varphi > \alpha$, show that there is an infinite family of formulas *with parameters in* M which each imply φ, are pairwise inconsistent and have Morley rank $\geq \alpha$.

EXERCISE 6.2.2. Let φ be a formula of Morley rank $\alpha < \infty$ and $\psi_0, \psi_1, \dots$ an infinite sequence of formulas. Assume that there is a number k such that the conjunction any k of the ψ_i has Morley rank smaller than α. Then $\mathrm{MR}(\varphi \wedge \psi_i) < \alpha$ for almost all i.

EXERCISE 6.2.3. Show that a totally transcendental group G has a *connected component*, i.e., a smallest definable subgroup G^0 of finite index. Show also that any finite normal subgroup of G^0 lies in the centre of G^0.

EXERCISE 6.2.4. If T is totally transcendental, then all types over ω-saturated models have Morley degree 1. (We will see in Corollary 8.5.12 that this is true without assuming ω-saturation.)

EXERCISE 6.2.5 (Lachlan). If a type p has Morley rank, then $\mathrm{MR}(p) < |T|^+$. Hence, if T is totally transcendental, we have $\mathrm{MR}(T) < |T|^+$.

EXERCISE 6.2.6. For a topological space X we define recursively on ordinals $X^0 := X, X^{\alpha+1} := X^\alpha \setminus \{x \mid x \text{ isolated in } X^\alpha\}$ and $X^\lambda := \bigcap_{\alpha<\lambda} X^\alpha$ if λ is a limit ordinal. The *Cantor–Bendixson rank* of $x \in X$ is equal to α if α is maximal with $x \in X^\alpha$.

Show that on $\mathrm{S}(\mathfrak{C})$ the Morley rank equals the Cantor–Bendixson rank. Note that $\mathrm{S}(\mathfrak{C})$ is not even a class. So, for this exercise we have to ignore set-theoretic subtleties.

EXERCISE 6.2.7. Call a function R which associates to every non-empty definable class an ordinal *dimensional* if $R(\varphi \vee \psi) = \max(R(\varphi), R(\psi))$. A function $\mathcal{R} : \mathrm{S}(\mathfrak{C}) \to \mathrm{On}$ is *continuous* if $\{p \mid \mathcal{R}(p) \geq \alpha\}$ is closed for every α. Show that

$$R(\varphi) = \max\{\mathcal{R}(p) \mid \varphi \in p\}$$
$$\mathcal{R}(p) = \min\{R(\varphi) \mid \varphi \in p\}$$

defines a bijection between dimensional R and continuous functions $\mathcal{R}$.

EXERCISE 6.2.8. If p is a type over $\mathrm{acl}(A)$, then p and $p \upharpoonright A$ have the same Morley rank.

6.3. Countable models of $\aleph_1$-categorical theories

Uncountable models of $\aleph_1$-categorical theories are determined up to isomorphism by their cardinality. Section 5.8 showed that this cardinality coincides with the dimension of a strongly minimal formula. We here extend this analysis in order to classify also the countable models of $\aleph_1$-categorical theories and show in Theorem 6.3.7 that for each possible dimension there is a unique model of the theory.

Throughout this section we fix a countable $\aleph_1$-categorical theory T. For models $M \prec N$ of T and $\varphi(x) \in L(M)$ a strongly minimal formula, we write $\dim_\varphi(N/M)$ for the φ-dimension of N over M.

THEOREM 6.3.1. *Let T be a countable $\aleph_1$-categorical theory, $M \prec N$ be models of T, $A \subseteq M$ and $\varphi(x) \in L(A)$ a strongly minimal formula.*

1. *If $b_1, \ldots, b_n \in \varphi(N)$ are independent over M and N is prime over $M \cup \{b_1, \ldots, b_n\}$, then*

$$\dim_\varphi(N/M) = n, \text{ and}$$

2. $\dim_\varphi(N) = \dim_\varphi(M) + \dim_\varphi(N/M)$.

PROOF. For ease of notation we assume $A = \emptyset$.

1: Let $c \in \varphi(N)$. We want to show that c is algebraic over $M \cup \{b_1, \dots, b_n\}$. Assume the contrary. Then $p(x) = \text{tp}(c/M \cup \{b_1, \dots, b_n\})$ is strongly minimal and is axiomatised by

$$\{\varphi(x)\} \cup \{\neg\varphi_i(x) \mid i \in I\},$$

where the φ_i range over all algebraic formulas defined over $M \cup \{b_1, \dots, b_n\}$. Since $\varphi(M)$ is infinite, any finite subset of $p(x)$ is realised by an element of M. Thus, $p(x)$ is not isolated. But all elements of the prime extension N are atomic over $M \cup \{b_1, \dots, b_n\}$ by Corollary 5.3.7, a contradiction.

2: This follows from Remark C.1.8, if we can show that a basis of $\varphi(N)$ over $\varphi(M)$ is also a basis of $\varphi(N)$ over M. So the proof is complete once we have established the following lemma. ⊣

LEMMA 6.3.2. *Let T be ω-stable, $M \prec N$ models of T, $\varphi(x)$ be strongly minimal and $b_i \in \varphi(N)$. If the b_i are independent over $\varphi(M)$, they are independent over M.*

PROOF. Assume that $b_1, \dots, b_n$ are algebraically independent over $\varphi(M)$ but dependent over $a \in M$. Put $\bar{b} = b_1 \dots b_n$. An argument as in the proof of Theorem 5.5.2 shows that we may assume that M is ω-saturated. Let p be the type of $\bar{b}$ over M. We choose a sequence $\bar{b}^0, \bar{b}^1, \dots$ in $\varphi(M)$ such that $\bar{b}^{2i}$ is an n-tuple of elements algebraically independent over $\bar{b}^0 \dots \bar{b}^{2i-1}$ and $\bar{b}^{2i+1}$ realises $p \restriction a\bar{b}^0 \dots \bar{b}^{2i}$. Let q be the type of a over the set B of elements of $(\bar{b}^i)$. Since the sequence $(\bar{b}^i)$ is indiscernible, every permutation π of ω defines a type $\pi(q)$ over B. If $\{i \mid \pi(2i) \text{ even}\} \neq \{i \mid \pi'(2i) \text{ even}\}$, we have $\pi(q) \neq \pi'(q)$. So there are uncountably many types over B and T is not ω-stable. ⊣

The previous lemma holds for arbitrary theories. This uses the fact that $\varphi(x)$ is stable in the sense of Exercise 8.3.5 and that by symmetry there are few types over parameter sets contained in $\varphi(\mathfrak{C})$ (see Definition 8.2.1).

COROLLARY 6.3.3. *The dimension*

$$\dim(N/M) = \dim_\varphi(N/M)$$

of N over M does not depend on φ: it is the maximal length of an elementary chain

$$M = N_0 \precneqq N_1 \precneqq \cdots \precneqq N_n = N$$

between M and N.

PROOF. This follows from the previous theorem since T has no Vaughtian pairs. ⊣

For the remainder of this section, we let M_0 denote the prime model of T. We also fix a strongly minimal formula $\varphi(x, \overline{a}_0) \in L(M_0)$ and put $p_0(x) = \mathrm{tp}(\overline{a}_0)$.

Note that the type $p_0(x)$ of $\overline{a}_0$ is isolated by Theorem 4.5.2, hence realised in every model of T. For any model M and realisation $\overline{a}$ of p_0 in M, let $\dim_{\varphi(x,\overline{a})}(M)$ denote the $\varphi(x, \overline{a})$-dimension of M over $\overline{a}$. To simplify notation we assume that $\overline{a}_0$ is some element a_0 rather than a tuple.

Since M_0 is ω-homogeneous by Corollary 4.5.4, the dimension

$$m_0 = \dim_{\varphi(x,a)}(M_0)$$

does not depend on the realisation a of p_0 in M_0. We will show in Lemma 6.3.6 that the same is true for *any* model of T.

LEMMA 6.3.4. *A countable model M is saturated if and only if its $\varphi(x, \overline{a})$-dimension is ω. Hence in this case, the dimension is independent of the realisation of $p_0(x)$ in M. In particular, T is $\aleph_0$-categorical if and only if $m_0 = \omega$.*

PROOF. In a saturated model the $\varphi(x, \overline{a})$-dimension is infinite. Since there exists a unique countable saturated model by Lemmas 5.2.8 and 5.2.9, the first claim follows. This obviously does not depend on the realisation of p_0. The last sentence now follows from Theorem 5.2.11. ⊣

We need the following observation.

LEMMA 6.3.5. *If M is prime over a finite set and $m_0 < \omega$, then $\dim_{\varphi(x,a)}(M)$ is finite.*

PROOF. Suppose M is prime over the finite set C. Let B be a basis of $\varphi(M, a)$ over M_0. Since M is the minimal prime extension of $M_0 \cup B$, C is atomic over $M_0 \cup B$. Thus there exists a finite subset B_0 of B such that C is contained in the prime extension N of $M_0 \cup B_0$. As M is prime over Ca, it suffices to show that $\dim_{\varphi(x,a)}(N)$ is finite and this follows by Theorem 6.3.1 from

$$\dim_{\varphi(x,a)}(N) = m_0 + |B_0|. \quad ⊣$$

The crucial lemma and promised converse to Corollary 5.8.3 is the following.

LEMMA 6.3.6. *The dimension $\dim_{\varphi(x,a)}(M)$ does not depend on the realisation a of p_0 in M.*

PROOF. The lemma is clear if M is uncountable and also if M is countable with infinite φ-dimension by Lemma 6.3.4. Therefore we may assume that M has finite φ-dimension, which implies that m_0 is finite.

For the proof we now introduce the following notion: let a_1 and a_2 realise p_0. Choose a model N of finite φ-dimension containing a_1 and a_2, which exists by Lemma 6.3.5, and put

$$\mathrm{diff}(a_1, a_2) = \dim_{\varphi(x,a_1)}(N) - \dim_{\varphi(x,a_2)}(N).$$

This definition does not depend on the model N: if $N' \prec N$ is prime over a_1, a_2, then by Theorem 6.3.1 we have for $i = 1, 2$

$$\dim_{\varphi(x,a_i)}(N) = \dim_{\varphi(x,a_i)}(N') + \dim(N/N'),$$

so

$$\operatorname{diff}(a_1, a_2) = \dim_{\varphi(x,a_1)}(N') - \dim_{\varphi(x,a_2)}(N').$$

Clearly we have

$$\operatorname{diff}(a_1, a_3) = \operatorname{diff}(a_1, a_2) + \operatorname{diff}(a_2, a_3).$$

This implies that $\operatorname{diff}(a_1, a_2)$ only depends on $\operatorname{tp}(a_1, a_2)$. We will show that $\operatorname{diff}(a_1, a_2) = 0$ for all realisations of p_0. This implies that $\dim_{\varphi(x,a_1)}(M) = \dim_{\varphi(x,a_2)}(M)$ for all models M which contain a_1 and a_2.

For the proof choose an infinite sequence $a_1, a_2, \ldots$ with

$$\operatorname{tp}(a_i, a_{i+1}) = \operatorname{tp}(a_1, a_2)$$

for all i.

Now we use the fact that $\aleph_1$-categorical theories are ω-stable, so the type p_0 has Morley rank and an extension q_0 to $\{a_1, a_2, \ldots\}$ of the same rank. Let b be a realisation of q_0. Then, by Corollary 6.2.13, there are at most $\mathrm{MD}(p_0)$ many different types of the form $\operatorname{tp}(ba_i)$. So let $i < j$ be such that $\operatorname{tp}(ba_i) = \operatorname{tp}(ba_j)$.

Then $\operatorname{diff}(a_i, b) = -\operatorname{diff}(b, a_j)$ and

$$(j - i)\operatorname{diff}(a_1, a_2) = \operatorname{diff}(a_i, a_j) = \operatorname{diff}(a_i, b) + \operatorname{diff}(b, a_j) = 0,$$

implying $\operatorname{diff}(a_1, a_2) = 0$. ⊣

Thanks to the previous lemma we obtain a complete account of the models of an uncountably categorical theory.

THEOREM 6.3.7 (Baldwin–Lachlan). *If T is uncountably categorical, then for any cardinal $m \geq m_0$ there is a unique model M of T with $\dim_{\varphi(x,a)}(M) = m$. These models are pairwise non-isomorphic.*

PROOF. If $m = m_0 + \beta$, choose M prime over $M_0 \cup \{b_i \mid i < \beta\}$ where the $b_i \in \varphi(\mathfrak{C}, a_0)$ are independent over M_0. Uniqueness follows from Corollary 5.8.3 and non-isomorphism from Lemma 6.3.6. ⊣

EXERCISE 6.3.1. All models of an $\aleph_1$-categorical theory are ω-homogeneous.

EXERCISE 6.3.2. Let T be strongly minimal and m_0 be the dimension of the prime model. Show that m_0 is the smallest number n such $\mathrm{S}_{n+1}(T)$ is infinite.

6.4. Computation of Morley rank

In this section we show that the Morley rank agrees with the dimension of the pregeometry on strongly minimal sets and give some examples of how to compute it in ω-stable fields. We start with some general computations and continue to assume that T is a countable complete theory with infinite models.

LEMMA 6.4.1. *If b is algebraic over aA, we have* $\mathrm{MR}(b/A) \leq \mathrm{MR}(a/A)$.

PROOF. Let $\mathrm{MR}(a/A) = \alpha$. We prove $\mathrm{MR}(b/A) \leq \alpha$ by induction on α. Let $d = \mathrm{MD}(b/Aa)$. Choose an $L(A)$-formula $\varphi(x, y)$ in $\mathrm{tp}(ab/A)$ such that $\mathrm{MR}(\exists y \varphi(x, y)) = \alpha$ and $|\varphi(a', \mathfrak{C})| \leq d$ for all a'.

We show that the Morley rank of $\chi(y) = \exists x \varphi(x, y)$ is bounded by α. For this consider an infinite family $\chi_i(\mathfrak{C})$ of disjoint subclasses of $\chi(\mathfrak{C})$ defined over some extension A' of A. Put $\psi_i(x) = \exists y(\varphi(x, y) \wedge \chi_i(y))$. Since any $d + 1$ of the ψ_i have empty intersection, some $\psi_i(x)$ has Morley rank $\beta < \alpha$. Let b' be any realisation of $\chi_i(y)$. Choose a' such that $\models \varphi(a', b')$. Then b' is algebraic over $a'A$ and since a' realises $\psi_i(x)$, we have $\mathrm{MR}(a'/A') \leq \beta$. So by induction we conclude $\mathrm{MR}(b'/A') \leq \beta$, which shows $\mathrm{MR}\,\chi_i \leq \beta$. So χ does not contain an infinite family of disjoint formulas of Morley rank greater or equal to α. So $\mathrm{MR}\,\chi \leq \alpha$. ⊣

THEOREM 6.4.2. *Let $\varphi(x)$ be a strongly minimal formula defined over B and $a_1, \dots, a_n$ a sequence of realisations. Then*

$$\mathrm{MR}(a_1, \dots, a_n/B) = \dim_\varphi(a_1, \dots, a_n/B).$$

PROOF. By the lemma we may assume that $a_1, \dots, a_n$ are independent over B. Let $a_1, \dots, a_n$ realise the $L(B)$-formula $\psi(x_1, \dots, x_n)$. By induction we have $\mathrm{MR}(a_1, a_2, \dots, a_n/Ba_1) = n - 1$. So the formula

$$\chi_{a_1}(\bar{x}) = \psi(x_1, \dots, x_n) \wedge x_1 \doteq a_1$$

has rank at least $n - 1$. The infinitely many conjugates of χ_{a_1} over B are disjoint and have rank $n - 1$ as well. This shows that $\mathrm{MR}\,\psi \geq n$.

Let $B' \supset B$ be an extension of B. By Corollary 5.7.7 there is only one type $p \in \mathrm{S}_n(B')$ which is realised by an B'-independent sequence of elements of $\varphi(\mathfrak{C})$. So by induction, there is only one n-type of elements of $\varphi(\mathfrak{C})$ of rank $\geq n$. This implies that $\varphi(x_1) \wedge \dots \wedge \varphi(x_n)$ has rank n. ⊣

COROLLARY 6.4.3. *Let $\varphi(x)$ be a strongly minimal formula and $\psi(x_1, \dots, x_n)$ be defined over B such that ψ implies $\varphi(x_i)$ for all i. Then*

$$\mathrm{MR}\,\psi = \max\{\dim_\varphi(\overline{a}/B) \mid \mathfrak{C} \models \psi(\overline{a})\}.$$ ⊣

On strongly minimal sets, *Morley rank is definable.*

COROLLARY 6.4.4. *For any strongly minimal formula $\varphi(x)$ and any formula $\psi(x_1, \dots, x_n, \overline{y})$ which implies $\varphi(x_i)$ for all i, we have that*

$$\{\overline{b} \mid \mathrm{MR}\,\psi(x_1, \dots, x_n, \overline{b}) = k\}$$

is a definable class for every k.

PROOF. We show that $\mathrm{MR}\,\psi(x_1, \dots, x_n, \overline{b}) \geq k$ is an elementary property of $\overline{b}$ by induction on n. The case $n = 1$ follows from the fact that $\mathrm{MR}\,\psi(x_1, \overline{b}) \geq 1$ is equivalent to $\exists^\infty x_1 \psi(x_1, \overline{b})$. This is an elementary property of $\overline{b}$ since φ is strongly minimal (see the discussion on page 82). For the induction step we conclude from Corollary 6.4.3 that MR $\psi(x_1, \ldots, x_n, \overline{b}) \geq k$ if and only if one of the following is true:

- there is an a_1 such that MR $\psi(a_1, x_2, \ldots, x_n, \bar{b}) \geq k$,
- there is an a_1 which is is not algebraic over $\overline{b}$ such that MR $\psi(a_1, x_2, \ldots, x_n, \overline{b}) \geq k - 1$.

The first part is an elementary property of $\overline{b}$ by induction. For the second part note that by induction MR $\psi(a_1, x_2, \ldots, x_n, \overline{b}) \geq k - 1$ can be expressed by a formula $\theta(a_1, \overline{b})$. The second condition is then equivalent to $\exists^{\infty} x_1 \theta(x_1, \overline{b})$. ⊣

For algebraically closed fields, these considerations translate into the following statement.

COROLLARY 6.4.5. *Let K be a subfield of a model of* ACF_p *and let a be a tuple of elements. Then the Morley rank of a over K equals the transcendence degree of $K(a)$ over K.* ⊣

Note that by quantifier elimination definable sets in algebraically closed fields are exactly the *constructible sets* in algebraic geometry. The previous corollary expresses the important fact that for a definable set in an algebraically closed field the Morley rank equals the dimension of its Zariski closure in the sense of algebraic geometry (see e.g., [51]).

We now turn to the theory of differentially closed fields of characteristic 0, DCF_0.

Let $K \subseteq F$ be an extension of differential fields and a an element of F. The *dimension* of a over K is defined as the transcendence degree of $K\{a\}$ over K. There is a unique quantifier-free type over K of infinite dimension.

REMARK 6.4.6. If the dimension of a over K equals n, then the type of a over K is determined by the *d-minimal polynomial* f of a over K: so f is irreducible in $K[x_0, \ldots, x_n]$ and $f(a, \ldots, d^n a) = 0$ (see Remark B.3.7). ⊣

COROLLARY 6.4.7. DCF_0 *is ω-stable. For a differential field K and elements a we have*

$$\mathrm{MR}(a/K) \leq \dim(a/K).$$

If a has infinite dimension, then the type of a over K has Morley rank ω.

PROOF. There are at most $|K|$ many d-minimal polynomials over K, so at most $|K|$ many 1-types. Thus, DCF_0 is ω-stable.

We may assume that K is $\aleph_1$-saturated (otherwise we take a extension of $\mathrm{tp}(a/K)$ to an $\aleph_1$-saturated field with the same Morley rank. Then the Morley rank stays the same and the dimension does not increase.) We show $\mathrm{MR}(a/K) \leq \dim(a/K)$ by induction on $\dim(a/K)$. If $\dim(a/K) = 0$, then a is algebraic over K and the Morley rank is 0. Let $\dim(a/K) = n$ and let f be the minimal polynomial of a over K. Apart from $\mathrm{tp}(a/K)$, all other types over K containing $f(x, \ldots, d^n x) \doteq 0$ have dimension, and hence Morley rank, less than n. Since K is sufficiently saturated, this implies that the Morley rank of $\mathrm{tp}(a/K)$ is at most n.

By the next remark there are types of rank n for every n. This implies that there must be 1-types of rank $\geq \omega$. Since there is only one type p_∞ of infinite dimension, it follows[4] that p_∞ has rank ω. ⊣

LEMMA 6.4.8. *If $a, \ldots, d^{n-1}a$ are algebraically independent over K and $d^n a \in K$, then* $\mathrm{MR}(a/K) = n$.

PROOF. We prove this by induction on n. Consider the formula $\varphi(x) = (d^n(x) \doteq d^n(a))$. If the claim is true for $n-1$, all formulas $\varphi_b(x) = (d^{n-1}(x) \doteq b)$ have rank $n-1$. For all constants c the $\varphi_{d^{n-1}(a)+c}(x)$ are contained in $\varphi(x)$. So $\varphi(x)$ has rank n. All a' which realise $\varphi(x)$ have either dimension at most $n-1$ or have the same type as a over K. This shows that a has Morley rank n over K. ⊣

Dimension in pregeometries is *additive*, i.e., we have $\dim(ab/B) = \dim(a/B) + \dim(b/aB)$. This translates into additivity of Morley rank for elements in the algebraic closure of strongly minimal sets.

PROPOSITION 6.4.9. *Let φ be a strongly minimal formula defined over B and a, b algebraic over $\varphi(\mathfrak{C}) \cup B$. Then* $\mathrm{MR}(ab/B) = \mathrm{MR}(a/B) + \mathrm{MR}(b/aB)$.

PROOF. Assume $B = \emptyset$ for notational simplicity. Then a is algebraic over some finite set of elements of $\varphi(\mathfrak{C})$. We can split this set into a sequence $\bar{f}$ of elements which are independent over a and a tuple $\bar{a}$ which is algebraic over $\bar{f}a$. By taking non-forking extensions if necessary (see Corollary 6.2.13 and the discussion thereafter), we may assume that $\bar{f}$ is independent from ab. Now $\bar{a}$ and a are interalgebraic over $\bar{f}$. In the same way we find tuples $\bar{g}$ and $\bar{b}$ in $\varphi(\mathfrak{C})$ such $F = \bar{f}\bar{g}$ is independent from ab and $\bar{b}$ is interalgebraic with b over aF. The claim now follows from

$$\begin{aligned}\mathrm{MR}(ab) &= \mathrm{MR}(ab/F)\\ &= \mathrm{MR}(\bar{a}\bar{b}/F) = \mathrm{MR}(\bar{a}/F) + \mathrm{MR}(\bar{b}/\bar{a}F)\\ &= \mathrm{MR}(a/F) + \mathrm{MR}(b/aF) = \mathrm{MR}(a) + \mathrm{MR}(b/a).\end{aligned}$$ ⊣

In fact, Exercise 6.4.6 shows that if F is any infinite B-independent subset F of $\varphi(\mathfrak{C})$ then every element of $\mathrm{acl}(\varphi(\mathfrak{C}) \cup B)$ is interalgebraic over FB with a tuple in $\varphi(\mathfrak{C})$.

EXERCISE 6.4.1. Let T be strongly minimal. Show that a finite tuple a is geometrically independent from B over C (in the sense of a pregeometry, see page 208 and Exercise C.1.1) if and only if $\mathrm{MR}(a/BC) = \mathrm{MR}(a/C)$.

EXERCISE 6.4.2. Let be ψ a formula without parameters. Assume that ψ is *almost strongly minimal*, i.e., that there is a strongly minimal formula φ defined over some set B such that all elements of $\psi(\mathfrak{C})$ are algebraic over

[4] This is immediate if K is ω-saturated, see Exercise 6.2.1.

$\varphi(\mathfrak{C}) \cup B$. Then for all a, b in $\psi(\mathfrak{C})$ and any set C we have $\mathrm{MR}(ab/C) = \mathrm{MR}(a/C) + \mathrm{MR}(b/aC)$.

The following exercise shows that for arbitrary totally transcendental theories the Morley rank need not be additive.

EXERCISE 6.4.3. Consider the following theory in a two-sorted language having sorts A and B and a function $f : B \to A$. Assume that sort A is split into infinitely many infinite predicates $A_1, A_2, \ldots$ such that any $a \in A_n$ has exactly n preimages under f. Let a be a *generic* element of A, i.e., an element such that $\mathrm{MR}(a)$ is maximal, and choose $f(b) = a$. Show that $\mathrm{MR}(ab) = \mathrm{MR}(a) = 2$, $\mathrm{MR}(b/a) = 1$.

EXERCISE 6.4.4. Let $f : \mathbb{B} \to \mathbb{A}$ be a definable map. Prove the following:

1. If f is surjective, then $\mathrm{MR}(\mathbb{A}) \leq \mathrm{MR}(\mathbb{B})$.
2. If f has finite fibres, then $\mathrm{MR}(\mathbb{B}) \leq \mathrm{MR}(\mathbb{A})$.
3. Let $\mathbb{A}$ have Morley rank α. Call a fibre $f^{-1}(a)$ generic if $\mathrm{MR}(a/C) = \alpha$ where C is a set of parameters over which $\mathbb{A}$, $\mathbb{B}$ and f are defined. Now assume that the rank of all fibres is bounded by $\beta > 0$ and the rank of the generic fibres is bounded by β_{gen}. Prove[5]
$$\mathrm{MR}(\mathbb{B}) \leq \beta \cdot \alpha + \beta_{\mathrm{gen}}.$$
(Hint: Use induction on β_{gen} and α. The slightly weaker inequality $\mathrm{MR}(\mathbb{B}) \leq \beta \cdot (\alpha + 1)$ is due to Shelah ([54], Thm. V 7.8) and Erimbetov [18].)

EXERCISE 6.4.5. A theory has the *definable multiplicity property* if for all $\varphi(x, y)$, n and k the class $\{b \mid \mathrm{MR}\,\varphi(x, b) = n,\ \mathrm{MD}\,\varphi(x, b) = k\}$ is definable. Find an example of a theory T which has definable Morley rank but not the definable multiplicity property.

EXERCISE 6.4.6. Let φ be a strongly minimal formula without parameters and F an infinite independent subset of $\varphi(\mathfrak{C})$. Then every element of $\mathrm{acl}\,\varphi(\mathfrak{C})$ is interalgebraic over F with a tuple in $\varphi(\mathfrak{C})$.

[5]We use here *ordinal addition and multiplication*: $\alpha + \beta$ is the order type of α followed by β and $\beta \cdot \alpha$ is the order type of the lexicographical ordering of $\alpha \times \beta$.

Chapter 7

SIMPLE THEORIES

So far, we have mainly studied totally transcendental theories, a small subclass of the class of stable theories, indeed the most stable ones. Before we turn to stable theories in general, we consider simple (but possibly unstable) theories, a generalisation which, after their first introduction by Shelah [56], gained new attention following the fundamental work of Kim and Pillay [34]. Interest in simple theories increased with Hrushovski's results on pseudo-finite fields, see [29]. The presentation given here owes much to Casanovas, see [14].

7.1. Dividing and forking

We will characterise simple theories by the existence of a well-behaved notion of independence, a relation on types satisfying certain properties. To this end we here define forking and dividing. In the context of totally transcendental theories, these concepts correspond to type extensions of smaller Morley rank. *Throughout this section, we work in a countable complete theory T with infinite models.*

We begin with a reformulation of the Standard Lemma 5.1.3 on indiscernibles.

LEMMA 7.1.1 (The Standard Lemma). *Let A be a set of parameters, $\mathcal{I}$ an infinite sequence of tuples and J a linear order. Then there is a sequence of indiscernibles over A of order type J realizing* $\mathrm{EM}(\mathcal{I}/A)$. $\dashv$

DEFINITION 7.1.2. We say $\varphi(x, b)$ *divides* over A (with respect to k) if there is a sequence $(b_i)_{i<\omega}$ of realisations of $\mathrm{tp}(b/A)$ such that $(\varphi(x, b_i))_{i<\omega}$ is k-inconsistent.[1] A set of formulas $\pi(x)$ divides over A if $\pi(x)$ implies some $\varphi(x, b)$ which divides over A. There is no harm in allowing $\varphi(x, y)$ to contain parameters from A.

If $\varphi(x, a)$ implies $\psi(x, a')$ and $\psi(x, a')$ divides over A, then $\varphi(x, a)$ divides over A. Thus φ divides over A if and only if $\{\varphi\}$ divides over A. Also a set

[1] A family $(\varphi_i(x))_{x\in I}$ is k-inconsistent if for every k-element subset K of I the set $\{\varphi_i \mid i \in K\}$ is inconsistent.

π divides over A if and only if a conjunction of formulas from π divides over A. Note that it makes sense to say that $\pi(\overline{x})$ divides over A for $\overline{x}$ an infinite sequence of variables as we may use dummy variables without changing the meaning of dividing.

EXAMPLE. In the theory DLO, the formula $b_1 < x < b_2$ divides over the empty set (for $k = 2$). The type $p = \{x > a \mid a \in \mathbb{Q}\}$ does not divide over the empty set for any k.

The following is easy to see.

REMARK 7.1.3. 1. If $a \notin \mathrm{acl}(A)$, then $\mathrm{tp}(a/Aa)$ divides over A.
2. If $\pi(x)$ is consistent and defined over $\mathrm{acl}(A)$, then $\pi(x)$ does not divide over A.

LEMMA 7.1.4. *The set $\pi(x, b)$ divides over A if and only if there is a sequence $(b_i)_{i<\omega}$ of indiscernibles over A with $\mathrm{tp}(b_0/A) = \mathrm{tp}(b/A)$ and $\bigcup_{i<\omega} \pi(x, b_i)$ inconsistent.*

We may replace ω by any infinite linear order. Note also that b may be a tuple of infinite length.

PROOF. If $(b_i)_{i<\omega}$ is a sequence of indiscernibles over A with $\mathrm{tp}(b_0/A) = \mathrm{tp}(b/A)$ and $\bigcup_{i<\omega} \pi(x, b_i)$ inconsistent there is a conjunction $\varphi(x, b)$ of formulas from $\pi(x, b)$ for which $\Sigma(x) = \{\varphi(x, b_i) \mid i < \omega\}$ is inconsistent. So Σ contains some k-element inconsistent subset. This implies that $(\varphi(x, b_i))_{i<\omega}$ is k-inconsistent.

Assume conversely that $\pi(x, b)$ divides over A. Then some finite conjunction $\varphi(x, b)$ of formulas from $\pi(x, b)$ divides. Let $(b_i)_{i<\omega}$ be a sequence of realisations of $\mathrm{tp}(b/A)$ such that $(\varphi(x, b_i) \mid i < \omega)$ is k-inconsistent. We may assume by Lemma 7.1.1 that $(b_i)_{i<\omega}$ is indiscernible over A. Clearly, $\bigcup_{i<\omega} \pi(x, b_i)$ is inconsistent. ⊣

COROLLARY 7.1.5. *The following are equivalent:*

1. $\mathrm{tp}(a/Ab)$ *does not divide over* A.
2. *For any infinite sequence of A-indiscernibles $\mathcal{I}$ containing b, there exists some a' with $\mathrm{tp}(a'/Ab) = \mathrm{tp}(a/Ab)$ and such that $\mathcal{I}$ is indiscernible over Aa'.*
3. *For any infinite sequence of A-indiscernibles $\mathcal{I}$ containing b, there exists $\mathcal{I}'$ with $\mathrm{tp}(\mathcal{I}'/Ab) = \mathrm{tp}(\mathcal{I}/Ab)$ and such that $\mathcal{I}'$ is indiscernible over Aa.*

PROOF. 2) ⇔ 3): this is clear by considering appropriate automorphisms. It is also easy to see that the conclusion of 2) and 3) is equivalent to:

(∗) *There exist a' and $\mathcal{I}'$ with $\mathrm{tp}(a'/Ab) = \mathrm{tp}(a/Ab)$ and $\mathrm{tp}(\mathcal{I}'/Ab) = \mathrm{tp}(\mathcal{I}/Ab)$ such that $\mathcal{I}'$ is indiscernible over Aa'.*

1) ⇒ (∗): Let $\mathcal{I} = (b_i)_{i \in I}$ be an infinite sequence of indiscernibles with $b_{i_0} = b$. Let $p(x, y) = \mathrm{tp}(ab/A)$. Then $\bigcup_{i \in I} p(x, b_i)$ is consistent by

Lemma 7.1.4. Let a' be a realisation. By Lemma 7.1.1, there is $\mathcal{I}'' = (b''_i)_{i \in I}$ indiscernible over Aa' and realising $\mathrm{EM}(\mathcal{I}/Aa')$. Since $\models p(a', b''_{i_0})$, there is an automorphism $\alpha \in \mathrm{Aut}(\mathfrak{C}/Aa')$ taking b''_{i_0} to b. Put $\mathcal{I}' = \alpha(\mathcal{I}'')$.

2) $\Rightarrow$ 1): Let $p(x, y) = \mathrm{tp}(ab/A)$ and let $(b_i)_{i<\omega}$ be a sequence of indiscernibles over A with $\mathrm{tp}(b_0/A) = \mathrm{tp}(b/A)$. We have to show that $\bigcup_{i<\omega} p(x, b_i)$ is consistent. By assumption there is a' with $\mathrm{tp}(a'/Ab) = \mathrm{tp}(a/Ab)$ such that $\mathcal{I}$ is indiscernible over Aa'. As $\models p(a', b)$, a' is a realisation of $\bigcup_{i<\omega} p(x, b_i)$. ⊣

The next proposition states a transitivity property of dividing. See Corollary 7.2.17, its proof and Exercise 7.2.5.

PROPOSITION 7.1.6. *If* $\mathrm{tp}(a/B)$ *does not divide over* $A \subseteq B$ *and* $\mathrm{tp}(c/Ba)$ *does not divide over* Aa, *then* $\mathrm{tp}(ac/B)$ *does not divide over* A.

PROOF. Let $b \in B$ be a finite tuple and $\mathcal{I}$ an infinite sequence of A-indiscernibles containing b. If $\mathrm{tp}(a/B)$ does not divide over A, there is some $\mathcal{I}'$ with $\mathrm{tp}(\mathcal{I}'/Ab) = \mathrm{tp}(\mathcal{I}, Ab)$ and indiscernible over Aa. If $\mathrm{tp}(c/Ba)$ does not divide over Aa, there is $\mathcal{I}''$ with $\mathrm{tp}(\mathcal{I}''/Aab) = \mathrm{tp}(\mathcal{I}'/Aab)$ and indiscernible over Aac proving the claim. ⊣

DEFINITION 7.1.7. The set of formulas $\pi(x)$ *forks* over A if $\pi(x)$ implies a disjunction $\bigvee_{\ell<d} \varphi_\ell(x)$ of formulas $\varphi_\ell(x)$ each dividing over A.

Thus, if $\pi(x)$ divides over A, it forks over A. The converse need not be true in general (see Exercise 7.1.6). By definition (and compactness), we immediately see the following.

REMARK 7.1.8 (Non-forking is closed). If $p \in \mathrm{S}(B)$ forks over A, there is some $\varphi(x) \in p$ such that any type in $\mathrm{S}(B)$ containing $\varphi(x)$ forks over A.

COROLLARY 7.1.9 (Finite character). *If* $p \in \mathrm{S}(B)$ *forks over* A, *there is a finite subset* $B_0 \subseteq B$ *such that* $p \upharpoonright AB_0$ *forks over* A. ⊣

LEMMA 7.1.10. *If* π *is finitely satisfiable in* A, *then* π *does not fork over* A.

PROOF. If $\pi(x)$ implies the disjunction $\bigvee_{\ell<d} \varphi_\ell(x, b)$, then some φ_ℓ has a realisation a in A. If the b_i, $i < \omega$, realise $\mathrm{tp}(b/A)$, then $\{\varphi_\ell(x, b_i) \mid i < \omega\}$ is realised by a. So φ_ℓ does not divide over A. ⊣

LEMMA 7.1.11. *Let* $A \subseteq B$ *and let* π *be a partial type over* B. *If* π *does not fork over* A, *it can be extended to some* $p \in \mathrm{S}(B)$ *which does not fork over* A.

PROOF. Let $p(x)$ be a maximal set of $L(B)$-formulas containing $\pi(x)$ which does not fork over A. Clearly, p is consistent. Let $\varphi(x) \in L(B)$. If neither φ nor $\neg\varphi$ belongs to p, then both $p \cup \{\varphi\}$ and $p \cup \{\neg\varphi\}$ fork over A, and hence p forks over A. Thus p is complete. ⊣

EXERCISE 7.1.1. If $\mathcal{I}$ is an infinite sequence of indiscernibles over A, then there is a model M extending A over which $\mathcal{I}$ is still indiscernible.

EXERCISE 7.1.2.
1. Let $\varphi(x)$ be a formula over A with Morley rank and let $\psi(x)$ define a subclass of $\varphi(\mathfrak{C})$. If ψ forks over A, it has smaller Morley rank than φ.
2. Let p be a type with Morley rank and q an extension of p. If q forks over A, it has smaller Morley rank than p.

We will see in Exercise 8.5.5 that in both statements the converse is also true.

EXERCISE 7.1.3. Let p be a type over the model M and $A \subseteq M$. Assume that M is $|A|^+$-saturated. Show that p forks over A if and only if p divides over A.

EXERCISE 7.1.4. A global type which is *A-invariant*, i.e., invariant under all $\alpha \in \mathrm{Aut}(\mathfrak{C}/A)$, does not fork over A.

EXERCISE 7.1.5. Let M be a κ-saturated and strongly κ-homogeneous model. If $p \in \mathrm{S}(M)$ forks over a subset A of cardinality smaller than κ, then p has κ many conjugates under $\mathrm{Aut}(M/A)$.

EXERCISE 7.1.6. Define the *cyclical order* on $\mathbb{Q}$ by

$$\mathrm{cyc}(a, b, c) \Leftrightarrow (a < b < c) \vee (b < c < a) \vee (c < a < b).$$

Show:

1. $(\mathbb{Q}, \mathrm{cyc})$ has quantifier elimination.
2. For $a \neq b$, $\mathrm{cyc}(a, x, b)$ divides over the empty set.
3. The unique type over the empty set forks (but of course does not divide) over the empty set.

EXERCISE 7.1.7. If $\mathrm{tp}(a/Ab)$ does not divide over A and $\varphi(x, b)$ divides over A with respect to k, then $\varphi(x, b)$ divides over Aa with respect to k.

7.2. Simplicity

In this section, we define simple theories and the notion of forking independence whose properties characterise such theories. By the absence of binary trees of consistent formulas, totally transcendental theories are simple. We will see in the next chapter that in fact all stable theories are simple. Recall that by our convention (see page 89), variables x and y may belong to different sorts representing n_x and n_y-tuples of elements, respectively. We continue to denote by T a countable complete theory with infinite models.

DEFINITION 7.2.1.
1. A formula $\varphi(x, y)$ has the *tree property* (with respect to k) if there is a tree of parameters $(a_s \mid \emptyset \neq s \in {}^{<\omega}\omega)$ such that:
   a) For all $s \in {}^{<\omega}\omega$, $(\varphi(x, a_{si}) \mid i < \omega)$ is k-inconsistent.
   b) For all $\sigma \in {}^{\omega}\omega$, $\{\varphi(x, a_s) \mid \emptyset \neq s \subseteq \sigma\}$ is consistent.
2. A theory T is *simple* if there is no formula $\varphi(x, y)$ with the tree property.

Clearly, for the notion of simplicity, it suffices to consider formulas $\varphi(x, y)$ without parameters.

REMARK 7.2.2. It is not hard to see that in totally transcendental theories no formula has the tree property. This is immediate for $k = 2$. The general case follows from Exercise 6.2.2.

DEFINITION 7.2.3. Let Δ be a finite set of formulas $\varphi(x, y)$ without parameters. A *Δ–k-dividing sequence* over A is a sequence $(\varphi_i(x, a_i) \mid i < \delta)$ such that:

1. $\varphi_i(x, y) \in \Delta$;
2. $\varphi_i(x, a_i)$ divides over $A \cup \{a_j \mid j < i\}$ with respect to k;
3. $\{\varphi_i(x, a_i) \mid i < \delta\}$ is consistent.

LEMMA 7.2.4. 1. *If φ has the tree property with respect to k, then for every A and μ there exists a φ–k-dividing sequence over A of length μ.*
2. *If no $\varphi \in \Delta$ has the tree property with respect to k, there is no infinite Δ–k-dividing sequence over $\emptyset$.*

PROOF. 1): Note first that we may assume that μ is a limit ordinal. A compactness argument shows that for every μ and κ there is a tree $(a_s \mid \emptyset \neq s \in {}^{<\mu}\kappa)$ such that all families $(\varphi(x, a_{si}) \mid i < \kappa)$ are k-inconsistent and for all $\sigma \in {}^{\mu}\kappa$, $\{\varphi(x, a_s) \mid \emptyset \neq s \subseteq \sigma\}$ is consistent. If κ is bigger than $2^{\max(|T|,|A|,\mu)}$, we recursively construct a path σ such that for all $s \in \sigma$, infinitely many a_{si} have the same type over $A \cup \{a_t \mid t \leq s\}$. Now $(\varphi(x, a_{\sigma \restriction i+1}) \mid i < \mu)$ is a φ–k-dividing sequence over A.

2): Suppose there is an infinite Δ–k-dividing sequence over $\emptyset$. If φ appears infinitely many times in this sequence, there is an infinite φ–k-dividing sequence $(\varphi(x, a_i) \mid i < \omega)$. For each i we choose a sequence $(a_i^n \mid n < \omega)$ with $\mathrm{tp}(a_i^n/\{a_j \mid j < i\}) = \mathrm{tp}(a_i/\{a_j \mid j < i\})$ such that $(\varphi(x, a_i^n) \mid n < \omega)$ is k-inconsistent. Then we find parameters b_s showing that φ has the tree property with respect to k as follows: assume $s \in {}^{i+1}\omega$ and $\vec{b} = (b_{s\restriction 1}, \ldots, b_{s \restriction i})$ have been defined such that $\mathrm{tp}(a_1, \ldots, a_{i-1}) = \mathrm{tp}(\vec{b})$. Choose $\alpha \in \mathrm{Aut}(\mathfrak{C})$ with $\alpha(a_1, \ldots, a_{i-1}) = \vec{b}$ and put $b_s = \alpha(a_i^{s(i)})$. ⊣

It is easy to see that in simple theories for every finite set Δ and all k there exists a finite bound on the possible lengths of Δ–k-dividing sequences.

PROPOSITION 7.2.5. *Let T be a complete theory. The following are equivalent.*
a) *T is simple.*
b) (Local Character) *For all $p \in \mathrm{S}_n(B)$ there is some $A \subseteq B$ with $|A| \leq |T|$ such that p does not divide over A.*
c) *There is some κ such that for all models M and $p \in \mathrm{S}_n(M)$ there is some $A \subseteq M$ with $|A| \leq \kappa$ such that p does not divide over A.*

PROOF. a) $\Rightarrow$ b): If b) does not hold, there is a sequence $(\varphi_i(x, b_i) \mid i < |T|^+)$ of formulas from $p(x)$ such that every $\varphi_i(x, b_i)$ divides over $\{b_j \mid j < i\}$

with respect to k_i. There is an infinite subsequence for which all $\varphi_i(x, y)$ equal $\varphi(x, y)$ and all $k_i = k$ yielding a φ–k-dividing sequence.

b) ⇒ c): Clear.

c) ⇒ a): If φ has the tree property, there are φ–k-dividing sequences $(\varphi(x, b_i) \mid i < \kappa^+)$. It is easy to construct an ascending sequence of models M_i, $(i < \kappa^+)$ such that $b_j \in M_i$ for $j < i$ and $\varphi(x, b_i)$ divides over M_i. Extend the set of $\varphi(x, b_i)$ to some type $p(x) \in S(M)$ where $M = \bigcup_{i<\kappa^+} M_i$. Then p divides over each M_i. ⊣

COROLLARY 7.2.6. *Let T be simple and $p \in S(A)$. Then p does not fork over A.*

PROOF. Suppose p forks over A, so p implies some disjunction $\bigvee_{l<d} \varphi_l(x, b)$ of formulas all of which divide over A with respect to k. Put $\Delta = \{\varphi_l(x, y) \mid l < d\}$.

We will show by induction that for all n there is a Δ–k-dividing sequence over A of length n. This contradicts the remark after Lemma 7.2.4. We will assume also that the dividing sequence is consistent with $p(x)$.

Suppose that $(\psi_i(x, a_i) \mid i < n)$ is a Δ–k-dividing sequence over A, consistent with $p(x)$. By Exercise 7.2.4 we can replace b with a conjugate b' over A such that $(\psi_i(x, a_i) \mid i < n)$ is a dividing sequence over Ab'. Now one of the formulas $\varphi_l(x, b')$, say $\varphi_0(x, b')$, is consistent with $p(x) \cup \{\psi_i(x, a_i) \mid i < n\}$. So $\varphi_0(x, b'), \psi_0(x, a_0), \dots, \psi_{n-1}(x, a_{n-1})$ is a Δ–k-dividing sequence over A and consistent with $p(x)$. ⊣

Let p be a type over A and q an extension of p. We call p a *forking* extension if q forks over A.

COROLLARY 7.2.7 (Existence). *If T is simple, every type over A has a non-forking extension to any B containing A.*

PROOF. This follows from Corollary 7.2.6 and Lemma 7.1.11. ⊣

DEFINITION 7.2.8. The set A is *independent* from B over C, written

$$A \underset{C}{\perp} B,$$

if for every finite tuple $\overline{a}$ from A, the type $\text{tp}(\overline{a}/BC)$ does not fork over C. If C is empty, we may omit it and write $A \perp B$.

This definition makes sense since forking of $\text{tp}(\overline{a}/BC)$ does not depend on the enumeration of $\overline{a}$ and since $\text{tp}(\overline{a}/BC)$ forks over C if the type of a subsequence of $\overline{a}$ forks over C. So this is the same as saying that $\text{tp}(A/BC)$ does not fork over C.

DEFINITION 7.2.9. Let I be a linear order. A sequence $(a_i)_{i \in I}$ is called

1. *independent* over A if $a_i \perp_A \{a_j \mid j < i\}$ for all i;
2. a *Morley sequence* over A if it is independent and indiscernible over A;

3. a *Morley sequence* in $p(x)$ over A if it is a Morley sequence over A consisting of realisations of p.

EXAMPLE 7.2.10. Let q be a global type invariant over A. Then any sequence $(b_i)_{i\in I}$ where each b_i realises $q \upharpoonright A\cup\{b_j \mid j<i\}$ is a Morley sequence over A.

PROOF. Let us call such sequences *good*. Clearly a subsequence of a good sequence is good again. So for indiscernibility it suffices to show that all finite good sequences $b_0\dots b_n$ and $b'_0\dots b'_n$ have the same type over A. Indeed, using induction, we may assume that $b_0\dots b_{n-1}$ and $b'_0\dots b'_{n-1}$ have the same type and so $\alpha(b_0\dots b_{n-1}) = b'_0\dots b'_{n-1}$ for some $\alpha\in \mathrm{Aut}(\mathfrak{C}/A)$. Then

$$\begin{aligned}\alpha(\mathrm{tp}(b_n/Ab_0\dots b_{n-1})) &= \alpha(q\upharpoonright Ab_0\dots b_{n-1}) = q\upharpoonright Ab'_0\dots b'_{n-1}\\ &= \mathrm{tp}(b'_n/Ab'_0\dots b'_{n-1}),\end{aligned}$$

which proves our claim. Independence follows from Exercise 7.1.4. ⊣

We call such a sequence $(b_i)_{i\in I}$ a *Morley sequence of q over A*. Note that our proof shows that the type of a Morley sequence of q over A is uniquely determined by its order type.

LEMMA 7.2.11. *If $(a_i)_{i\in I}$ is independent over A and $J<K$ are subsets of I, then $\mathrm{tp}((a_k)_{k\in K}/A\{a_j\mid j\in J\})$ does not divide over A.*

PROOF. We may assume that K is finite. The claim now follows from Proposition 7.1.6 by induction on $|K|$. ⊣

LEMMA 7.2.12 (Shelah). *For all A there is some λ such that for any linear order I of cardinality λ and any family $(a_i)_{i\in I}$ there exists an A-indiscernible sequence $(b_j)_{j<\omega}$ such that for all $j_1<\dots<j_n<\omega$ there is a sequence $i_1<\dots<i_n$ in I with $\mathrm{tp}(a_{i_1}\dots a_{i_n}/A)=\mathrm{tp}(b_{j_1}\dots b_{j_n}/A)$.*

PROOF. We only need that λ satisfies the following. Let $\tau=\sup_{n<\omega}|\mathrm{S}_n(A)|$.

1. $\mathrm{cf}(\lambda)>\tau$
2. For all $\kappa<\lambda$ and all $n<\omega$ there is some $\kappa'<\lambda$ with $\kappa'\to(\kappa)^n_\tau$ (see Definition C.3.1).

By Erdős–Rado (see Theorem C.3.2) we may take $\lambda=\beth_{\tau^+}$.

We now construct a sequence of types $p_1(x_1)\subseteq p_2(x_1,x_2)\subseteq\cdots$ with $p_n\in \mathrm{S}_n(A)$ such that for all $\kappa<\lambda$ there is some $I'\subseteq I$ with $|I'|=\kappa$ such that $\mathrm{tp}(a_{i_1}\dots a_{i_n})=p_n$ for all $i_1<\dots<i_n$ from I'.

Then we can choose the $(b_i)_{i<\omega}$ as a realisation of $\bigcup_{i<\omega}p_i$.

If p_{n-1} has been constructed and we are given $\kappa<\lambda$, we choose $\kappa'<\lambda$ with $\kappa'\to(\kappa)^n_\tau$ and some $I'\subseteq I$ with $|I'|=\kappa'$ such that $\mathrm{tp}(a_{i_1}\dots a_{i_{n-1}}/A)=p_{n-1}$ for all $i_1<\dots<i_{n-1}$ from I'. Thus there are $I''\subseteq I'$ and p_n^κ with $\mathrm{tp}(a_{i_1}\dots a_{i_n})=p_n^\kappa$ for all $i_1<\dots<i_n$ from I''. Since $\mathrm{cf}(\lambda)>\tau$, there is some p_n with $p_n^\kappa=p_n$ for cofinally many κ. ⊣

The existence of a Ramsey cardinal $\kappa > \tau$ (see p. 210) would directly imply that any sequence of order type κ contains a countable indiscernible subsequence (in fact even an indiscernible subsequence of size κ).

LEMMA 7.2.13. *If $p \in S(B)$ does not fork over A, there is an infinite Morley sequence in p over A which is indiscernible over B. In particular, if T is simple, for every $p \in S(A)$, there is an infinite Morley sequence in p over A.*

PROOF. Let a_0 be a realisation of p. By Lemma 7.1.11 there is a non-forking extension p' of p to Ba_0. Let a_1 be a realisation of p'. Continuing in this way we obtain a sequence $(a_i)_{i<\lambda}$ with $a_i \mathop{\smile\!\!\!\!\!\!|}_A B(a_j)_{j<i}$ for arbitrary λ. By Lemma 7.2.12 we obtain a sequence of length ω with the same property and indiscernible over B. The last sentence is immediate by Corollary 7.2.6. $\dashv$

PROPOSITION 7.2.14. *Let T be simple and $\pi(x, y)$ be a partial type over A. Let $(b_i)_{i<\omega}$ be an infinite Morley sequence over A and $\bigcup_{i<\omega} \pi(x, b_i)$ consistent. Then $\pi(x, b_0)$ does not divide over A.*

PROOF. By Lemma 7.1.1, for every linear order I there is a Morley sequence $(b_i)_{i\in I}$ in $\text{tp}(b_0/A)$ over A such that $\Sigma(x) = \bigcup_{i\in I} \pi(x, b_i)$ is consistent. Choose I having the inverse order type of $|T|^+$. Let c be a realisation of Σ. By Proposition 7.2.5(b) there is some i_0 such that $\text{tp}(c/A \cup \{b_i \mid i \in I\})$ does not divide over $A \cup \{b_i \mid i > i_0\}$. This implies that $\text{tp}(c/A \cup \{b_i \mid i \geq i_0\})$ does not divide over $A \cup \{b_i \mid i > i_0\}$. By Lemma 7.2.11, $\text{tp}((b_i \mid i > i_0)/Ab_{i_0})$ does not divide over A. Hence $\text{tp}(c\ (b_i \mid i > i_0)/Ab_{i_0})$ does not divide over A by Proposition 7.1.6. This implies that $\pi(x, b_{i_0})$ does not divide over A. $\dashv$

PROPOSITION 7.2.15. *Let T be simple. Then $\pi(x, b)$ divides over A if and only if it forks over A.*

PROOF. By definition, if $\pi(x, b)$ divides over A, it forks over A. For the converse assume $\pi(x, b)$ does not divide over A. So if $\psi(x, b) = \bigvee_{l<d} \varphi_l(x, b)$ is implied by $\pi(x, b)$, it does not divide over A. Let $(b_i)_{i<\omega}$ be a Morley sequence in $\text{tp}(b/A)$ over A, which exists since T is simple. So $\{\psi(x, b_i) \mid i \in \omega\}$ is consistent. By the pigeon-hole principle there must be some l and some infinite $I \subseteq \omega$ such that $\{\varphi_l(x, b_i) \mid i \in I\}$ is consistent. By Proposition 7.2.14, $\varphi_l(x, b)$ does not divide over A. Hence $\pi(x, b)$ does not fork over A. $\dashv$

PROPOSITION 7.2.16 (Symmetry). *In simple theories, independence is symmetric.*

PROOF. Assume $A \mathop{\smile\!\!\!\!\!\!|}_C B$ and consider finite tuples $a \in A$ and $b \in B$. Since $a \mathop{\smile\!\!\!\!\!\!|}_C b$, Lemma 7.2.13 gives an infinite Morley sequence $(a_i)_{i<\omega}$ in $\text{tp}(a/Cb)$ over C, indiscernible over Cb. Let $p(x, y) = \text{tp}(ab/C)$. Then $\bigcup_{i<\omega} p(a_i, y)$ is consistent because it is realised by b. Thus, by Proposition 7.2.14, $p(a, y)$ does not divide over C. This proves $b \mathop{\smile\!\!\!\!\!\!|}_C a$. Since this holds for all $a \in A$, $b \in B$, it follows $B \mathop{\smile\!\!\!\!\!\!|}_C A$ by Finite Character. $\dashv$

COROLLARY 7.2.17 (Monotonicity and Transitivity). *Let T be simple, $B \subseteq C \subseteq D$. Then we have $A \mathop{\downarrow}_B D$ if and only if $A \mathop{\downarrow}_B C$ and $A \mathop{\downarrow}_C D$.*

PROOF. One direction of this equivalence, *Monotonicity*, holds for arbitrary theories and follows easily from the definition. For *Transitivity*, the other direction, note that by Proposition 7.2.15 we may read Proposition 7.1.6 after replacing finite tuples by infinite ones as

$$A' \mathop{\downarrow}_A B \text{ and } C \mathop{\downarrow}_{AA'} B \quad \Rightarrow \quad CA' \mathop{\downarrow}_A B.$$

Swapping the left and the right hand sides, this is exactly the transitivity. Hence the claim follows from Proposition 7.2.16. ⊣

COROLLARY 7.2.18. *That $(a_i)_{i \in I}$ is independent over A does not depend on the ordering of I.*

PROOF. Let i be an element of I and J, K two subsets such that $J < i < K$. Write $a_J = \{a_j \mid j \in J\}$ and $a_K = \{a_k \mid k \in K\}$. We have to show that $a_i \mathop{\downarrow}_A a_J a_K$. Now by Lemma 7.2.11 we have $a_K \mathop{\downarrow}_A a_J a_i$. Monotonicity yields $a_K \mathop{\downarrow}_{Aa_J} a_i$ and by Symmetry we have $a_i \mathop{\downarrow}_{Aa_J} a_K$. The claim follows now from $a_i \mathop{\downarrow}_A a_J$ and Transitivity. ⊣

So we can define a *family* $(a_i \mid i \in I)$ to be independent over A if it is independent for some ordering of I. Clearly $(a_i)_{i \in I}$ is independent over A if and only if $a_i \mathop{\downarrow}_A \{a_j \mid j \neq i\}$ for all i. One calls a *set* B independent over A if $b \mathop{\downarrow}_A (B \setminus \{b\})$ for all $b \in B$.

The following lemma is a generalisation of Proposition 7.2.14.

LEMMA 7.2.19. *Let T be simple and $\mathcal{I}$ be an infinite Morley sequence over A. If $\mathcal{I}$ is indiscernible over Ac, then $c \mathop{\downarrow}_A \mathcal{I}$.*

PROOF. We may assume $\mathcal{I} = (a_i)_{i<\omega}$. Consider any $\varphi(x, a_0, \dots, a_{n-1}) \in \mathrm{tp}(c/A\mathcal{I})$. Put $b_i = (a_{ni}, \dots, a_{ni+n-1})$. Then by Lemma 7.2.11 $(b_i)_{i<\omega}$ is again a Morley sequence over A and $\{\varphi(x, b_i) \mid i \in \omega\}$ is consistent since realised by c. We see from Proposition 7.2.14 that $\varphi(x, a_0, \dots, a_{n-1})$ does not fork over A. ⊣

EXERCISE 7.2.1. If T is simple, there does not exist an ascending chain $(p_\alpha)_{\alpha \in |T|^+}$ of forking extensions. Hence, there do not exist an A-independent sequence $(b_\alpha)_{\alpha \in |T|^+}$ and a finite tuple c such that $b_\alpha \not\mathop{\downarrow}_A c$ for all α.

EXERCISE 7.2.2 (Diamond Lemma). Assume T to be simple and $p \in \mathrm{S}(A)$. Let q be a non-forking extension of p and r any extension of p. Then there is an A-conjugate r' of r with a non-forking extension s which also extends q.

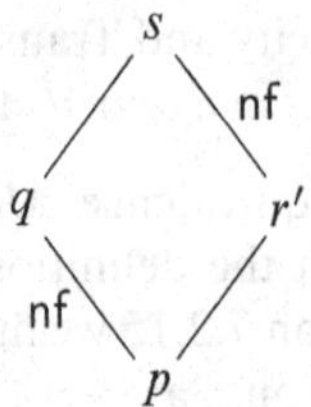

We can choose r' in such a way that the domains of r' and q are independent over A.

EXERCISE 7.2.3. If T is simple and $(a_i)_{i \in I}$ is an A-independent sequence, then

$$a_X \underset{Aa_{X\cap Y}}{\perp} a_Y$$

for all $X, Y \subseteq I$ where $a_X = \{a_i \mid i \in X\}$.

EXERCISE 7.2.4. If $\varphi(x, b)$ divides over A and $A \subseteq B$, there is some A-conjugate B' of B such that $\varphi(x, b)$ divides over B'.

EXERCISE 7.2.5. If T is simple, then

$$ab \underset{A}{\perp} B \Longleftrightarrow a \underset{A}{\perp} B \text{ and } b \underset{Aa}{\perp} B.$$

EXERCISE 7.2.6. Assume that T simple and $b_1 \dots b_n \perp_A C$. Then the sequence $b_1, \dots, b_n$ is independent over A if and only if it is independent over AC.

EXERCISE 7.2.7. If T is simple and $Aa \perp_B C$, then $a \in \operatorname{acl}(ABC)$ implies $a \in \operatorname{acl}(AB)$.

EXERCISE 7.2.8. Use Proposition 7.1.6 to show that T is simple if no formula $\varphi(x, y)$ for a single variable x has the tree property or, equivalently, if every 1-type does not divide over a set of cardinality at most $|T|$.

7.3. The independence theorem

The core of this section is the characterisation of simple theories in terms of a suitable notion of independence. This is due to Kim and Pillay and will be applied to pseudo-finite fields in Section 7.5. We will later specialise this characterisation to stable theories. *Unless explicitly stated otherwise, we assume throughout this section that T is a simple theory.*

DEFINITION 7.3.1. For any set A we write $\mathrm{nc}_A(a, b)$ if a and b start an infinite sequence of indiscernibles over A.

A formula $\theta(x, y)$ is called *thick* if there are no infinite antichains, i.e., sequences $(c_i)_{i<\omega}$ where $\neg\theta(c_i, c_j)$ for all $i < j < \omega$. By compactness this says that there is a bound $k < \omega$ on the length of finite antichains. See Exercise 7.3.2 for an explanation of the terminology.

LEMMA 7.3.2. (T *arbitrary.*) *For any set A and n-tuples a, b the following are equivalent*:

a) $\mathrm{nc}_A(a, b)$.

b) $\models \theta(a, b)$ *for all thick* $\theta(x, y)$ *defined over* A.

In particular, nc_A *is type-definable.*

PROOF. Let $p(x, y) = \mathrm{tp}(ab/A)$. By 5.1.3, a) and b) start an infinite sequence of indiscernibles if and only if there is a sequence $(c_i)_{i<\omega}$ with $\models p(c_i, c_j)$ for $i < j$ if and only if for all $\varphi \in p$ the complement of $\varphi(\mathfrak{C})$ contains arbitrarily long antichains, and so

$$\not\models \psi(a, b) \;\Rightarrow\; \psi \text{ is not thick.} \qquad \dashv$$

COROLLARY 7.3.3. (T *arbitrary.*) *If a and b have the same type over a model M, there is some c such that* $\mathrm{nc}_M(a, c)$ *and* $\mathrm{nc}_M(c, b)$.

PROOF. We have to show $\models \exists z(\varphi(a, z) \wedge \varphi(z, b))$ for every thick formula $\varphi(x, y) \in L(M)$. We may assume that φ is symmetric.[2] Since M is a model, there is a maximal antichain $a_0, \ldots, a_{k-1}$ of φ in M. Thus for some i we have $\models \varphi(a, a_i)$ and hence $\models \varphi(b, a_i)$. $\dashv$

In Exercise 8.1.2 below we give a different proof of the corollary, independent of Lemma 7.3.2.

LEMMA 7.3.4. (T *arbitrary.*) *Let $(b_i)_{i<\omega}$ be indiscernible over A and $(b_i)_{1\leq i<\omega}$ indiscernible over Aa_0b_0. Then there is some a_1 such that* $\mathrm{nc}_A(a_0b_0, a_1b_1)$.

PROOF. Choose a_i with $\mathrm{tp}(a_ib_ib_{i+1}\ldots/A) = \mathrm{tp}(a_0b_0b_1\ldots/A)$. Let $a_0'b_0$, $a_1'b_1, \ldots$ be a sequence of indiscernibles realizing the EM-type of $a_0b_0, a_1b_1, \ldots$ over A. Since $(b_i)_{1\leq i<\omega}$ is indiscernible over Aa_0b_0, we have

$$\mathrm{tp}(a_{i_1}', b_{i_1}, b_{i_2}, \ldots, b_{i_n}/A) = \mathrm{tp}(a_0, b_0, b_1, \ldots, b_n/A)$$

for all $i_1 < \cdots < i_n$. So $\mathrm{tp}(a_0', b_0, b_1, \ldots/A) = \mathrm{tp}(a_0, b_0, b_1, \ldots/A)$ and we may assume that $a_0b_0, a_1b_1, \ldots$ is indiscernible over A. $\dashv$

LEMMA 7.3.5. *Let $\mathcal{I}$ be indiscernible over A and $\mathcal{J}$ an infinite initial segment without last element. Then $\mathcal{I} \setminus \mathcal{J}$ is a Morley sequence over $A\mathcal{J}$.*

PROOF. Let $\mathcal{I} = (a_i)_{i\in I}$ and $\mathcal{J} = (a_i)_{i\in J}$. It suffices to show $a_i \mathop{\smile\!\!\!\!\!\!\mid}_{A\mathcal{J}} a_X$ for all $i \in I \setminus J$ and all finite $X \subseteq I$ with $X < i$. But this follows from Lemma 7.1.10 as $\mathrm{tp}(a_X/A\mathcal{J}a_i)$ is finitely satisfiable in $A\mathcal{J}$. $\dashv$

PROPOSITION 7.3.6. *If $\varphi(x, a)$ does not fork over A and* $\mathrm{nc}_A(a, b)$, *then* $\varphi(x, a) \wedge \varphi(x, b)$ *does not fork over* A.

[2]That is, $\models \forall x, y\, (\theta(x, y) \rightarrow \theta(y, x))$.

PROOF. Let $\mathcal{I}$ be an infinite sequence of indiscernibles over A containing a and b. We extend $\mathcal{I}$ by an infinite initial segment $\mathcal{J}$ without last element. Let c be a realisation of $\varphi(x,a)$ independent from $\mathcal{J}a$ over A. By Corollary 7.1.5 we may assume $\mathcal{I}$ to be indiscernible over $A\mathcal{J}c$.

It follows from Lemma 7.3.5 that $\mathcal{I}$ is a Morley sequence over $A\mathcal{J}$. So by Lemma 7.2.19 we have $c \mathop{\smile\!\!\!\!\!|}_{A\mathcal{J}} \mathcal{I}$. Transitivity now implies $c \mathop{\smile\!\!\!\!\!|}_{A} \mathcal{J}\mathcal{I}$ and hence the claim. ⊣

LEMMA 7.3.7. *Let* $\mathrm{nc}_A(b,b')$ *and* $a \mathop{\smile\!\!\!\!\!|}_{Ab} b'$. *Then there is some* a' *with* $\mathrm{nc}_A(ab,a'b')$.

PROOF. Let $(b_i)_{i<\omega}$ indiscernible over A, $b=b_0$ and $b'=b_1$. By Corollary 7.1.5 we may assume $(b_i)_{1\leq i<\omega}$ to be indiscernible over Aab. The claim now follows from Lemma 7.3.4. ⊣

PROPOSITION 7.3.8. *If* $\varphi(x,a)\wedge\psi(x,b)$ *does not fork over* A, $\mathrm{nc}_A(b,b')$ *and* $a \mathop{\smile\!\!\!\!\!|}_{Ab} b'$, *then neither does* $\varphi(x,a)\wedge\psi(x,b')$ *fork over* A.

PROOF. By Lemma 7.3.7 there is some a' such that $\mathrm{nc}_A(ab,a'b')$. Proposition 7.3.6 implies that $\varphi(x,a)\wedge\psi(x,b)\wedge\varphi(x,a')\wedge\psi(x,b')$ does not fork over A. ⊣

COROLLARY 7.3.9. *Assume that* $\varphi(x,a)\wedge\psi(x,b)$ *does not fork over* A, $a \mathop{\smile\!\!\!\!\!|}_{A} b'b$ *and that* b *and* b' *have the same type over some model containing* A. *Then* $\varphi(x,a)\wedge\psi(x,b')$ *does not fork over* A.

PROOF. By Corollary 7.3.3 there is some c such that $\mathrm{nc}_A(b,c)$ and $\mathrm{nc}_A(c,b')$. By replacing a, if necessary, by a realisation of a non-forking extension of $\mathrm{tp}(a/Abb')$ to $Abb'c$, which exists by Corollary 7.2.7, we may assume that $a \mathop{\smile\!\!\!\!\!|}_{A} bb'c$. Proposition 7.3.8 yields now first that $\varphi(x,a)\wedge\psi(x,c)$ does not fork over A and then that $\varphi(x,a)\wedge\psi(x,b')$ does not fork over A. ⊣

COROLLARY 7.3.10. *Let* $a \mathop{\smile\!\!\!\!\!|}_{M} b$, $a' \mathop{\smile\!\!\!\!\!|}_{M} a$, $b' \mathop{\smile\!\!\!\!\!|}_{M} b$, $\models\varphi(a',a)\wedge\psi(b',b)$ *and assume that* a' *and* b' *have the same type over* M. *Then* $\varphi(x,a)\wedge\psi(x,b)$ *does not fork over* M.

PROOF. Choose a'' such that $\mathrm{tp}(a''a'/M)=\mathrm{tp}(bb'/M)$ and $a'' \mathop{\smile\!\!\!\!\!|}_{Ma'} abb'$. Then by Transitivity we have

$$a'' \mathop{\smile\!\!\!\!\!|}_{M} aa'bb'.$$

It follows that the sequences a,a',a'' and a,b,a'' are both independent over A. This implies

$$a' \mathop{\smile\!\!\!\!\!|}_{M} aa'' \tag{7.1}$$

and

$$a \mathop{\smile\!\!\!\!\!|}_{M} a''b \tag{7.2}$$

Since $\models \psi(a', a'')$, (7.1) implies that $\varphi(x, a) \wedge \psi(x, a'')$ does not fork over M. So $\varphi(x,a) \wedge \psi(x,b)$ does not fork over M by (7.2) and Corollary 7.3.9. ⊣

THEOREM 7.3.11 (Independence Theorem). *Suppose that b and c have the same type over the model M and suppose that*

$$B \underset{M}{\perp} C,\ b \underset{M}{\perp} B \text{ and } c \underset{M}{\perp} C.$$

Then there exists some d with $\mathrm{tp}(d/B) = \mathrm{tp}(b/B)$ *and* $\mathrm{tp}(d/C) = \mathrm{tp}(c/C)$ *and such that*

$$d \underset{M}{\perp} BC.$$

PROOF. By Corollary 7.3.10, $\mathrm{tp}(b/B) \cup \mathrm{tp}(c/C)$ does not fork over M. So we find some d such that $d \perp_M BC$, $\mathrm{tp}(d/B) = \mathrm{tp}(b/B)$ and $\mathrm{tp}(d/C) = \mathrm{tp}(c/C)$. ⊣

COROLLARY 7.3.12. *Let* $B_i, i \in I$, *be independent over M and let* b_i *be such that* $b_i \perp_M B_i$ *all* b_i *having the same type over M. Then there is some d with* $\mathrm{tp}(d/B_i) = \mathrm{tp}(b_i/B_i)$ *for all i and*

$$d \underset{M}{\perp} \{B_i \mid i \in I\}.$$

PROOF. Well-order I and show the existence of d by recursively constructing $p_i = \mathrm{tp}(d/\{B_i \mid i \in I\})$. The details are left as an exercise. ⊣

THEOREM 7.3.13 (Kim–Pillay [34]). *Let T be a complete theory and* $a \perp^0_A B$ *a relation between finite tuples a and sets A and B invariant under automorphisms and having the following properties:*

a) (MONOTONICITY AND TRANSITIVITY) $a \perp^0_A BC$ *if and only if* $a \perp^0_A B$ *and* $a \perp^0_{AB} C$.
b) (SYMMETRY) $a \perp^0_A b$[3] *if and only if* $b \perp^0_A a$.
c) (FINITE CHARACTER) $a \perp^0_A B$ *if* $a \perp^0_A b$ *for all finite tuples* $b \in B$.
d) (LOCAL CHARACTER) *There is a cardinal κ, such that for all a and B there exists* $B_0 \subseteq B$ *of cardinality less than κ such that* $a \perp^0_{B_0} B$
e) (EXISTENCE) *For all a, A and C there is a′ such that* $\mathrm{tp}(a'/A) = \mathrm{tp}(a/A)$ *and* $a' \perp^0_A C$.
f) (INDEPENDENCE OVER MODELS) *Let M be a model,* $\mathrm{tp}(a'/M) = \mathrm{tp}(b'/M)$ *and*

$$a \underset{M}{\perp^0} b,\ a' \underset{M}{\perp^0} a,\ b' \underset{M}{\perp^0} b.$$

Then there is some c such that $\mathrm{tp}(c/Ma) = \mathrm{tp}(a'/Ma)$, $\mathrm{tp}(c/Mb) = \mathrm{tp}(b'/Mb)$ *and* $c \perp^0_M ab$.

Then T is simple and $\perp^0 = \perp$.

[3]We here consider a tuple as a finite set.

We have seen that in simple theories the relation $\mathop{\smile\!\!\!\!\!|}$ satisfies the properties of the previous theorem (for $\kappa = |T|^+$).

PROOF. We may assume κ to be a regular cardinal, otherwise just replace κ by κ^+. Assume now $a \mathop{\smile\!\!\!\!\!|}^0_A b$. We will use Lemma 7.1.4 to show that $\mathrm{tp}(a/Ab)$ does not divide over A. So, let $(b_i)_{i<\omega}$ be a sequence of A-indiscernibles starting with $b = b_0$.

CLAIM. *We can find a model M containing A such that $(b_i)_{i<\omega}$ is indiscernible over M and $b_i \mathop{\smile\!\!\!\!\!|}^0_M \{b_j \mid j < i\}$ for all i.*

PROOF OF CLAIM. By Lemma 7.1.1 we can extend the sequence to $(b_i)_{i\leq\kappa}$. Furthermore, it is easy to construct an ascending sequence of models $A \subseteq M_0 \subseteq M_1 \subseteq \cdots$ such that for all $i < \kappa$ all b_j, $(j < i)$ is contained in M_i and $(b_j)_{i\leq j\leq\kappa}$ is indiscernible over M_i. By LOCAL CHARACTER there is some i_0 such that $b_\kappa \mathop{\smile\!\!\!\!\!|}^0_{M_{i_0}} \{b_j \mid i_0 \leq j < \kappa\}$. From the Indiscernibility it now follows that $b_i \mathop{\smile\!\!\!\!\!|}^0_{M_{i_0}} \{b_j \mid i_0 \leq j < i\}$ for all i. We can take $M = M_{i_0}$ and the sequence $b_{i_0}, b_{i_0+1}, \ldots$. ⊣

CLAIM. *We may assume $a \mathop{\smile\!\!\!\!\!|}^0_M b$.*

PROOF OF CLAIM. By EXISTENCE we may replace a by a' with $\mathrm{tp}(a'/Ab) = \mathrm{tp}(a/Ab)$ and $a' \mathop{\smile\!\!\!\!\!|}^0_{Ab} M$ and then apply MONOTONICITY AND TRANSITIVITY. ⊣

We now find elements $a = a_0, a_1, \ldots$ such that $a_i \mathop{\smile\!\!\!\!\!|}^0_M \{b_j \mid j \leq i\}$, $q_i(x) = \mathrm{tp}(a_{i+1}/M\{b_j \mid j \leq i\}) = \mathrm{tp}(a_i/M\{b_j \mid j \leq i\})$ and $\mathrm{tp}(a_ib_i/M) = \mathrm{tp}(ab/M)$: given $a_0, \ldots, a_i$, choose a' with $\mathrm{tp}(a'b_{i+1}/M) = \mathrm{tp}(ab/M)$. Now apply INDEPENDENCE OVER MODELS to

$$\{b_j \mid j \leq i\} \mathop{\smile\!\!\!\!\!|}^0_M b_{i+1},\ a_i \mathop{\smile\!\!\!\!\!|}^0_M \{b_j \mid j \leq i\} \text{ and } a' \mathop{\smile\!\!\!\!\!|}^0_M b_{i+1}$$

to find a_{i+1}.

This implies that $\bigcup_{i<\omega} q_i(x)$ is consistent and contains all $p(x, b_i)$ where $p(x, y) = \mathrm{tp}(ab/M)$. By Lemma 7.1.4, $\mathrm{tp}(a/Ab)$ does not divide over A. Simplicity of T now follows from LOCAL CHARACTER and Proposition 7.2.5.

It remains to show $a \mathop{\smile\!\!\!\!\!|}^0_A b$ if $\mathrm{tp}(a/Ab)$ does not divide over A. Using EXISTENCE we construct for any λ a sequence $(b_i)_{i<\lambda}$ which is $\mathop{\smile\!\!\!\!\!|}^0_A$-independent and for which $\mathrm{tp}(b_i/A) = \mathrm{tp}(b/A)$. If this sequence is sufficiently long, the same argument used for Lemma 7.2.13 (but now using MONOTONICITY and FINITE CHARACTER) yields an A-indiscernible sequence $(b'_i)_{i<\kappa}$ which is $\mathop{\smile\!\!\!\!\!|}^0_A$-independent as well and satisfies $\mathrm{tp}(b'_i/A) = \mathrm{tp}(b/A)$ and $b = b'_0$. We now apply Corollary 7.1.5 and obtain a' such that $\mathrm{tp}(a'/Ab) = \mathrm{tp}(a/Ab)$ and so that $(b'_i)_{i<\kappa}$ is indiscernible over Aa'. LOCAL CHARACTER and MONOTONICITY yield the existence of i_0 with

$$a' \mathop{\smile\!\!\!\!\!|}^0_{A\{b'_i \mid i<i_0\}} b'_{i_0}.$$

Since

$$b'_{i_0} \mathop{\smile\!\!\!\!\!\!\;|}^{0}_{A} \{b'_i \mid i < i_0\}$$

we get

$$a' \mathop{\smile\!\!\!\!\!\!\;|}^{0}_{A} b'_{i_0} \text{ and hence } a \mathop{\smile\!\!\!\!\!\!\;|}^{0}_{A} b$$

from SYMMETRY and TRANSITIVITY using that $\mathrm{tp}(a'b'_{i_0}/A) = \mathrm{tp}(a'b'_0/A) = \mathrm{tp}(ab/A)$. ⊣

COROLLARY 7.3.14. *The theory of the random graph is simple.*

PROOF. Define $A \mathop{\smile\!\!\!\!\!\!\;|}^{0}_{B} C$ by $A \cap C \subseteq B$ and apply Theorem 7.3.13, ⊣

EXERCISE 7.3.1. Let T be simple. Assume that the partial type $\pi(x, b)$ does not fork over A and that $\mathcal{I}$ is an infinite sequence of indiscernibles over A containing b. Show that there is a realisation c of $\pi(x, b)$ such that $c \mathop{\smile\!\!\!\!\!\!\;|}_{A} \mathcal{I}$ and $\mathcal{I}$ is indiscernible over Ac.

EXERCISE 7.3.2. A symmetric formula is thick if and only if there is *no infinite anti-clique*, i.e., a sequence $(c_i)_{i<\omega}$, where $\models \neg\theta(x, y)$ for all $i \neq j$. This explains the notation nc_A. Prove the following, without using Lemma 7.3.2:

1. The conjunction of two thick formulas is thick.
2. If $\theta(x, y)$ is thick, then $\theta^{\sim}(x, y) = \theta(y, x)$ is thick.
3. A formula is thick if and only if it is implied by a symmetric thick formula.

EXERCISE 7.3.3. Let T be a simple theory and A a set of parameters. Assume that there is an element b which is algebraic over A but not definable over A. Then the Independence Theorem does not hold if in its formulation the model M is replaced by A.

EXERCISE 7.3.4. Fill in the details of the proof of Corollary 7.3.12.

EXERCISE 7.3.5. Prove directly from the axioms in Theorem 7.3.13 that

$$a \mathop{\smile\!\!\!\!\!\!\;|}^{0}_{A} B \Leftrightarrow a \mathop{\smile\!\!\!\!\!\!\;|}^{0}_{A} AB.$$

EXERCISE 7.3.6. Let T be simple and p be a type over a model. Then p has either exactly one non-forking extension to $\mathfrak{C}$ (p is *stationary*) or arbitrarily many.

7.4. Lascar strong types

In this section we will prove a version of the Independence Theorem 7.3.11 over arbitrary parameter sets A. For this we have to strengthen the assumption that b and c have the same type over A to having the same *Lascar strong type* over A. In what follows T is an arbitrary complete theory.

DEFINITION 7.4.1. Let A be any set of parameters. The group $\mathrm{Aut}_f(\mathfrak{C}/A)$ of *Lascar strong automorphisms* of $\mathfrak{C}$ over A is the group generated by all $\mathrm{Aut}(\mathfrak{C}/M)$ where the M are models containing A. Two tuples a and b have the same *Lascar strong type* over A if $\alpha(a) = b$ for some $\alpha \in \mathrm{Aut}_f(\mathfrak{C}/A)$. We denote this by $\mathrm{Lstp}(a/A) = \mathrm{Lstp}(b/A)$.

It is easy to see that tuples a and b have the same Lascar strong type over A if and only if there is a sequence $a = b_0, b_1, \ldots, b_n = b$ such that for all for all $i < n$, b_i and b_{i+1} have the same type over some model containing A.

LEMMA 7.4.2. *Assume that T is simple. If $\varphi(x, a) \wedge \psi(x, b)$ does not fork over A, and if* $\mathrm{Lstp}(b/A) = \mathrm{Lstp}(b'/A)$ *and $a \mathop{\smile\!\!\!\!\!\mid}_A b'b$, then $\varphi(x, a) \wedge \psi(x, b')$ does not fork over A.*

PROOF. Choose a sequence $b = b_0, b_1, \ldots, b_n = b'$ such that for each $i < n$, b_i and b_{i+1} have the same type over some model containing A. By the properties of forking we may assume that $a \mathop{\smile\!\!\!\!\!\mid}_A b_0 b_1 \ldots b_n$. We thus always have $a \mathop{\smile\!\!\!\!\!\mid}_A b_i b_{i+1}$ and the claim follows by induction from Corollary 7.3.9. ⊣

As in the previous proof we will repeatedly use EXISTENCE (Corollary 7.2.7) to assume that we have realisations of types which are independent from other sets. This is also crucial in the following lemma.

LEMMA 7.4.3. *Assume T to be simple. For all a, A and B there is some a' such that* $\mathrm{Lstp}(a'/A) = \mathrm{Lstp}(a/A)$ *and $a' \mathop{\smile\!\!\!\!\!\mid}_A B$.*

PROOF. By EXISTENCE and SYMMETRY, we can choose a model $M \supset A$ with $a \mathop{\smile\!\!\!\!\!\mid}_A M$. By EXISTENCE again we can find a' such that $\mathrm{tp}(a'/M) = \mathrm{tp}(a/M)$ and $a' \mathop{\smile\!\!\!\!\!\mid}_M B$, so the claim now follows from Transitivity. ⊣

A stronger statement will be proved in Exercise 7.4.4.

COROLLARY 7.4.4. *Let* $\mathrm{Lstp}(a/A) = \mathrm{Lstp}(b/A)$. *For all a', B there exists b' such that* $\mathrm{Lstp}(aa'/A) = \mathrm{Lstp}(bb'/A)$ *and $b' \mathop{\smile\!\!\!\!\!\mid}_{Ab} B$.*

PROOF. Choose bb'' such that $\mathrm{Lstp}(aa'/A) = \mathrm{Lstp}(bb''/A)$ and by the Lemma there is some b' such that $b' \mathop{\smile\!\!\!\!\!\mid}_{Ab} B$ and $\mathrm{Lstp}(b''/Ab) = \mathrm{Lstp}(b'/Ab)$. It is easy to see that this implies $\mathrm{Lstp}(bb''/A) = \mathrm{Lstp}(bb'/A)$. ⊣

COROLLARY 7.4.5. *Let $a \mathop{\smile\!\!\!\!\!\mid}_A b$, $a' \mathop{\smile\!\!\!\!\!\mid}_A a$, $b' \mathop{\smile\!\!\!\!\!\mid}_A b$,* $\mathrm{Lstp}(a'/A) = \mathrm{Lstp}(b'/A)$ *and $\models \varphi(a', a) \wedge \psi(b', b)$. Then $\varphi(x, a) \wedge \psi(x, b)$ does not fork over A.*

PROOF. Choose a'' by Corollary 7.4.4 such that $\mathrm{Lstp}(a''a'/A) = \mathrm{Lstp}(bb'/A)$ and $a'' \mathop{\smile\!\!\!\!\!\mid}_{Aa'} abb'$. Then proceed as in the proof of Corollary 7.3.10, but use Lemma 7.4.2 instead of Corollary 7.3.9. ⊣

This yields the following generalisation of Theorem 7.3.11.

THEOREM 7.4.6 (Independence Theorem). *Let T be simple and suppose* $\mathrm{Lstp}(b/A) = \mathrm{Lstp}(c/A)$,

$$B \mathop{\smile\!\!\!\!\!\mid}_A C, \; b \mathop{\smile\!\!\!\!\!\mid}_A B \text{ and } c \mathop{\smile\!\!\!\!\!\mid}_A C.$$

Then there exists some d such that $d \mathop{\smile\!\!\!\!\!\mid}_A BC$, $\mathrm{Lstp}(d/B) = \mathrm{Lstp}(b/B)$ and $\mathrm{Lstp}(d/C) = \mathrm{Lstp}(c/C)$.

PROOF. By Corollary 7.4.5, $\mathrm{tp}(b/B) \cup \mathrm{tp}(c/C)$ does not fork over A. So we find some d such that $d \mathop{\smile\!\!\!\!\!\mid}_A BC$, $\mathrm{tp}(d/B) = \mathrm{tp}(b/B)$ and $\mathrm{tp}(d/C) = \mathrm{tp}(c/C)$. The stronger claim about Lascar strong types is left as Exercise 7.4.3. ⊣

COROLLARY 7.4.7. *Assume T to be simple and let B_i, $i \in I$, be independent over A and b_i such that $b_i \mathop{\smile\!\!\!\!\!\mid}_A B_i$ with all b_i having the same Lascar strong type over A. Then there is some d such that $d \mathop{\smile\!\!\!\!\!\mid}_A \{B_i \mid i \in I\}$ and $\mathrm{Lstp}(d/B_i) = \mathrm{Lstp}(b_i/B_i)$ for all i.*

PROOF. Assume the B_i are models containing A. The proof goes then as the proof of Corollary 7.3.12. ⊣

LEMMA 7.4.8. *Let T be simple, $a \mathop{\smile\!\!\!\!\!\mid}_A b$ and $\mathrm{Lstp}(a/A) = \mathrm{Lstp}(b/A)$, there is an infinite Morley sequence over A containing a and b.*

PROOF. Consider $p(x, y) = \mathrm{tp}(ab/A)$. Starting from $a_0 = a$ and $a_1 = b$ we recursively construct a long independent sequence (a_i) of elements all having the same Lascar type over A. If $(a_i \mid i < \alpha)$ is given, the $p(a_i, y)$ are realised by elements b_i with $b_i \mathop{\smile\!\!\!\!\!\mid}_A a_i$ and $\mathrm{Lstp}(b_i/A) = \mathrm{Lstp}(a/A)$. By Corollary 7.4.7 there is some a_α with $a_\alpha \mathop{\smile\!\!\!\!\!\mid}_A \{a_i \mid i < \alpha\}$, $\models p(a_i, a_\alpha)$ for all $i < \alpha$ and $\mathrm{Lstp}(a_\alpha/A) = \mathrm{Lstp}(a/A)$. If the sequence is sufficiently long, then by Lemma 7.2.12 there is an A-indiscernible sequence $(a'_i)_{i<\omega}$ such that $\models p(a'_i, a'_j)$ for all $i < j$ and furthermore the sequence is independent over A because all types $(a'_j/A\{a'_i \mid i < j\})$ appear in (a_i). Since $\mathrm{tp}(a'_1 a'_0/A) = \mathrm{tp}(ba/A)$, we may assume $a'_0 = a$ and $a'_1 = b$. ⊣

COROLLARY 7.4.9. *We have $\mathrm{Lstp}(a/A) = \mathrm{Lstp}(b/A)$ if and only if $\mathrm{nc}^2_A(a, b)$. In particular, the equivalence relation $\mathrm{E}^{\mathrm{L}}_A(a, b)$ defined as $\mathrm{Lstp}(a/A) = \mathrm{Lstp}(b/A)$ is type-definable.*

PROOF. Choose c with $c \mathop{\smile\!\!\!\!\!\mid}_A ab$ and $\mathrm{Lstp}(c/A) = \mathrm{Lstp}(a/A) = \mathrm{Lstp}(b/A)$. By Lemma 7.4.8, we have $\mathrm{nc}_A(c, a)$ and $\mathrm{nc}_A(c, b)$. Hence

$$\mathrm{E}^{\mathrm{L}}_A(x, y) \Leftrightarrow \exists z\, (\mathrm{nc}_A(x, z) \wedge \mathrm{nc}_A(z, y))$$

and this is type-definable by Lemma 7.3.2. ⊣

It is an open problem whether in simple theories Lascar strong types are the same as strong types (see Exercise 8.4.9).

EXERCISE 7.4.1. Show that an automorphism of $\mathfrak{C}$ is Lascar strong if and only if it preserves the Lascar strong type of any tuple of length $|T|$.

EXERCISE 7.4.2. Show that $\mathrm{nc}_A(a, b)$ implies $\mathrm{nc}_{\mathrm{acl}(A)}(a, b)$.

EXERCISE 7.4.3. Deduce Theorem 7.4.6 from the weaker version which claims only the equalities $\mathrm{tp}(d/B) = \mathrm{tp}(b/B)$ and $\mathrm{tp}(d/C) = \mathrm{tp}(c/C)$.

EXERCISE 7.4.4. Show that in arbitrary theories $\mathrm{nc}_A(x, a)$ does not divide over A. Use this to prove a stronger version of Lemma 7.4.3: *Assume that T is simple. For all a, A, B there is some a' such that $\mathrm{nc}_A(a, a')$ and $a' \mathop{\smile\!\!\!\!\!\mid}_A B$.*

EXERCISE 7.4.5. If $\mathrm{nc}_A(a, b)$, there is some model M containing A such that $\mathrm{tp}(a/M) = \mathrm{tp}(b/M)$. Conclude that $\mathrm{E}^{\mathrm{L}}_A$ is the transitive closure of nc_A.

EXERCISE 7.4.6. A relation $R \subseteq \mathfrak{C}^n \times \mathfrak{C}^n$ is called bounded if there are no arbitrarily long antichains, i.e., sequences $(c_\alpha \mid \alpha < \kappa)$ with $\neg R(c_\alpha, c_\beta)$ for all $\alpha < \beta < \kappa$. Show that the intersection of a family R_i, $(i \in I)$, of bounded relations is again bounded.

EXERCISE 7.4.7. We call a relation *A-invariant* if it is invariant under all automorphisms in $\mathrm{Aut}(\mathfrak{C}/A)$. Show that nc_A is the smallest bounded A-invariant relation.

EXERCISE 7.4.8. Show that $\mathrm{E}^{\mathrm{L}}_A$ is the smallest bounded A-invariant equivalence relation.

7.5. Example: pseudo-finite fields

We now turn to an important example of simple theories, namely those of *pseudo-finite fields*. A perfect field K is called pseudo-finite if it is *pseudo-algebraically closed*, i.e., if every absolutely irreducible affine variety defined over K has a K-rational point and if its absolute Galois group is $\widehat{\mathbb{Z}}$, i.e., K has a unique extension of degree n for each $n \geq 1$. Equivalently, a field is pseudo-finite if it is elementarily equivalent to an infinite ultraproduct of finite fields, see Exercise 7.5.2 or [1], Theorem 8. For background on pseudo-finite fields and profinite groups see Section B.4.

PROPOSITION 7.5.1. *Let L_1 and L_2 be regular procyclic extensions of a field K and let L_2 be pseudo-finite. Then L_1 can be regularly embedded over K into an elementary extension of L_2.*

PROOF. By Lemma B.4.16, we may assume that N is a common regular procyclic extension of the L_i. As a regular procyclic extension of L_2, N is 1-free (by Corollary B.4.14). Since L_2 is existentially closed in N (see Lemma B.4.2), N is embeddable over L_2 into an elementary extension L_2' of L_2. Let N' denote the image of this embedding. By B.4.14, L_2'/N' is regular. $\dashv$

THEOREM 7.5.2. *Let L_1 and L_2 be regular pseudo-finite extensions of K. Then L_1 and L_2 are elementarily equivalent over K.*

PROOF. By Proposition 7.5.1 we obtain an alternating elementary chain. Its union is an elementary extension of both L_1 and L_2. $\dashv$

The *absolute part* $\mathrm{Abs}(L)$ of a field L is the relative algebraic closure in L of its prime field. Since a perfect field L is a regular extension of $\mathrm{Abs}(L)$ (see Proposition B.4.13) we obtain.

COROLLARY 7.5.3. *The elementary theory of a pseudo-finite field L is determined by the isomorphism type of* $\mathrm{Abs}(L)$. *A field K algebraic over its prime field is the absolute part of some pseudo-finite field if and only if it is procyclic. (This is always true in finite characteristic.)*

We now fix the complete theory of a pseudo-finite field and work in its monster model $\mathfrak{C}$.

COROLLARY 7.5.4. *Let K be a subfield of $\mathfrak{C}$, and a and b tuples of elements $\mathfrak{C}$. Then a and b have the same type over K if and only if the relative algebraic closures of $K(a)$ and $K(b)$ in $\mathfrak{C}$ are isomorphic over K via some isomorphism taking a to b.*

PROOF. Let A and B be the relative algebraic closures of $K(a)$ and $K(b)$, respectively. If a and b have the same type over K, then A and B are isomorphic in the required way by Lemma 5.6.4. Conversely, if A and B are isomorphic over K by such an isomorphism, the claim follows immediately from Theorem 7.5.2. ⊣

THEOREM 7.5.5. *In pseudo-finite fields, algebraic independence has all the properties of forking listed in Theorem* 7.3.13.

PROOF. We keep working in $\mathfrak{C}$. All properties are clear except (EXISTENCE) and (INDEPENDENCE OVER MODELS).

(EXISTENCE): Let K be a subfield of $\mathfrak{C}$ and L and H two extensions of K. We may assume that all three fields are relatively algebraically closed in $\mathfrak{C}$. By Lemma B.4.16, there is a procyclic extension C of H (not necessarily contained in $\mathfrak{C}$) containing a copy L' of L/K independent from H over K and such that C/L' is regular. By Proposition 7.5.1, C can be regularly over H embedded into $\mathfrak{C}$. Let L'' denote the image of L' in $\mathfrak{C}$. Then L'' and H are independent over K and L'' and L have the same type over K (see p. 23).

(INDEPENDENCE OVER MODELS): Let M be an elementary submodel of $\mathfrak{C}$ and let K and L be field extensions independent over M. Assume further that we are given extensions K' and L' so that K and K' as well as L and L' are independent over M and such that K' and L' have the same type over M.

We may assume that all these fields are relatively algebraically closed in $\mathfrak{C}$. Then, if σ is a generator of $\mathrm{G}(\mathfrak{C})$, the relative algebraic closures of KK' and LL' in $\mathfrak{C}$ are the fixed fields of $\kappa' = \sigma \upharpoonright (KK')^{\mathrm{alg}}$ in $(KK')^{\mathrm{alg}}$ and of $\lambda' = \sigma \upharpoonright (LL')^{\mathrm{alg}}$, respectively.

We now take another field extension H/M (possibly outside $\mathfrak{C}$) isomorphic to K'/M and L'/M and independent of KL over M. Then KK' and KH, and LL' and LH, respectively, are isomorphic over M and the isomorphisms are compatible with the given isomorphism between K' and L'. We transport κ'

and λ' to $(KH)^{\mathrm{alg}}$ and $(LH)^{\mathrm{alg}}$ via these isomorphism and call the transported automorphisms κ and λ. Clearly κ, λ and $\mu = \sigma \restriction (KL)^{\mathrm{alg}}$ agree on K^{alg} and L^{alg}. They also agree on H^{alg}, since $\kappa \restriction H^{\mathrm{alg}}$ and $\lambda \restriction H^{\mathrm{alg}}$ are both the unique extension of $\sigma \restriction M^{\mathrm{alg}}$ to $\mathrm{G}(H)$.

Since $(KH)^{\mathrm{alg}}$ and $(LH)^{\mathrm{alg}}$ are independent over H^{alg}, it follows that κ and λ extend to an automorphism μ' of $(KH)^{\mathrm{alg}}(LH)^{\mathrm{alg}}$ which agrees with μ on $K^{\mathrm{alg}}L^{\mathrm{alg}}$. We will see in Corollary 8.1.8 that

$$((KH)^{\mathrm{alg}}(LH)^{\mathrm{alg}}) \cap (KL)^{\mathrm{alg}} = KL.$$

So μ' and μ have a common extension to some automorphism of

$$(KH)^{\mathrm{alg}}(LH)^{\mathrm{alg}}(KL)^{\mathrm{alg}}$$

which again can be extended to some automorphism τ of $(KLH)^{\mathrm{alg}}$. Let C be the fixed field of τ, so C is procyclic. The relative algebraic closures of KL, KH and LH in C are isomorphic to the relative algebraic closures of KL, KK' and LL' in $\mathfrak{C}$. Let N be the relative algebraic closure of KL in $\mathfrak{C}$. Then C is a regular extension of N. So, by Proposition 7.5.1 we find a regular embedding of C into $\mathfrak{C}$ over N. The image of H has the required properties.

Note that we did not make use of the fact that M is a model, but only that M is 1-free and $\mathfrak{C}/M$ is regular. ⊣

We have now proved

COROLLARY 7.5.6. *Pseudo-finite fields are simple. Forking independence agrees with algebraic independence.*

EXERCISE 7.5.1. Let K be a procyclic field which is algebraic over its prime field. It is shown in [1] that K is the absolute part of an infinite ultraproduct of finite fields. While the characteristic 0 case uses Čebotarev's Density Theorem, the proof of the characteristic p is an easy exercise.

EXERCISE 7.5.2.
1. Use the previous exercise to show that every pseudo-finite field is elementarily equivalent to an infinite ultraproduct of finite fields.
2. Show that pseudo-finite fields are exactly the infinite models of the theory of all finite fields (see Exercise 2.1.2.)

EXERCISE 7.5.3. Show that in pseudo-finite fields every formula $\varphi(\bar{x})$ is equivalent to a Boolean combination of formulas of the form $\exists y f(\bar{x}, y) \doteq 0$, where $f(\bar{X}, Y)$ is a polynomial over $\mathbb{Z}$. (Hint: Use Lemma 3.1.1, Theorem 7.5.2 and Lemma B.3.13.)

Chapter 8

STABLE THEORIES

Recall from Section 5.2, that a theory is κ-stable if there are only κ-many types over any parameter set of size κ. A theory is stable if it is κ-stable for some κ. This is equivalent to the definition given in Section 8.2: a theory is stable if no formula has the *order property*; the equivalence will be proven in Exercise 8.2.7. In order to apply the results of the previous chapter to stable theories, we will eventually show that stable theories are simple and then specialise the characterisation given in Theorem 7.3.13 to stable theories. But before that we will introduce some of the classical notions of stability theory, all essentially describing forking in stable theories.

8.1. Heirs and coheirs

In this section we fix an arbitrary complete theory T. For types over models we here define some special extensions to supersets, viz. heirs, coheirs, and definable type extensions. All these extensions have in common that they do not add too much new information to the given type. For stable theories, we will see in Section 8.3 that these extensions coincide with the non-forking extension (and this is in fact unique).

DEFINITION 8.1.1. Let p be a type over a model M of T and $q \in \mathrm{S}(B)$ an extension of p to $B \supset M$.

1. We call q an *heir* of p if for every $L(M)$-formula $\varphi(x, y)$ such that $\varphi(x, b) \in q$ for some $b \in B$ there is some $m \in m$ with $\varphi(x, m) \in p$.
2. We call q a *coheir* of p if q is finitely satisfiable in M.

It is easy to see that $\mathrm{tp}(a/Mb)$ is an heir of $\mathrm{tp}(a/M)$ if and only if $\mathrm{tp}(b/Ma)$ is a coheir of $\mathrm{tp}(b/M)$.

The following observation is trivial, but used frequently.

REMARK 8.1.2. Suppose q is an heir of $p \in \mathrm{S}(M)$. If $\varphi(x, b) \in q$ and $\models \sigma(b)$, then there is some $m \in M$ with $\models \sigma(m)$ and $\varphi(x, m) \in p$.

LEMMA 8.1.3. *Let $q \in S(B)$ be a (co)heir of $p \in S(M)$ and C an extension of B. Then q can be extended to a type $r \in S(C)$ which is again a (co)heir of p.*

PROOF. Suppose q is an heir of p. We have to show that

$$s(x) = q(x) \cup \{\varphi(x, \overline{c}) \mid \overline{c} \in C,\ \varphi(x, \overline{y}) \in L(M), \\ \varphi(x, \overline{m}) \in p \text{ for all } \overline{m} \in M\}$$

is consistent. If there are formulas $\varphi(x, \overline{b}), \varphi_1(x, \overline{c}_1), \ldots, \varphi_n(x, \overline{c}_n) \in s(x)$ with $\varphi(x, \overline{b}) \in q(x)$ whose conjunction is inconsistent, then as M is a model and q is an heir of p there would be $\overline{m}, \overline{m}_1, \ldots, \overline{m}_n \in M$ with $\varphi(x, \bar{m}) \in p$ and its conjunction with $\varphi_1(x, \overline{m}_1), \ldots, \varphi_n(x, \overline{m}_n)$ inconsistent. Since $\varphi_i(x, \overline{m}_i) \in p$, this is impossible. Any type $r(x) \in S(C)$ containing $s(x)$ is then an heir of $p(x)$.

If q is a coheir of p, let r be a maximal set of $L(C)$-formulas containing q which is finitely satisfiable in M. Clearly, r is consistent. Let $\varphi(x) \in L(C)$. If neither φ nor $\neg\varphi$ belongs to r, then both $r \cup \{\varphi\}$ and $r \cup \{\neg\varphi\}$ are not finitely satisfied in M and so neither is r (see the proof of Lemma 2.2.2). $\dashv$

DEFINITION 8.1.4. A type $p(\overline{x}) \in S_n(B)$ is *definable* over C if the following holds: for any L-formula $\varphi(\overline{x}, \overline{y})$ there is an $L(C)$-formula $\psi(\overline{y})$ such that for all $\overline{b} \in B$

$$\varphi(\overline{x}, \overline{b}) \in p \text{ if and only if } \models \psi(\overline{b}).$$

We say p is *definable* if it is definable over its domain B.

We write $\psi(\overline{y})$ as $\mathrm{d}_p\,\overline{x}\varphi(\overline{x}, \overline{y})$ to indicate the dependence on p, $\varphi(\overline{x}, \overline{y})$ and the choice of the variable tuple $\overline{x}$. (So d_p has the syntax of a generalised quantifier, see [59].) Thus, we have

$$\varphi(\overline{x}, \overline{b}) \in p \text{ if and only if } \models \mathrm{d}_p\,\overline{x}\varphi(\overline{x}, \overline{b}).$$

Note that $\mathrm{d}_p\,\overline{x}\varphi(\overline{x}, \overline{y})$ is also meaningful for formulas φ with parameters in B.

EXAMPLE. In strongly minimal theories all types $p \in S(A)$ are definable. To see this fix $\varphi_0 \in p$ of minimal Morley rank k and minimal degree and consider a formula $\psi(x, y)$ without parameters. The discussion on page 98 shows that $\psi(x, a) \in p$ if and only if $\mathrm{MR}(\varphi_0(x) \wedge \neg\psi(x, a)) < k$. By Corollary 6.4.4 this is an A-definable property of a.

LEMMA 8.1.5. *A definable type $p \in S(M)$ has a unique extension $q \in S(B)$ definable over M for any set $B \supset M$, namely*

$$\{\varphi(x, \overline{b}) \mid \varphi(x, \overline{y}) \in L,\ \overline{b} \in B,\ \mathfrak{C} \models \mathrm{d}_p\, x\varphi(x, \overline{b})\},$$

and q is the only heir of p.

PROOF. The fact that the $d_p\, x\varphi(x,\overline{y})$ define a type is a first-order property expressible in M and is hence true in any elementary extension of M. This proves existence. On the other hand, if q is a definable extension of p, then $d_q\, x\varphi(x,\overline{y})$ and $d_p\, x\varphi(x,\overline{y})$ agree on M and hence in all elementary extensions, proving uniqueness. Clearly, q is an heir of p. If $q' \in S(B)$ is different from q, then for some $\varphi(x,\overline{b}) \in q'$ we have $\not\models d_p x\varphi(x,b)$. But there is no $\overline{m}$ with $\varphi(x,\overline{m}) \wedge \neg d_p x\varphi(x,\overline{m}) \in p$, so q' is not an heir of p. ⊣

LEMMA 8.1.6. *A global type which is a coheir of its restriction to a model M is invariant over M.*

PROOF. Let $q \in S(\mathfrak{C})$ be finitely satisfiable in M and $\alpha \in \operatorname{Aut}(\mathfrak{C}/M)$. Consider a formula $\varphi(x,c)$. Since c and $\alpha(c)$ have the same type over M, we have $\varphi(x,c) \wedge \neg\varphi(x,\alpha(c))$ is not satisfiable in M. So $\varphi(x,c) \in q$ implies $\varphi(x,\alpha(c)) \in q$. ⊣

We conclude by exhibiting coheirs in strongly minimal theories.

PROPOSITION 8.1.7. *Let T be strongly minimal, M a model and B an extension of M. Then* $\operatorname{tp}(a/B)$ *is an heir of* $\operatorname{tp}(a/M)$ *if and only if* $\operatorname{MR}(a/B) = \operatorname{MR}(a/M)$.

Note that in strongly minimal theories $\operatorname{MR}(a/B) = \operatorname{MR}(a/M)$ is equivalent to a and B being geometrically independent over M (see Exercise 6.4.1). This is a symmetric notion, which implies that in strongly minimal theories heirs and coheirs coincide. We will later see in Corollary 8.3.7 (see also Corollary 8.5.11) that this is actually true for all *stable* theories. Note also that this implies that in strongly minimal theories types over models have a unique extension of the same Morley rank, i.e., they have Morley degree 1. This is true in all totally transcendental theories (Corollary 8.5.12, see also Corollary 8.5.4.)

PROOF. Let k be the Morley rank of $p = \operatorname{tp}(a/M)$. Choose a formula $\varphi_0 \in p$ of same rank and degree as p. We saw in the Example on page 130 that the unique heir q of p on B is given by

$$\{\psi(x)\ L(B)\text{-formula} \mid \operatorname{MR}(\varphi_0(x) \wedge \neg\psi(x,a)) < k\}.$$

On the other hand this set of formulas must be contained in all extensions of p to B having rank k. So q is also the unique extension of p of rank k. ⊣

COROLLARY 8.1.8 (Hrushovski–Chatzidakis). *Let K, L, H be algebraically closed extensions of an algebraically closed field M. If H algebraically independent from KL over M, then*

$$\big((KH)^{\text{alg}}(LH)^{\text{alg}}\big) \cap (KL)^{\text{alg}} = KL.$$

PROOF. We work in the monster model $\mathfrak{C}$. Let c be an element of the left hand side. So there are tuples $a \in (KH)^{\text{alg}}$ and $b \in (LH)^{\text{alg}}$ such that $c \in \operatorname{dcl}(a,b)$, witnessed by, say, $\models \varphi(a,b,c)$. Furthermore, there are tuples $a' \in K$ and $h_1 \in H$ such that a is algebraic over $a'h_1$ witnessed by, say,

$\models \varphi_1(a, a', h_1)$. Similarly, we find $\models \varphi_2(b, b', h_2)$ for b. By independence and the heir property there are $h_1', h_2' \in M$ such that

$$\models \exists x, y\, \varphi(a, x, y) \wedge \varphi_1(x, a', h_1') \wedge \varphi_2(y, b', h_2').$$

Since K and L are algebraically closed, this implies $c \in KL$. $\dashv$

EXERCISE 8.1.1. Let X be a compact topological space and $\mathcal{F}$ an ultrafilter on I. Then every family $(x_i)_{i \in I}$ has a unique $\mathcal{F}$-*limit*, which is the unique x such that $\{i \in I \mid x_i \in N\}$ belongs to $\mathcal{F}$ for every neighbourhood N of x.

Let A be an extension of the model M and $p \in M$. Show that one can construct all coheir extensions of $p \in \mathrm{S}(M)$ to A as follows: choose an ultrafilter $\mathcal{F}$ on M such that p is the $\mathcal{F}$-limit of $(\mathrm{tp}(m/M))_{m \in M}$, then the $\mathcal{F}$-limit of $(\mathrm{tp}(m/A))_{m \in M}$ in $\mathrm{S}(A)$ is a coheir of p.

EXERCISE 8.1.2. Use Example 7.2.10 and Lemma 8.1.6 to give an alternative proof of Corollary 7.3.3.

EXERCISE 8.1.3. Show that a formula which is satisfiable in every model extending A does not divide over A.

EXERCISE 8.1.4. Let T be a complete theory, M an ω-saturated model.

1. Let $\psi(x)$ be a formula over M with Morley rank, and φ a formula over arbitrary parameters with the same Morley rank and $\varphi(\mathfrak{C}) \subseteq \psi(\mathfrak{C})$. Show that φ is realized in M.
2. Let B an extension of M and $\mathrm{MR}(a/B) = \mathrm{MR}(a/M) < \infty$. Show that $\mathrm{tp}(a/B)$ is a coheir of $\mathrm{tp}(a/M)$.

It will follow from Corollaries 8.3.7 and 8.5.11 that in totally transcendental theories this is true for arbitrary M. In fact this holds for arbitrary theories, see Exercise 8.5.5.

EXERCISE 8.1.5 (Hrushovski–Pillay). Let $p(x)$ and $q(y)$ be global types, and suppose that $p(x)$ is A-invariant. We define a global type $p(x) \otimes q(y)$ by setting $(p \otimes q) \restriction B = \mathrm{tp}(ab/B)$ for any $B \supset A$ where b realizes $q(y) \restriction B$ and a realizes $p \restriction Bb$. Show that $p(x) \otimes q(y)$ is well defined, and A-invariant if both $p(x)$ and $q(y)$ are.

8.2. Stability

By analogy with simple theories we here define stable theories via (several equivalent) properties of their formulas and note that this definition fits well with the definition of κ-stability given in Section 5.2.

In this section, let T be a complete (possibly uncountable) theory. For a formula $\varphi(x, y)$ let $\mathrm{S}_\varphi(B)$ denote the set of all φ-types over B; these are maximal consistent sets of formulas of the form $\varphi(x, b)$ or $\neg\varphi(x, b)$ where

$b \in B$. Recall that the variables x and y may have different sorts representing n_x and n_y-tuples of elements, respectively.

DEFINITION 8.2.1. Let $\varphi(x, y)$ be a formula in the language of T.

1. The formula φ is *stable* if there is an infinite cardinal λ such that $|\,\mathrm{S}_\varphi(B)| \le \lambda$ whenever $|B| \le \lambda$. The theory T is *stable* if all its formulas are stable.
2. The formula φ has the *order property* if there are elements $a_0, a_1, \ldots$ and $b_0, b_1, \ldots$ such that for all $i, j \in \omega$
$$\models \varphi(a_i, b_j) \text{ if and only if } i < j.$$
3. The formula $\varphi(x, y)$ has the *binary tree property* if there is a binary tree $(b_s \mid s \in {}^{<\omega}2)$ of parameters such that for all $\sigma \in {}^{\omega}2$, the set
$$\{\varphi^{\sigma(n)}(x, b_{\sigma\restriction n}) \mid n < \omega\}$$
is consistent. (We use the notation $\varphi^0 = \neg\varphi$ and $\varphi^1 = \varphi$.)

It is important to note that T is stable if and only if it is κ-stable for some κ, see Exercise 8.2.7.

REMARK 8.2.2. The notion of $\varphi(x, y)$ having the order property is symmetrical in x and y. This means that if $\varphi(x, y)$ has the order property, then there are elements $a_0, a_1, \ldots$ and $b_0, b_1, \ldots$ such that $\models \varphi(a_i, b_j)$ if and only if $j < i$.

PROOF. Apply Lemma 7.1.1 to $\mathcal{I} = (a_i b_i)_{i<\omega}$ and $J = (\omega, >)$. ⊣

THEOREM 8.2.3. *For a formula $\varphi(x, y)$ the following are equivalent:*

a) *φ is stable.*
b) *$|\,\mathrm{S}_\varphi(B)| \le |B|$ for any infinite set B.*
c) *φ does not have the order property.*
d) *φ does not have the binary tree property.*

PROOF. a) $\Rightarrow$ d): Let μ be minimal such that $2^\mu > \lambda$. Then the tree $I = {}^{<\mu}2$ has cardinality at most λ. If $\varphi(x, y)$ has the binary tree property, by compactness there are parameters b_s, $(s \in I)$, such that for all $\sigma \in {}^{\mu}2$, $q_\sigma = \{\varphi^{\sigma(\alpha)}(x, b_{\sigma\restriction\alpha}) \mid \alpha < \mu\}$ is consistent. Complete every q_σ to a φ-type p_σ over $B = \{b_s \mid s \in I\}$. Since the p_σ are pairwise different, we have $|B| \le \lambda < 2^\mu \le |\,\mathrm{S}_\varphi(B)|$.

d) $\Rightarrow$ c): Choose a linear ordering of $I = {}^{\le\omega}2$ such that for all $\sigma \in {}^{\omega}2$ and $n < \omega$
$$\sigma < \sigma \restriction n \Leftrightarrow \sigma(n) = 1.$$
If $\varphi(x, y)$ has the order property, then by Lemma 7.1.1 one can find a_i and b_i indexed by I such that
$$\models \varphi(a_i, b_j) \text{ if and only if } i < j.$$
Now the tree $\varphi(x, b_s)$, $s \in {}^{<\omega}2$, shows that φ has the binary tree property.

c) $\Rightarrow$ b): Let B be an infinite set of parameters and $|S_\varphi(B)| > |B|$. For any a the φ-type of a over B is given by $S_a = \{b \in B^n \mid \models \varphi(a, b)\}$. Since $|B| = |B^n|$ we may assume for simplicity that $n = 1$ and so $S_a \subseteq B$. Applying the Erdős–Makkai Theorem C.2.1 to B and $\mathcal{S} = \{S_a \mid a \in \mathfrak{C}\}$, we obtain a sequence $(b_i \mid i < \omega)$ of elements of B and a sequence $(a_i)_{i<\omega}$ such that either $b_i \in S_{a_j} \Leftrightarrow j < i$ or $b_i \in S_{a_j} \Leftrightarrow i < j$ for all i, j. In the first case φ has the order property by definition. In the second case $\varphi(x, y)$ has the order property by Remark 8.2.2.

b) $\Rightarrow$ a): Clear. $\dashv$

If φ has the binary tree property witnessed, say, by $(b_s \mid s \in {}^{<\omega}2)$, then the family

$$\varphi_s = \bigwedge_{n<|s|} \varphi^{s(n)}(x, b_{s\restriction n}),\ s \in {}^{<\omega}2,$$

is a binary tree of consistent formulas. This shows that totally transcendental theories are stable. We will see below (Corollary 8.3.6, also Exercise 8.2.11)) that stable theories are simple.

REMARK 8.2.4. By Example 8.6.6 and Exercise 8.2.7 the theory of any R-module is stable (but not necessarily totally transcendental) providing a rich class of examples for stable theories. Note that the theory of the random graph is simple by Corollary 7.3.14 but not stable (see Exercise 8.2.3).

EXERCISE 8.2.1. The theory T is unstable if and only if there is an L-formula $\psi(\overline{x}, \overline{y})$ and elements $\overline{a}_0, \overline{a}_1, \ldots$, ordered by ψ; i.e., such that

$$\models \psi(\overline{a}_i, \overline{a}_j) \Longleftrightarrow i < j :$$

ψ may contain parameters.

EXERCISE 8.2.2. A formula $\varphi(x, y)$ is said to have the *independence property* (IP) if there are $a_i, i \in \omega$, such that for each $A \subseteq \omega$ the set $\{\varphi(x, a_i) \mid i \in A\} \cup \{\neg\varphi(x, a_i) \mid i \notin A\}$ is consistent. Show that T is unstable if it contains a formula with the independence property.

EXERCISE 8.2.3. Show that the theory of the random graph is not stable.

EXERCISE 8.2.4. A formula $\varphi(x, y)$ is said to have the *strict order property* (SOP) if there is a sequence $(a_i)_{i<\omega}$ such that

$$\models \forall y(\varphi(a_i, y) \to \varphi(a_j, y)) \Leftrightarrow i \leq j.$$

The theory T has the strict order property if there is a formula in T with the strict order property. Show that T has the SOP if and only if there is a partial ordering with infinite chains definable in T^{eq}. (For the definition of T^{eq} see p. 140.)

EXERCISE 8.2.5. Show that a theory with SOP is not simple.

EXERCISE 8.2.6. (Shelah) If T is unstable, either there is a formula having the IP or a formula having the SOP.

EXERCISE 8.2.7. The following are equivalent:

a) T is stable.
b) T is λ-stable for all λ such that $\lambda^{|T|} = \lambda$.
c) T is λ-stable for some λ.

It follows from this and Lemma 5.2.2 that T is stable if and only if all $\varphi(x, y)$ for a single variable x are stable.

EXERCISE 8.2.8. Show that for any infinite λ there is a linear order of cardinality greater than λ with a dense subset of size λ.

EXERCISE 8.2.9. Show that for all tuples of variables x, y the set of stable formulas $\varphi(x, y)$ is closed under Boolean combinations, i.e., conjunction (disjunction) and negation. Use this to show that the theory Tree, defined on page 60, is stable.

EXERCISE 8.2.10. Fix an L-formula $\varphi(x, y)$. Let Φ denote the class of Boolean combinations of formulas of the form $\varphi(x, b)$. Define the *φ-rank* R_φ as the smallest function from formulas $\psi(x)$ to $\{-\infty\} \cup \mathrm{On} \cup \{\infty\}$ (here On is the class of ordinals) such that

$\mathrm{R}_\varphi(\psi) \geq 0$ if ψ is consistent;

$\mathrm{R}_\varphi(\psi) \geq \beta + 1$ if there are infinitely many $\delta_i \in \Phi$ which are pairwise inconsistent and such that $\mathrm{R}_\varphi(\psi \wedge \delta_i) \geq \beta$ for all i.

Prove:

1. $\mathrm{R}_\varphi(\psi) < \infty$ if and only if $\psi(x) \wedge \varphi(x, y)$ is stable.
2. If $\mathrm{R}_\varphi(\psi) < \infty$, then $\mathrm{R}_\varphi(\psi) < \omega$.

EXERCISE 8.2.11. If φ has the tree property, it is unstable.

This shows that stable theories are simple. We will give a different proof in Corollary 8.3.6.

8.3. Definable types

Definability of types turns out to be a crucial feature of stable theories. We show here that in stable theories the extensions of a type over a model given by its definition agree with the non-forking extensions (and with heirs and coheirs). We continue to assume that T is a complete theory.

THEOREM 8.3.1. *The formula $\varphi(x, y)$ is stable if and only if all φ-types are definable.*

PROOF. Let A be a set of parameters of size $\geq |T|$. If all φ-types over A are definable, there exists no more φ-types over A than there are defining formulas, i.e., at most $|A|$ many. So φ is stable.

For the converse assume that $\varphi(x, y)$ is stable. Define for any formula $\theta(x)$ the degree $\mathrm{D}_\varphi(\theta)$ to be the largest n for which there is a finite tree $(b_s \mid s \in {}^{<n}2)$ of parameters such that for every $\sigma \in {}^{n}2$ the set $\{\theta(x)\} \cup \{\varphi^{\sigma(i)}(x, b_{\sigma\restriction i}) \mid i < n\}$ is consistent. This is well defined since φ does not have the binary tree property. Now, let p be a φ-type over B. Let θ be a conjunction of formulas in p with $n = \mathrm{D}_\varphi(\theta)$ minimal. Then $\varphi(x, b)$ belongs to p if and only if $n = \mathrm{D}_\varphi(\theta(x) \wedge \varphi(x, b))$. This shows that p is definable. ⊣

COROLLARY 8.3.2. *The theory T is stable if and only if all types are definable.* ⊣

Observe that the proof of Theorem 8.3.1 applies also to a proper class of parameters. From this we obtain the following important corollary.

COROLLARY 8.3.3 (Separation of variables). *Let T be stable and let $\mathbb{F}$ be a 0-definable class. Then any definable subclass of $\mathbb{F}^n$ is definable using parameters from $\mathbb{F}$.*

PROOF. Let $\psi(a, \mathfrak{C})$ be a definable subclass of $\mathbb{F}^n$. The type $q = \mathrm{tp}(a/\mathbb{F})$ is definable over a subset of $\mathbb{F}$ by Corollary 8.3.2. Thus,

$$\psi(a, \mathfrak{C}) = \{f \in \mathbb{F}^n \mid \models \mathrm{d}_q\, x \psi(x, f)\}.$$ ⊣

If the property in the conclusion of Corollary 8.3.3 holds for a 0-definable class $\mathbb{F}$ (in a not necessarily stable theory T), then $\mathbb{F}$ is called *stably embedded.* For equivalent definitions see Exercise 10.1.5.

At first glance, the next lemma looks mysterious. In essence it states that in stable theories heirs and coheirs coincide. We need it in the proof of Corollary 8.5.3.

LEMMA 8.3.4 (Harrington). *Let T be stable and let $p(x)$ and $q(y)$ be global types. Then for every formula $\varphi(x, y)$ with parameters*

$$\mathrm{d}_p\, x\varphi(x, y) \in q(y) \;\Leftrightarrow\; \mathrm{d}_q\, y\varphi(x, y) \in p(x).$$

PROOF. Let p, q and φ be definable over A. We recursively define sequences a_i and b_i, $i \in \omega$: if $a_0, \ldots, a_{n-1}$ and $b_0, \ldots, b_{n-1}$ have been defined, let b_n be a realisation of $q \restriction Aa_0, \ldots, a_{n-1}$ and a_n a realisation of $p \restriction Ab_0, \ldots, b_n$. Then we have for $i < j$

$$\models \varphi(a_i, b_j) \Leftrightarrow \models \mathrm{d}_q\, y\varphi(a_i, y) \;\Leftrightarrow\; \mathrm{d}_q\, y\varphi(x, y) \in p(x)$$

and for $j \leq i$

$$\models \varphi(a_i, b_j) \Leftrightarrow \models \mathrm{d}_p\, x\varphi(x, b_j) \;\Leftrightarrow\; \mathrm{d}_p\, x\varphi(x, y) \in q(y).$$

Since φ does not have the order property, the claim follows. ⊣

LEMMA 8.3.5. *Let $p \in S(\mathfrak{C})$ be a global type.*

1. *If p is definable over A, then p does not divide over A.*
2. *If T is stable and p does not divide over the model M, then p is definable over M.*

Note that for global types dividing and forking coincide (Exercise 7.1.3).

PROOF. 1): Consider a formula $\varphi(x, m) \in p$ and an infinite sequence of indiscernibles $m = m_0, m_1, \ldots$ over A. If p is definable over A, all $\varphi(x, m_i)$ belong to p. So $\varphi(x, m)$ does not divide over A by Lemma 7.1.4.

2): Now assume that T is stable and p does not divide over the model M. We will show that p is an heir of $p \upharpoonright M$. By Corollary 8.3.2 and Lemma 8.1.5 this implies that p is definable over M. So assume that $\varphi(x, b) \in p$, we want to show that $\varphi(x, b') \in p$ for some $b' \in M$.

Let $\mathcal{I} = (b_i)_{i<\omega}$ be a Morley sequence of a global coheir extension of $\mathrm{tp}(b/M)$ over M starting with $b_0 = b$ (see Example 7.2.10 and Lemma 8.1.6). Since $\mathrm{tp}(a/Mb)$ does not divide over M, Lemma 7.1.5 implies that we may assume that $\mathcal{I}$ is indiscernible over Ma. So we have $\models \varphi(a, b_i)$ for all i. By Corollary 8.3.2, the type $q = \mathrm{tp}(a/M\{b_i \mid i < \omega\})$ is definable. Assume that the parameters of $\mathrm{d}_q\, x\varphi(x, y)$ are in $M\{b_0, \ldots, b_{n-1}\}$. Since $\mathrm{tp}(b_n/M\{b_0, \ldots, b_{n-1}\})$ is a coheir of $\mathrm{tp}(b/M)$, and since $\models \mathrm{d}_q\, x\varphi(x, b_n)$, there is a $b' \in M$ with $\models \mathrm{d}_q\, x\varphi(x, b')$. This implies $\models \varphi(a, b')$ and so $\varphi(x, b') \in \mathrm{tp}(a/M) = p \upharpoonright M$. ⊣

COROLLARY 8.3.6. *Stable theories are simple.*

PROOF. Let p be a type over a model M. Then p is definable over some $A \subseteq M$ of cardinality $\leq |T|$. Let p' be the global extension of p given by the definition over A. By Lemma 8.3.5(1), p' and hence also p does not divide over A. Proposition 7.2.5 implies that T is simple. ⊣

This implies in particular that forking and dividing coincide in stable theories (see Proposition 7.2.15).

COROLLARY 8.3.7. *Let T be a stable theory, p a type over a model M and A an extension of M. Then p has a unique extension $q \in S(A)$ with the following equivalent properties:*

a) *q does not fork over M.*
b) *q is definable over M.*
c) *q is an heir of p.*
d) *q is a coheir of p.*

PROOF. By Lemma 8.3.5, q does not fork over M if and only if it is definable over M. Since p is definable, we know by Lemma 8.1.5 that there is a unique extension q which is definable over M, and which is also the unique heir of p.

To prove the equivalence with d) we may assume that $A = M \cup \{a\}$ for a finite tuple a. Fix a realisation b of q. Then $q = \mathrm{tp}(b/Ma)$ is a coheir of $p = \mathrm{tp}(b/M)$ if and only if $\mathrm{tp}(a/Mb)$ is an heir and hence, by the first part of the proof, a non-forking extension of $\operatorname{tp}(a/M)$. Now forking symmetry and the first part of the proof imply the desired. ⊣

EXERCISE 8.3.1. Find a theory T and a type p over the empty set such that no definition of p defines a global type. (A definition which defines a global type is called a *good* definition of p, see Theorem 8.5.1).

EXERCISE 8.3.2. Let T be an arbitrary complete theory and M be a model. Consider the following four properties of a global type p:

(D) p is definable over M.
(C) p is a coheir of $p \upharpoonright M$.
(I) p is M-invariant.
(H) p is an heir of $p \upharpoonright M$.

Use the example $T = \mathsf{DLO}$ and $M = \mathbb{Q}$ to show that D→I, C→I, D→H are the only logical relations between these notions.

EXERCISE 8.3.3. Let $p(x) \in S(\mathfrak{C})$ be definable over B. Then, for any n, the map

$$r(y_1, \ldots, y_n) \mapsto \{\varphi(x, y_1, \ldots, y_n) \mid \mathrm{d}_p\, x\varphi \in r\}$$

defines a continuous section $\pi_n\colon S_n(B) \to S_{n+1}(B)$. Show that this defines a bijection between all types definable over B and all "coherent" families (π_n) of continuous sections $S_n(B) \to S_{n+1}(B)$.

EXERCISE 8.3.4. Let $\varphi(x)$ be a formula without parameters and let M be a model of T. Show that $\varphi(M)$ is *stably embedded in M* (i.e., every M-definable relation of $\varphi(M)$ is definable over $\varphi(M)$) if and only if for all n, every $p(x_1, \ldots, x_n) \in S_n(\varphi(M))$ which contains $\varphi(x_1) \wedge \cdots \wedge \varphi(x_n)$ has a unique extension $p' \in S_n(M)$. If φ is (absolutely) stably embedded and p is definable, show that p' is definable over $\varphi(M)$.

EXERCISE 8.3.5. Call a formula $\psi(x)$ in *one* variable (though possibly representing a tuple) *stable* if $\psi(x) \wedge \varphi(x, y)$ is stable for all $\varphi(x, y)$ according to Definition 8.2.1. We call a type stable if it contains a stable formula. Prove:

1. Types with Morley rank are stable.
2. Stable types are definable.
3. Stable formulas are stably embedded.

EXERCISE 8.3.6. Let T be stable, and $p \in S(A)$. Show that p is definable over C if p is finitely satisfiable in C. Furthermore for every $\varphi(x, y)$, $\mathrm{d}_p\, x\varphi(x, y)$ is a positive Boolean combination of formulas $\varphi(c, y)$, $c \in C$.

EXERCISE 8.3.7. If T is stable, then for any formula $\varphi(x, y)$, there is a formula $\Delta(y, z)$ such that for every set A and every type $p(x)$ over A there is a tuple b in A such that $\{a \in A \mid \varphi(x, a) \in p(x)\} = \{a \in A \mid \models \Delta(a, b)\}$.

EXERCISE 8.3.8. We call q a *weak heir* of $p \in S(M)$ if the heir property holds for all $\varphi(x, y)$ without parameters. Show that in stable theories, weak heirs are in fact heirs.

EXERCISE 8.3.9. In Corollary 8.3.7 prove the equivalence of c) and d) directly from Lemma 8.3.4.

EXERCISE 8.3.10. Show that in a stable theory a formula does not fork over A if and only if it is realized in every model which contains A.

8.4. Elimination of imaginaries and T^{eq}

This section is an excursion outside the realm of stable theories: for a model M of an *arbitrary complete theory* T and any 0-definable equivalence relation $E(\overline{x}, \overline{y})$ on n-tuples, we now consider the equivalence classes of M^n/E as elements of a new sort of so-called *imaginary elements*. Adding these imaginaries makes many arguments more convenient. For certain theories, these imaginaries are already coded in the original structure. However, if this is not already the case, then adding imaginaries leads to a new theory T^{eq} which does have this property so that we do not run into an infinite regression.

DEFINITION 8.4.1. A finite tuple $d \subseteq \mathfrak{C}$ is called a *canonical parameter* for a definable class $\mathbb{D}$ in $\mathfrak{C}^n$ if d is fixed by the same automorphisms of $\mathfrak{C}$ which leave $\mathbb{D}$ invariant. A *canonical base* for a type $p \in S(\mathfrak{C})$ is a set B which is pointwise fixed by the same automorphisms which leave p invariant.

Lemma 6.1.10 implies that $\mathbb{D}$ is definable over d, and by Corollary 6.1.12(1) d is determined by $\mathbb{D}$ up to interdefinability. We write $d = \ulcorner \mathbb{D} \urcorner$, or $d = \ulcorner \varphi(x) \urcorner$ if $\mathbb{D} = \varphi(\mathfrak{C})$. Note that the empty tuple is a canonical parameter for every 0-definable class.

DEFINITION 8.4.2. The theory T *eliminates imaginaries* if any class e/E of a 0-definable equivalence relation E on $\mathfrak{C}^n$ has a canonical parameter $d \subseteq \mathfrak{C}$.

THEOREM 8.4.3. *If T eliminates imaginaries, then the following hold*:

1. *Every definable class $\mathbb{D} \subseteq \mathfrak{C}^n$ has a canonical parameter c.*
2. *Every definable type $p \in S(\mathfrak{C})$ has a canonical base.*

PROOF. Write $\mathbb{D} = \varphi(\mathfrak{C}, e)$. Define the equivalence relation E by

$$y_1 E y_2 \iff \forall x\ \varphi(x, y_1) \leftrightarrow \varphi(x, y_2)$$

and let d be a canonical parameter of e/E. Then d is a canonical parameter of $\mathbb{D}$.

If d_p is a definition of p, the set $B = \{\ulcorner \mathrm{d}_p\, x \varphi(x, y) \urcorner \mid \varphi(x, y)\ L\text{-formula}\}$ is a canonical base of p. ⊣

LEMMA 8.4.4. *Assume that T eliminates imaginaries. Let A be a set of parameters and $\mathbb{D}$ a definable class. Then the following are equivalent*:

a) $\mathbb{D}$ *is* $\operatorname{acl}(A)$*-definable.*
b) $\mathbb{D}$ *has only finitely many conjugates over* A.
c) $\mathbb{D}$ *is the union of equivalence classes of an A-definable equivalence relation with finitely many classes* (*a* finite *equivalence relation*).

PROOF. Let d be a canonical parameter of $\mathbb{D}$. Then $\mathbb{D}$ is definable over $\operatorname{acl}(A)$ if and only if d belongs to $\operatorname{acl}(A)$. On the other hand $\mathbb{D}$ has as many conjugates over A as d. So a) and b) are equivalent.

For the equivalence of b) and c), first notice that any class of an A-definable finite equivalence relation has only finitely many conjugates over A, which yields c) $\Rightarrow$ b). For the converse, let $\mathbb{D} = \mathbb{D}_1, \ldots, \mathbb{D}_n$ be the conjugates of $\mathbb{D}$ over A. Consider the finite equivalence relation $E(c, c')$ defined by

$$c \in \mathbb{D}_i \text{ if and only if } c' \in \mathbb{D}_i \text{ for all } i.$$

Clearly $\mathbb{D}$ is a union of E-classes. Also E is definable and since it is invariant under all A-automorphisms of $\mathfrak{C}$, it is in fact definable over A.

Note that elimination of imaginaries was only used for b) $\Rightarrow$ a). ⊣

The previous results show why it is convenient to work in a theory eliminating imaginaries. It is easy to see that a theory eliminates imaginaries if every 0-definable equivalence relation arises from fibres of a 0-definable function. While not all theories have this property (e.g., the theory of an equivalence relation with infinitely many infinite classes), we now show how to extend any complete theory T to a theory T^{eq} (in a corresponding language L^{eq}) which does.

Let $E_i(\overline{x}_1, \overline{x}_2)$, $(i \in I)$, be a list of all 0-definable equivalence relations on n_i-tuples. For any model M of T we consider the many-sorted structure

$$M^{\mathrm{eq}} = (M, M^{n_i}/E_i)_{i \in I},$$

which carries the *home sort* M and for every i the natural projection

$$\pi_i : M^{n_i} \to M^{n_i}/E_i.$$

The elements of the sorts $S_i = M^{n_i}/E_i$ are called *imaginary* elements, the elements of the home sort are *real* elements.

The M^{eq} form an elementary class axiomatised by the (complete) theory T^{eq} which, in in the appropriate many-sorted language L^{eq}, is axiomatised by the axioms of T and for each $i \in I$ by

$$\forall y \, \exists \overline{x} \; \pi_i(\bar{x}) \doteq y \qquad (y \text{ a variable of sort } S_i)$$

and

$$\forall \overline{x}_1, \overline{x}_2 \; (\pi_i(\overline{x}_1) \doteq \pi_i(\bar{x}_2) \leftrightarrow E_i(\overline{x}_1, \overline{x}_2)).$$

The algebraic (definable, respectively) closure of in M^{eq} is denoted by acl^{eq} (dcl^{eq}, respectively).

The first two statements of the following proposition explain why we consider T^{eq} as an inessential expansion of T.

PROPOSITION 8.4.5. 1. *Elements of $\mathfrak{C}^{eq}$ are definable over $\mathfrak{C}$ in a uniform way.*

2. *The 0-definable relations on the home sort of C^{eq} are exactly the same as those in $\mathfrak{C}$.*
3. *The theory T^{eq} eliminates imaginaries,*

PROOF. 1: Every element of sort S_i has the form $\pi_i(\overline{a})$ for an n_i-tuple $\overline{a}$ from $\mathfrak{C}$.

2: We show that every L^{eq}-formula $\varphi(\overline{x})$ with free variables from the home-sort is equivalent to an L-formula $\varphi^*(\overline{x})$ by induction on the complexity of φ. If φ is atomic, it is either an L-formula or of the form $\pi_i(\bar{x}_1) \doteq \pi_i(\overline{x}_2)$, in which case we set $\varphi^* = E_i(\bar{x}_1, \overline{x}_2)$. We let $*$ commute with negations, conjunctions and quantification over home-sort variables. Finally, if y is a variable of sort S_i, we set

$$\left(\exists y \psi(\overline{x}, y)\right)^* = \exists \overline{x}' \psi(\overline{x}, \pi_i(\overline{x}'))^*.$$

3: We observe first that $\mathfrak{C}^{eq}$ is the monster model of T^{eq}, i.e., that every type $p(y)$ over a set A is realized in $\mathfrak{C}^{eq}$. By Part 1 we may assume that $A \subseteq \mathfrak{C}$. If y is of sort S_i, the set $\Sigma(\overline{x}) = p(\pi_i(\overline{x}))$ is finitely satisfiable. By Part 2 Σ is equivalent to a set Σ^* of L-formulas; this set has a realisation $\overline{b}$, which gives us a realisation $\pi_i(\overline{b})$ of p.

It is now clear that $\pi_i(e)$ is a canonical parameter of the class e/E_i. By the proof of Theorem 8.4.3 this implies that every relation in $\mathfrak{C}^{eq}$ that is definable with parameters from $\mathfrak{C}$ has a canonical parameter in $\mathfrak{C}^{eq}$. On the other hand, by Part 1, every definable relation in $\mathfrak{C}^{eq}$ is definable in $\mathfrak{C}$. ⊣

COROLLARY 8.4.6. *The theory T eliminates imaginaries if and only if in T^{eq} every imaginary is interdefinable with a real tuple.*

PROOF. Since every automorphism of $\mathfrak{C}$ extends (uniquely) to an automorphism of $\mathfrak{C}^{eq}$, a real tuple d is a canonical parameter of e/E_i in the sense of T if and only if it is a canonical parameter in the sense of T^{eq}. But this is equivalent to d being interdefinable with $\pi_i(e)$. ⊣

The proof of the following criterion for elimination of imaginaries shows how T^{eq} can be used.

LEMMA 8.4.7. *The following are equivalent:*

a) *The theory T eliminates imaginaries and has at least two 0-definable elements.*

b) *Every 0-definable equivalence relation on $\mathfrak{C}^n$ is the fibration of a 0-definable function $f : \mathfrak{C}^n \to \mathfrak{C}^m$.*

PROOF. b) $\Rightarrow$ a): If E is the fibration of a 0-definable function $f : \mathfrak{C}^n \to \mathfrak{C}^m$, we have $\ulcorner e/E \urcorner = f(e)$. To see that there are at least two 0-definable elements look at the following equivalence relation on $\mathfrak{C}^2$:

$$x_1 x_2 \mathrel{E} y_1 y_2 \Leftrightarrow (x_1 = x_2 \leftrightarrow y_1 = y_2).$$

This has two classes, which are both 0-definable. If E is the fibration of a 0-definable $\pi \colon \mathfrak{C}^2 \to \mathfrak{C}^m$, the two images of π are two different 0-definable m-tuples.

a) $\Rightarrow$ b): Let E be a 0-definable equivalence relation on $\mathfrak{C}^n$. Every e/E is interdefinable with an element of some power $\mathfrak{C}^{m_e}$. So by Exercise 6.1.12, e/E belongs to a 0-definable $\mathbb{D} \subseteq \mathfrak{C}^n/E$ with a 0-definable injection $f \colon \mathbb{D} \to \mathfrak{C}^{m_e}$. A compactness argument shows that we can cover $\mathfrak{C}^n/E$ by finitely many 0-definable classes $\mathbb{D}_1, \dots, \mathbb{D}_k$ with 0-definable injections $f_i \colon \mathbb{D}_i \to \mathfrak{C}^{m_i}$. We may assume that the $\mathbb{D}_i$ are pairwise disjoint, otherwise we replace $\mathbb{D}_i$ by $\mathbb{D}_i \setminus (\mathbb{D}_0 \cup \dots \cup \mathbb{D}_{i-1})$. Now, using the two 0-definable elements, we can find, for some big m, 0-definable injections $g_i \colon \mathfrak{C}^{m_i} \to \mathfrak{C}^m$ with pairwise disjoint images. The union of the $g_i f_i$ is a 0-definable injection from $\mathfrak{C}^n/E$ into $\mathfrak{C}^m$. $\dashv$

Using parts 1 and 2 of the previous proposition, one can see that in general all properties of T which concern us here are preserved when going from T to T^{eq}. Here are some examples.

LEMMA 8.4.8. 1. *The theory T is $\aleph_1$-categorical if and only if T^{eq} is $\aleph_1$-categorical.*
2. *T is λ-stable if and only if T^{eq} is λ-stable.*
3. *T is stable if and only if T^{eq} is stable.*

PROOF. Part 1 is clear.

For Part 2 let A be a set of parameters in T^{eq} of cardinality λ. This set A is contained in the definable closure of some set B of cardinality λ of the home sort. For any $p \in \mathrm{S}(B)$ we may first take the unique extension of p to $\mathrm{dcl}^{\mathrm{eq}}(B)$ and then its restriction to A. This defines a surjection $\mathrm{S}(B) \to \mathrm{S}(A)$. Notice that we now have to specify not only the number of variables but also the sorts for the variables in the types.

If $\mathrm{S}'(A)$ consists of types of elements of the sort $\mathfrak{C}^n/E$, $\mathrm{S}_n(A)$ denotes the n-types of the home sort and π the projection $\mathfrak{C}^n \to \mathfrak{C}^n/E$, then $\mathrm{tp}(\bar{b}/A) \mapsto \mathrm{tp}(\pi(\bar{b})/A)$ defines a surjection $\mathrm{S}_n(A) \to \mathrm{S}'(A)$. This shows that T^{eq} is stable if T is.

Of course Part 3 follows from 2. and Exercise 8.2.7. Still we give a direct proof as an example of how to translate between T and T^{eq}. Let $\varphi(y_1, y_2)$ be a formula in T^{eq} with the order property. If y_1 and y_2 belong to $\mathfrak{C}^{n_1}/E_1$ and $\mathfrak{C}^{n_2}/E_2$, respectively, there are tuples $\bar{a}_i$ of the home sort such that

$$\models \varphi(\pi_1(\bar{a}_1), \pi_2(\bar{a}_2)) \text{ if and only if } i < j.$$

By Proposition 8.4.5 the formula $\varphi(\pi_1(\bar{x}_1), \pi_2(\bar{x}_2))$ is equivalent to some L-formula $\psi(\bar{x}_1, \bar{x}_2)$, which has the order property in T. ⊣

For applications the following special cases are often useful.

DEFINITION 8.4.9. 1. The theory T *eliminates finite imaginaries* if every finite set of n-tuples has a canonical parameter.

2. T has *weak elimination of imaginaries* if for every imaginary e there is a real tuple c such that $e \in \mathrm{dcl}^{\mathrm{eq}}(c)$ and $c \in \mathrm{acl}(e)$.

LEMMA 8.4.10. *The theory T eliminates imaginaries if and only if it has weak elimination of imaginaries and eliminates finite imaginaries*

PROOF. This follows from the observation that T has weak elimination of imaginaries if and only if every imaginary e is interdefinable with the canonical parameter of a finite set of real n-tuples. Indeed, if $e \in \mathrm{dcl}^{\mathrm{eq}}(c)$ and $c \in \mathrm{acl}(e)$, and $\{c_1, \ldots, c_m\}$ are the conjugates of c over e, then e is interdefinable with $\ulcorner\{c_1, \ldots, c_m\}\urcorner$. If conversely e is interdefinable with $\ulcorner\{c_1, \ldots, c_m\}\urcorner$, then $e \in \mathrm{dcl}^{\mathrm{eq}}(c_1 \ldots c_m)$ and $c_1 \ldots c_m x \in \mathrm{acl}(e)$. ⊣

LEMMA 8.4.11 (Lascar–Pillay). *Let T be strongly minimal and* $\mathrm{acl}(\emptyset)$ *infinite. Then T has weak elimination of imaginaries.*

PROOF. Let $e = c/E$ be an imaginary. It suffices to show that c/E contains an element algebraic over e or, more generally, that every non-empty definable $X \subseteq \mathfrak{C}^n$ contains an element of $A = \mathrm{acl}(\ulcorner X\urcorner)$. We proceed by induction on n. For $n = 1$ there are two cases: if X is finite, it is a subset of A; if X is infinite, almost all elements of $\mathrm{acl}(\emptyset)$ belong to X. If $n > 1$, consider the projection Y of X to the first coordinate. Such a Y contains an element a of $\mathrm{acl}(\ulcorner Y\urcorner)$, which is a subset of A. By induction the fibre X_a contains an element b of $\mathrm{acl}(\ulcorner X_a\urcorner)$, which is also a subset of A. So (a, b) is in $X \cap A$. ⊣

COROLLARY 8.4.12. *The theory* ACF_p *of algebraically closed fields of characteristic p eliminates imaginaries.*

PROOF. By the preceding lemmas it suffices to show that every theory of fields eliminates finite imaginaries. Let $S = \{c_0, \ldots, c_{k-1}\}$ be a set of n-tuples $c_i = (c_{i,j})_{j<n}$. Consider the polynomial

$$p(X, Y_0, \ldots, Y_{n-1}) = \prod_{i<k} \big(X - \sum_{j<n} c_{i,j} Y_j\big).$$

An automorphism leaves p fixed if and only if it permutes S. So the coefficients of p serve as a canonical parameter of S. ⊣

LEMMA 8.4.13. *A totally transcendental theory in which every global type has a canonical base in* $\mathfrak{C}$ *has weak elimination of imaginaries.*

PROOF. Let $e = c/E$ be an imaginary and α the Morley rank of the class c/E. Let p be a global type of Morley rank α which contains $E(x, c)$. By assumption p has a canonical base $d \subseteq \mathfrak{C}$. Since there are only finitely many such p, d is algebraic over e. Also e is definable from d since for an automorphism α fixing p, c/E and $\alpha c/E$ cannot be disjoint, so they must be equal. Clearly we may assume that d is a finite tuple (see also Exercises 8.4.1 and 8.4.7). ⊣

COROLLARY 8.4.14. DCF$_0$ *eliminates imaginaries.*

PROOF. By quantifier elimination every global type $p(x)$ is axiomatised by its quantifier-free part $p_{qf}(x)$, which is equivalent to a union of $q_i(x, dx, \dots, d^i)$, $i = 0, 1, \dots$, where the $q_i(x_0, \dots, x_i)$ are quantifier-free pure field-theoretic types. If C_i is the canonical base of q_i in the sense of ACF$_0$, then $C_0 \cup C_1 \cup \cdots$ is a canonical base of p. ⊣

EXERCISE 8.4.1. Let $\mathbb{D}$ be a definable class. Assume that there is a set D which is fixed by the same automorphisms which leave $\mathbb{D}$ invariant. Show that D contains a canonical parameter of $\mathbb{D}$.

EXERCISE 8.4.2. A theory T has weak elimination of imaginaries if and only if for every definable class $\mathbb{D}$ there is a smallest algebraically closed set over which $\mathbb{D}$ is definable.

EXERCISE 8.4.3. Use Exercise 8.4.2 to prove that the theories Infset and DLO (not easy) have weak elimination of imaginaries. Show also that DLO has elimination of imaginaries, but Infset does not.

EXERCISE 8.4.4. Show that all extensions of $p \in S(A)$ to acl(A) are conjugate over A. More generally this remains true for every *normal* extension B of A. These are sets which are invariant under all $\alpha \in \mathrm{Aut}(\mathfrak{C}/A)$. Note that normal extensions must be subsets of acl(A).

EXERCISE 8.4.5. An algebraic type over A has a *good* definition (see Exercise 8.3.1 or p. 145) over $B \subseteq A$ if and only if it is realised in dcl(B).

EXERCISE 8.4.6. Let d be a canonical parameter of $\mathbb{D}$. Then d is 0-definable in the $L \cup \{P\}$-structure $(\mathfrak{C}, \mathbb{D})$.

EXERCISE 8.4.7. Let T be totally transcendental and p a global type.

1. Show that p has a finite canonical base in $\mathfrak{C}^{\mathrm{eq}}$.
2. If p has a canonical base $D \subseteq \mathfrak{C}$, then it has a finite base $d \subseteq \mathfrak{C}$.

EXERCISE 8.4.8. Show that Lemma 8.4.13 is true for stable theories. (Hint: In the proof of 8.4.13 replace p by a suitable $E(x, y)$-type.)

EXERCISE 8.4.9. Define the *strong type* of a over A as $\mathrm{stp}(a/A) = \mathrm{tp}(a/\mathrm{acl}^{\mathrm{eq}}(A))$. Show that stp$(a/A)$ is axiomatised by

$$\{E(x, a) \mid E(x, y)\ A\text{-definable finite equivalence relation}\}.$$

EXERCISE 8.4.10 (Poizat). Let T be a complete theory with elimination of imaginaries. Consider the group $G = \mathrm{Aut}(\mathrm{acl}(\emptyset))$ of elementary permutations of $\operatorname{acl}(\emptyset)$. This G is a topological group if we use the stabilisers of finite sets as a basis of neighbourhoods of 1. Show that there is a Galois correspondence between the closed subgroups H of G and the definably closed subsets A of $\operatorname{acl}(\emptyset)$.

8.5. Properties of forking in stable theories

Except in Theorem 8.5.10 *we assume throughout this section that T is stable.* We now collect the crucial properties of forking in stable theories. As with simple theories, we will see in Theorem 8.5.10 that these properties characterise stable theories and forking. Since we already know that stable theories are simple, some of these properties are immediate. For completeness and reference we restate them in the context of stable theories.

Let $p \in S(B)$ be defined by $L(B)$-formulas $\mathrm{d}_p\, x\varphi$. We call the definition *good* if it defines a global type (or, equivalently, if it defines a type over some model containing B).

THEOREM 8.5.1. *Let T be stable. A type $p \in S(B)$ does not fork over $A \subseteq B$ if and only if p has a good definition over* $\operatorname{acl}^{\mathrm{eq}}(A)$.

PROOF. If p does not fork over A, p has a global extension p' which does not fork over A. Let M be any model which contains A. By Lemma 8.3.5, p' is definable over M, so the canonical base of p' belongs to M^{eq}. By Exercise 6.1.2 the canonical base belongs to $\operatorname{acl}^{\mathrm{eq}}(A)$.

If conversely p has a good definition over $\operatorname{acl}^{\mathrm{eq}}(A)$, p does not fork over $\operatorname{acl}^{\mathrm{eq}}(A)$ and therefore does not fork over A. ⊣

DEFINITION 8.5.2. A type is *stationary* if and only if it has a unique non-forking extension to any superset.

COROLLARY 8.5.3 (Uniqueness). *If T is stable, any type over* $\operatorname{acl}^{\mathrm{eq}}(A)$ *is stationary.*

If T eliminates imaginaries, this just says that any type over an algebraically closed set is stationary.

PROOF. Let $A = \operatorname{acl}^{\mathrm{eq}}(A)$. Let p' and p'' two global non-forking extensions of $p \in S(A)$. Consider any formula $\varphi(x, b)$, and let $q(y)$ be a global non-forking extension of $\operatorname{tp}(b/A)$. By Theorem 8.5.1, p', p'' and q are definable over A. Now we apply Harrington's Lemma 8.3.4:

$$\begin{aligned}\varphi(x,b) \in p' &\Leftrightarrow \mathrm{d}_{p'}\, x\varphi(x,y) \in q \Leftrightarrow \mathrm{d}_q\, y\varphi(y,x) \in p \\ &\Leftrightarrow \mathrm{d}_{p''}\, x\varphi(x,y) \in q \Leftrightarrow \varphi(x,b) \in p''.\end{aligned}$$

⊣

COROLLARY 8.5.4. *In a stable theory, types over models are stationary.*

PROOF. This is immediate by the above proof since we can replace $\mathrm{acl}^{\mathrm{eq}}(A)$ everywhere by M. It follows also formally from Corollary 8.5.3 since $M^{\mathrm{eq}} = \mathrm{dcl}^{\mathrm{eq}}(M)$ is an elementary substructure of $\mathfrak{C}^{\mathrm{eq}}$ and so algebraically closed. $\dashv$

For the remainder of this section we also assume that T eliminates imaginaries. In view of T^{eq} (see Section 8.4) this is a harmless assumption.

As stable theories are simple we may first collect some of the properties of forking established in Section 7.2.

We keep using

$$a \mathop{\smile\!\!\!\!\!\!|}_{C} B$$

to express that $\mathrm{tp}(a/BC)$ does not fork over C.

THEOREM 8.5.5. *If T is stable, forking independence has the following properties*:

1. (Monotonicity and Transitivity) *Let* $A \subseteq B \subseteq C$ *and* $q \in \mathrm{S}(C)$. *Then* q *does not fork over* A *if and only if* q *does not fork over* B *and* $q \upharpoonright B$ *does not fork over* A.
2. (Symmetry) $a \mathop{\smile\!\!\!\!\!\!|}_{A} b \Longrightarrow b \mathop{\smile\!\!\!\!\!\!|}_{A} a$
3. (Finite Character) *If* $p \in \mathrm{S}(B)$ *forks over* A, *there is a finite subset* $B_0 \subseteq B$ *such that* $p \upharpoonright AB_0$ *forks over* A.
4. (Local Character) *For* $p \in \mathrm{S}(A)$ *there is some* $A_0 \subseteq A$ *of cardinality at most* $|T|$ *such that* p *does not fork over* A_0
5. (Existence) *Every type* $p \in \mathrm{S}(A)$ *has a non-forking extension to any set containing* A.
6. (Algebraic Closure)
   (a) $p \in \mathrm{S}(\mathrm{acl}(A))$ *does not fork over* A.
   (b) *If* $\mathrm{tp}(a/Aa)$ *does not fork over* A, *then* a *is algebraic over* A.

PROOF. This is contained in 7.2.17, 7.2.16, 7.1.9, 7.2.5, 7.2.7 and 7.1.3. $\dashv$

The following properties do not hold in arbitrary simple theories.

THEOREM 8.5.6. *Assume T is stable.*

1. (Conjugacy) *If* $A \subseteq M$ *and* M *is strongly* κ*-homogeneous for some* $\kappa > \max(|T|, |A|)$, *then all non-forking extensions of* $p \in \mathrm{S}(A)$ *to* M *are conjugate over* A.
2. (Boundedness) *Any* $p \in \mathrm{S}(A)$ *has at most* $2^{|T|}$ *non-forking extensions for every* $B \supset A$.

PROOF. For Part 1 let q_1 and q_2 be non-forking extensions of p to M. Any A-automorphism of M which takes $q_1 \upharpoonright \mathrm{acl}(A)$ to $q_2 \upharpoonright \mathrm{acl}(A)$ (see Exercise 8.4.4) takes q_1 to q_2. Since types over algebraically closed sets are stationary by Corollary 8.5.3, the claim now follows.

To prove Part 2 let A_0 be a subset of A of cardinality at most $|T|$ such that p does not fork over A_0. Then p has at most as many non-forking extensions as $p \restriction A_0$ has extensions to $\operatorname{acl}(A_0)$. ⊣

COROLLARY 8.5.7. *Let T be stable and $p \in S(A)$. Then p is stationary if and only if it has a good definition over A.*

PROOF. Assume first that p is stationary and let q be the global non-forking extension. So q is definable and invariant under all automorphisms over A, hence definable over A by Lemma 6.1.10. This shows that p has a good definition over A. For the converse assume that p has a good definition over A. So p has a non-forking global extension p', definable over A by 8.5.1. Since all global non-forking extensions of p are conjugate over A, and p' is fixed by all automorphisms over A, p' is the only global non-forking extension of p. ⊣

Let $p \in S(A)$ be a stationary type. The *canonical base* $\operatorname{Cb}(p)$ of p is the canonical base of the non-forking global extension of p. We call p *based* on B if p is *parallel* to some stationary type q defined over B, i.e., if p and q have the same global non-forking extension (see Exercise 9.1.4). Note that parallel types necessarily have the same free variables.

LEMMA 8.5.8. *A stationary type $p \in S(A)$ is based on B if and only if $\operatorname{Cb}(p) \subseteq \operatorname{dcl}(B)$. So p does not fork over $B \subseteq A$ if and only if $\operatorname{Cb}(p) \subseteq \operatorname{acl}(B)$.*

PROOF. Let r be the global non-forking extension of p and $q = r \restriction B$. Assume that p is based on B. Then q is stationary and r the unique non-forking extension of q. By Corollary 8.5.7, q has a good definition over B, which also defines r. So r is definable over B, which means $\operatorname{Cb}(p) \subseteq \operatorname{dcl}(B)$.

If, conversely, r is definable over B, we know by Theorem 8.5.1 that r does not fork over B and that q is stationary by Corollary 8.5.7.

The last statement follows from the easy fact that p does not fork over B if and only if p is based on $\operatorname{acl}(B)$. ⊣

For $A \subseteq B$ let $N(B/A)$ be the set of all types over B that do not fork over A. By Remark 7.1.8, $N(B/A)$ is closed in $S(B)$. For future reference we record the following useful fact.

THEOREM 8.5.9. (Open mapping) *The restriction map $\pi : N(B/A) \longrightarrow S(A)$ is open.*

PROOF. It is easy to see that we may replace B by $\mathfrak{C}$. If $\pi(q) = \pi(q')$ for some $q, q' \in N(\mathfrak{C}/A)$, then q and q' are conjugate. So if O is a (relative) open subset of $N(\mathfrak{C}/A)$, then

$$O' = \pi^{-1}(\pi(O)) = \bigcup\{\alpha(O) \mid \alpha \in \operatorname{Aut}(\mathfrak{C}/A)\}$$

is again open. So

$$S(A) \setminus \pi(O) = \pi(N(\mathfrak{C}/A) \setminus O')$$

is closed since it is the image of a closed set. ⊣

THEOREM 8.5.10 (Characterisation of Forking). *Let T be a complete theory and $n > 0$. Then T is stable if and only if there is a special class of extensions of n-types, which we denote by $p \sqsubset q$, with the following properties.*

a) (INVARIANCE) *$\sqsubset$ is invariant under* $\mathrm{Aut}(\mathfrak{C})$,
b) (LOCAL CHARACTER) *There is a cardinal κ such that for $q \in \mathrm{S}_n(C)$ there is $C_0 \subseteq C$ of cardinality at most κ such that $q \restriction C_0 \sqsubset q$.*
c) (WEAK BOUNDEDNESS) *For all $p \in \mathrm{S}_n(A)$ there is a cardinal μ such that p has, for any $B \supset A$, at most μ extensions $q \in \mathrm{S}_n(B)$ with $p \sqsubset q$.*

If $\sqsubset$ satisfies in addition

d) (EXISTENCE) *For all $p \in \mathrm{S}_n(A)$ and $A \subseteq B$, there is $q \in \mathrm{S}_n(B)$ such that $p \sqsubset q$,*
e) (TRANSITIVITY) *$p \sqsubset q \sqsubset r$ implies $p \sqsubset r$,*
f) (WEAK MONOTONICITY) *$p \sqsubset r$ and $p \subseteq q \subseteq r$ implies $p \sqsubset q$,*

then $\sqsubset$ coincides with the non-forking relation.

PROOF. In a stable theory, non-forking extensions satisfy properties a), b) and c) (and d), e), f)) by Theorems 8.5.5 and 8.5.6.

Assume conversely that properties a), b) and c) hold. Then a) and c) allow us to find a sufficiently large cardinal μ' so that for all A_0 of cardinality at most κ all n-types over A_0 have at most μ' $\sqsubset$-extensions to any superset.

Let A be a set of parameters. Then the number of n-types over A is bounded by the product of the number of subsets A_0 of A of cardinality at most κ, *times* a bound for the number of types p over A_0, *times* a bound for the number of $\sqsubset$-extensions of $p \in \mathrm{S}_n(A_0)$ to A. So we have

$$|\mathrm{S}_n(A)| \leq |A|^{\kappa} \cdot 2^{\max(\kappa,|T|)} \cdot \mu'$$

and it follows that T is λ-stable if $\lambda^{\kappa} = \lambda$ and $\lambda \geq \max(2^{|T|}, \mu')$, hence stable by Exercise 8.2.7.

Now let $\sqsubset$ have the properties a) to f). Consider a type $p \in \mathrm{S}_n(A)$ with an extension $q \in \mathrm{S}_n(B)$.

Assume first that $p \sqsubset q$. Let μ be the cardinal given BOUNDEDNESS applied to p. By Exercise 7.1.5 there is an extension M of B such that every $r \in \mathrm{S}_n(M)$ which forks over A has more than μ conjugates over A. By EXISTENCE and TRANSITIVITY q has an extension r to M such that $p \sqsubset r$. By INVARIANCE we have $p \sqsubset r'$ for all conjugates r'. So r has no more than μ conjugates, which implies that r does not fork over A and that q is a non-forking extension of p.

Now assume that q is a non-forking extension of p. Choose an extension M of B which is sufficiently saturated in the sense of Theorem 8.5.6(1). Let $r \in \mathrm{S}_n(M)$ be a non-forking extension of q and $r' \in \mathrm{S}_n(M)$ such that $p \sqsubset r'$. By the above r' is a non-forking extension of p. So r and r' are conjugate over A. This implies $p \sqsubset r$ and $p \sqsubset q$ by WEAK MONOTONICITY. ⊣

COROLLARY 8.5.11. *Let T be totally transcendental, $p \in \mathrm{S}(A)$ and q an extension of p to some superset of A. Then q is a non-forking extension if and only if $\mathrm{MR}(p) = \mathrm{MR}(q)$. Hence p is stationary if and only if it has Morley degree 1.*

PROOF. In a totally transcendental theory, extensions having the same Morley rank satisfy the conditions of Theorem 8.5.10 (see Section 6.2). $\dashv$

The same is true for types with Morley rank in stable theories (see Exercise 8.5.5). It follows in particular that in totally transcendental theories for any type $p \in \mathrm{S}(A)$ there is a finite set $A_0 \subseteq A$ such that p does not fork over A_0. Stable theories with this property are called *superstable*: see Section 8.6.

COROLLARY 8.5.12. *In a totally transcendental theory, types over models have Morley degree 1.*

PROOF. This follows from Corollaries 8.5.4 and 8.5.11. $\dashv$

COROLLARY 8.5.13. *If T is strongly minimal, we have $A \mathop{\smile\!\!\!\!\!\!\mid}_B C$ if and only if A and C are algebraically independent over B in the pregeometry sense, i.e., if $\dim(a/B) = \dim(a/BC)$ for all finite tuples $a \in A$.*

PROOF. This follows from Theorem 6.4.2 and Corollary 8.5.11. $\dashv$

COROLLARY 8.5.14. *Let $K \subseteq F_1, F_2$ be differential fields contained in a model of DCF_0. Then $F_1 \mathop{\smile\!\!\!\!\!\!\mid}_K F_2$ if and only if F_1 and F_2 are algebraically independent over K.*

PROOF. By Exercises 3.3.2 and 7.2.7, $F_1 \mathop{\smile\!\!\!\!\!\!\mid}_K F_2$ implies algebraic independence. For the converse we may assume that K is algebraically closed. So let F_1 and F_2 be algebraically independent over K. By the existence of non-forking extensions choose a copy F' of F_1 satisfying the same type over K and forking independent of F_2 over K. Then F' is algebraically independent of F_2 over K. Since K is algebraically closed, F' and F_1 satisfy the same type over F_2 in the sense of field theory. Since F', F_1 and F_2 are d-closed and F' and F_1 are isomorphic as d-fields, we conclude that F' and F_1 have the same type over F_2. Thus, $F_1 \mathop{\smile\!\!\!\!\!\!\mid}_K F_2$. $\dashv$

EXERCISE 8.5.1. Show that in ACF_p the type of a finite tuple a over a field K is stationary if and only if $K(a)$ and K^{sep} are linearly disjoint over K.

We continue assuming that T eliminates imaginaries.

EXERCISE 8.5.2 (Finite Equivalence Relation Theorem). Let $A \subseteq B$ and let $\mathrm{tp}(a/B) \neq \mathrm{tp}(b/B)$ be types which do not fork over A. Then there is an A-definable finite equivalence relation E with $q_1(x) \cup q_2(y) \vdash \neg E(x, y)$.

EXERCISE 8.5.3. If a and b are independent realisations of the same type over $\mathrm{acl}(A)$, then $\mathrm{tp}(a/Ab)$ is stationary. Hence the canonical base of $\mathrm{tp}(b/\mathrm{acl}(A))$ is contained in $\mathrm{dcl}(bA)$.

EXERCISE 8.5.4. Prove the following.

1. Let $p \in S(A)$ be stable and q an extension of p. Then q does not fork over A if and only if q has a good definition over $\operatorname{acl}(A)$.
2. Stable types over algebraically closed sets are stationary.

EXERCISE 8.5.5.

1. Let T be stable, $p \in S(A)$ a type with Morley rank and q an extension of p to some superset of A. Then q is a non-forking extension if and only if $\mathrm{MR}(p) = \mathrm{MR}(q)$. It follows that a type with Morley rank is stationary if and only if it has Morley degree 1.
2. Show that the same is true for an arbitrary theory T.

EXERCISE 8.5.6. Assume that T is stable. For any $p \in S(A)$ there is some $A_0 \subseteq A$ of cardinality at most $|T|$ such that p is the unique non-forking extension of $p \upharpoonright A_0$ to A. If p has Morley rank, A_0 can be chosen as a finite set.

EXERCISE 8.5.7 (Forking multiplicity). (T stable) Define the *multiplicity* of a type p as the number $\operatorname{mult}(p)$ of its global non-forking extensions. Show:

1. If p is algebraic, $\operatorname{mult}(p)$ equals $\deg(p)$, the number of realisations of p. (See page 79 and Remark 5.6.3.)
2. If T is countable, then $\operatorname{mult}(p)$ is either finite or $2^{\aleph_0}$.
3. If p has Morley rank, show that $\operatorname{mult}(p) = \mathrm{MD}(p)$.

EXERCISE 8.5.8. Let G be a totally transcendental group. Show:

1. If a and b are independent elements, then $\mathrm{MR}(a \cdot b) \geq \mathrm{MR}(a)$. Equality holds if and only if $a \cdot b$ and b are independent.
2. Assume that G is ω-saturated. Then $a \in G$ is *generic*, i.e., $\mathrm{MR}(a) = \mathrm{MR}(G)$, if and only if for all $b \in G$

$$a \perp b \Rightarrow a \cdot b \perp b.$$

It can be shown that all ω-saturated stable groups contain *generic* elements, i.e., elements satisfying property 2, see Poizat [46], or Wagner [61].

EXERCISE 8.5.9 (Group configuration). Let G be a totally transcendental group, and let $a_1, a_2, a_3 \in G$ be independent generic elements, i.e., elements of maximal rank in $\mathrm{Th}(G)$. Put $b_1 = a_1 \cdot a_2$, $b_2 = a_1 \cdot a_2 \cdot a_3$ and $b_3 = a_2 \cdot a_3$. We consider these six elements as the points of a geometry with "lines" $A_0 = \{a_1, b_1, a_2\}$, $A_1 = \{a_2, b_3, a_3\}$, $A_2 = \{a_1, b_2, b_3\}$ and $A_3 = \{b_1, b_2, a_3\}$. It is easy to see that every permutation of the four lines gives rise to an automorphism of this geometry.

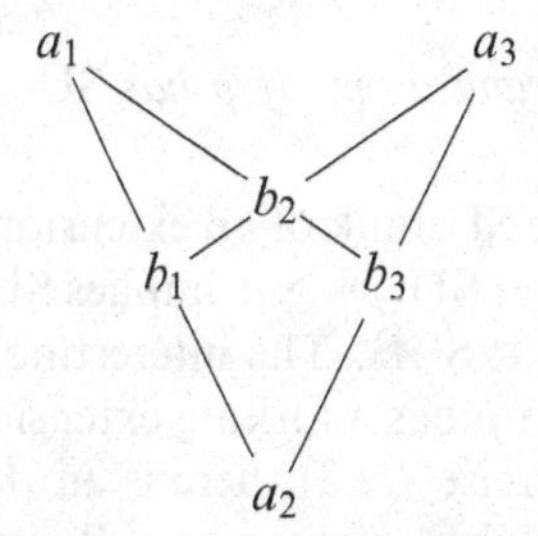

Show:

1. Each point on a line is algebraic over the other two points on the line.
2. Any three non-collinear points are independent.

Any family of points $a_1, a_2, a_3, b_1, b_2, b_3$ with these properties[1] is called a *group configuration*. Hrushovski proved that whenever a totally transcendental structure contains a group configuration, there is a group definable in this structure whose Morley rank equals the Morley rank of any of the points. For more details see Bouscaren [11], Wagner [61] or Pillay [44].

EXERCISE 8.5.10. Let T be an arbitrary complete theory, not necessarily stable. For any set of parameters A the map $\mathrm{S}(\mathrm{acl}(A)) \longrightarrow \mathrm{S}(A)$ is open. (For stable theories, this is just the Open Mapping Theorem.)

8.6. SU-rank and the stability spectrum

We saw that in totally transcendental theories forking is governed by the Morley rank. The SU-rank, which we define here, generalises this to *superstable* theories. We use it to show that the *stability spectrum* of countable theories is rather restrictive: there are only four possibilities for the class of cardinals in which a countable theory is stable.

DEFINITION 8.6.1. Let T be a simple theory. We define $\mathrm{SU}(p) \geq \alpha$ for a type p by recursion on α:

$\mathrm{SU}(p) \geq 0$ for all types p;

$\mathrm{SU}(p) \geq \beta+1$ if p has a forking extension q with $\mathrm{SU}(q) \geq \beta$;

$\mathrm{SU}(p) \geq \lambda$ (for a limit ordinal λ) if $\mathrm{SU}(p) \geq \beta$ for all $\beta < \lambda$.

and the SU-rank $\mathrm{SU}(p)$ of p as the maximal α such that $\mathrm{SU}(p) \geq \alpha$. If there is no maximum, we set $\mathrm{SU}(p) = \infty$.

LEMMA 8.6.2. *Assume T to be simple. Let p have ordinal valued* SU*-rank and let q be an extension of p. Then q is a non-forking extension of p if and*

[1] It is easy to see that Part 2 can equivalently be replaced by: 2a *Any two points on a line are independent* and 2b *Any two lines are independent over their intersection.*

only if q has the same SU*-rank as p. If p has* SU*-rank ∞, then so does any non-forking extension.* ⊣

PROOF. It is clear that the SU-rank of an extension cannot increase. So it is enough to show for all α that $\mathrm{SU}(p) \geq \alpha$ implies $\mathrm{SU}(q) \geq \alpha$ whenever q is a non-forking extension of $p \in \mathrm{S}(A)$. The interesting case is where $\alpha = \beta + 1$ is a successor ordinal. Then p has a forking extension r with $\mathrm{SU}(r) \geq \beta$. By the Diamond Lemma (Exercise 7.2.2) there is an A-conjugate r' of r with a non-forking extension s which also extends q. By induction $\mathrm{SU}(s) \geq \beta$. But s is a forking extension of q, so $\mathrm{SU}(q) \geq \beta$. ⊣

Since every type does not fork over some set of cardinality at most $|T|$, there are at most $2^{|T|}$ different SU-ranks. Since they form an initial segment of the ordinals, all ordinal ranks are smaller than $(2^{|T|})^+$. (Actually one can prove that they are smaller than $|T|^+$.) It follows that every type of SU-rank ∞ has a forking extension of SU-rank ∞.

DEFINITION 8.6.3. A simple theory is *supersimple* if every type does not fork over some finite subset of its domain. A stable, supersimple theory is called *superstable*.

Note that totally transcendental theories are superstable.

LEMMA 8.6.4. *The theory T is supersimple if and only if every type has* SU*-rank $< \infty$.*

PROOF. If $\mathrm{SU}(p) = \infty$, there is an infinite sequence $p = p_0 \subseteq p_1 \subseteq \cdots$ of forking extensions of SU-rank ∞. The union of the p_i forks over every finite subset of its domain. If $p \in \mathrm{S}(A)$ forks over every finite subset of A, there is an infinite sequence $A_0 \subseteq A_1 \subseteq \cdots$ of finite subsets of A such that $p \restriction A_{i+1}$ forks over A_i. This shows that $p \restriction \emptyset$ has SU-rank ∞. ⊣

Let T be a complete theory. The *stability spectrum* $\mathrm{Spec}(T)$ of T is the class of all infinite cardinals in which T is stable.

THEOREM 8.6.5. *Let T be a countable complete theory. There are four cases:*

1. *T is totally transcendental. Then* $\mathrm{Spec}(T) = \{\kappa \mid \kappa \geq \aleph_0\}$.
2. *T is* superstable *but not totally transcendental. Then* $\mathrm{Spec}(T) = \{\kappa \mid \kappa \geq 2^{\aleph_0}\}$.
3. *T is stable but not superstable. Then* $\mathrm{Spec}(T) = \{\kappa \mid \kappa^{\aleph_0} = \kappa\}$.
4. *T is unstable. Then* $\mathrm{Spec}(T)$ *is empty.*

Note that these are really four different possibilities: it follows from Theorem A.3.6 that $\kappa^{\aleph_0} > \kappa$ for all κ with countable cofinality, e.g., for all $\kappa = \beth_\omega(\mu)$.

PROOF. 1: This follows from Theorem 5.2.6.

2: Let T be superstable and $|A| = \kappa$. Since every type over A does not fork over a finite subset of A, an upper bound for the size of $\mathrm{S}(A)$ can be computed as the product of

- the number of finite subsets E of A,
- the number of types $p \in \mathrm{S}(E)$,
- the number of non-forking extensions of p to A.

So we have $|\,\mathrm{S}(A)| \leq \kappa \cdot 2^{\aleph_0} \cdot 2^{\aleph_0} = \max(2^{\aleph_0}, \kappa)$. If T is not totally transcendental, the proof of Theorem 5.2.6 shows that T cannot be stable in cardinals smaller than $2^{\aleph_0}$.

3: If T is stable, then T is κ-stable whenever $\kappa^{\aleph_0} = \kappa$ by Exercise 8.2.7. If T is not superstable, the proof of Lemma 8.6.4 shows that there is a type p over the empty set of infinite SU-rank with a forking extension p' of infinite SU-rank. Let q be a non-forking global extension of p' and let $\kappa \geq \aleph_0$. By Exercise 7.1.5 q has κ many different conjugates q_α, $(\alpha < \kappa)$. Choose A_0 of size κ such that all $p_\alpha = q_\alpha \restriction A_0$ are different. By Lemma 8.6.2 the p_α have again infinite SU-rank. Continuing in this manner we get a sequence $A_0 \subseteq A_1 \subseteq \cdots$ of parameter sets and a tree of types $p_{\alpha_0,\dots,\alpha_n} \in \mathrm{S}(A_{n+1})$, $(n < \omega, \alpha_i < \kappa)$. We may assume that all A_i have cardinality κ. Each path through this tree defines a type over $A = \bigcup_{n<\omega} A_n$. So we have $|\,\mathrm{S}(A)| \geq \kappa^{\aleph_0}$.

4: Clear from the definition. ⊣

The spectrum of uncountable theories is more difficult to describe, see [54, Chapter III].

EXAMPLE 8.6.6 (Modules). For any R-module M, the $\mathrm{L}_{\mathrm{Mod}}(R)$-theory of M is κ-stable if $\kappa = \kappa^{|R|+\aleph_0}$.

See [47] or [63] for more on the model theory of modules.

PROOF. Let B be a subset of some model N of $\mathrm{Th}(M)$, $|B| \leq \kappa$. Let $\mathrm{S}_N(B)$ denote the set of all complete types over B which are realized in N. Every type $\mathrm{tp}(a/B)$ is axiomatised by

$$\mathrm{tp}^{\pm}(a/B) = \mathrm{tp}^{+}(a/B) \cup \mathrm{tp}^{-}(a/B),$$

where

$$\mathrm{tp}^{+}(a/B) = \{\varphi(x, b) \mid \varphi \text{ pp-formula}, \ b \in B, \ N \models \varphi(a, b)\}$$

and

$$\mathrm{tp}^{-}(a/B) = \{\neg\varphi(x, b) \mid \varphi \text{ pp-formula}, \ b \in B, \ N \models \neg\varphi(a, b)\}.$$

Clearly, tp^{-} is determined by tp^{+}.

By Corollary 3.3.8, $\mathrm{tp}^{+}(a/B)$ contains – up to equivalence – at most one formula $\varphi(x, b)$ for any pp-formula $\varphi(x, y)$. Hence $\mathrm{tp}^{+}(a/B)$ is determined by a partial map f from the set of pp-formulas to the set of finite tuples in B in the sense that it is axiomatised by $\{\varphi(x, f(\varphi)) \mid \varphi \text{ pp-formula}\}$. Hence we have

$$|\,\mathrm{S}_N(B)| \leq (|B| + \aleph_0)^{|R|+\aleph_0}.$$

Thus T is κ-stable in every κ with $\kappa = \kappa^{|R|+\aleph_0}$. ⊣

EXAMPLE 8.6.7 (Separably closed fields). The theory $\mathsf{SCF}_{p,e}$ of separably closed fields is stable for all κ with $\kappa = \kappa^{\aleph_0}$.

PROOF. Let L be a model of $\mathsf{SCF}_{p,e}$. Fix a p-basis $b_1, \ldots, b_e$ and consider the corresponding λ-functions λ_ν. Now let K be a subfield of cardinality κ. By Theorem 3.3.20 every type $tp(a/K)$ is axiomatised by Boolean combinations of equations $t(x) \doteq 0$ where the $t(x)$ are $L(c_1, \ldots, c_e, \lambda_\nu)_{\nu \in p^e}$-terms with parameters from K. To compute an upper bound for the number of types over K we may assume that K contains the p-basis and is closed under the λ-functions. It is now easy to see that every $t(x)$ is equivalent to a term $p(\lambda_{\bar\nu_1}(x), \ldots, \lambda_{\bar\nu_k}(x))$ where $p(X_1, \ldots, X_k) \in K[X_1, \ldots, X_k]$ and the $\lambda_{\bar\nu_i}$ are iterated λ-functions:

$$\lambda_{\bar\nu} = \lambda_{\nu^1 \ldots \nu^m} = \lambda_{\nu^1} \circ \cdots \circ \lambda_{\nu^m}.$$

So, if $\lambda_{\bar\nu_0}, \lambda_{\bar\nu_1}, \ldots$ is a list of all iterated λ-functions, the type of a over K is determined by the sequence of the quantifier-free L-types of the tuples $(\lambda_{\bar\nu_0}(a), \ldots, \lambda_{\bar\nu_n}(a))$ over K. By Example 5.2.3 for each n there are only κ many such types. So we we can bound the number of types over K by $\kappa^{\aleph_0}$. ⊣

EXERCISE 8.6.1. Show that the types of SU-rank 0 are exactly the algebraic types. Show also that a type is minimal if and only if it is stationary and has SU-rank 1.

EXERCISE 8.6.2. Let T be an arbitrary theory. Define the U-rank (or *Lascar rank*) $\mathrm{U}(p)$ of a type $p \in \mathrm{S}(A)$ as its SU-rank, except for the clause

$\mathrm{U}(p) \geq \beta + 1$ if for any κ there is a set $B \supset A$ to which p has at least κ many extensions q with $\mathrm{U}(q) \geq \beta$.

Show that in stable theories U-rank and SU-rank coincide.

EXERCISE 8.6.3. Use Exercise 7.2.5 to show that a simple theory is supersimple if every 1-type has SU-rank $< \infty$.

EXERCISE 8.6.4. Show that in simple theories $\mathrm{SU}(p) \leq \mathrm{MR}(p)$.

EXERCISE 8.6.5 (Lascar inequality). Let T be simple, $\mathrm{SU}(a/C) = \alpha$ and $\mathrm{SU}(b/aC) = \beta$. Prove[2]

$$\beta + \alpha \leq \mathrm{SU}(ab/C) \leq \beta \oplus \alpha.$$

If a and b are independent over C, we have $\mathrm{SU}(ab/C) = \beta \oplus \alpha$.

EXERCISE 8.6.6. Let $\varphi(x, y)$ be a formula without parameters and k natural number. Define the rank $\mathrm{D}(p, \varphi, k)$ by

$\mathrm{D}(p, \varphi, k) \geq \beta + 1$ if p has an extension q with $\mathrm{D}(q, \varphi, k) \geq \beta$ and which contains a formula $\varphi(x, b)$ which divides over the domain of p with respect to k.

---

[2]Ordinal addition was defined in Exercise 6.4.4. The *strong* sum $\oplus$ is the smallest function $\mathrm{On} \times \mathrm{On} \to \mathrm{On}$ which is strictly monotonous in both arguments.

Show for simple T:

1. That $D(p, \varphi, k)$ is bounded by a natural number which depends only on φ and k.
2. Let q be an extension of p. Then p is a non-forking extension of p if and only if $D(p, \varphi, k) = D(q, \varphi, k)$ for all φ and k.

EXERCISE 8.6.7. A countable theory is ω-stable if and only if it is superstable, small and if every type (over a finite set) has finite multiplicity.

EXERCISE 8.6.8 (Lachlan). Show that in an $\aleph_0$-categorical theory there are only finitely many strong 1-types over a finite set. Conclude that an $\aleph_0$-categorical superstable theory is ω-stable.

Note that there are stable $\aleph_0$-categorical theories which are not ω-stable (see [26]).

EXERCISE 8.6.9. Let T be a simple theory. Assume that there is a sequence of definable equivalence relations $E_0 \subseteq E_1 \subseteq \cdots$ such that every E_i-class contains infinitely many E_{i+1}-classes. Show that T is not supersimple.

EXERCISE 8.6.10. Use Exercise 8.6.9 to show that a superstable group has no infinite descending sequence of definable subgroups $G = G_0 \geq G_1 \geq G_2 \geq G_3 \geq \cdots$ each of infinite index in the previous one. Conclude from this that $\mathsf{SCF}_{p,e}$ is not superstable if $e > 0$.

EXERCISE 8.6.11. Prove that a module M is totally transcendental if and only if it has the dcc on pp-definable subgroups of M. A module M is superstable if and only if there is no infinite descending sequence of pp-definable subgroups each of which is of infinite index in its predecessor.

Chapter 9

PRIME EXTENSIONS

In this chapter we return to questions around the uniqueness of prime extensions. We will now prove their uniqueness for totally transcendental theories and for countable stable theories having prime extensions.

9.1. Indiscernibles in stable theories

We assume throughout this section that T is complete, stable and eliminates imaginaries. Indiscernibles in stable theories are in fact indiscernible for every ordering of the underlying set. More importantly, we show that they form a Morley sequence in some appropriate stationary type.

A family $\mathcal{I} = (a_i)_{i \in I}$ of tuples is *totally indiscernible* over A, if

$$\mathfrak{C} \models \varphi(a_{i_1}, \dots, a_{i_k}) \leftrightarrow \varphi(a_{j_1}, \dots, a_{j_k})$$

for all $L(A)$-formulas φ and sequences $i_1, \dots, i_k, j_1, \dots, j_k$ of pairwise distinct indices.

LEMMA 9.1.1. *If T is stable, indiscernibles are totally indiscernible.*

PROOF. Assume that $\mathcal{I} = (a_i)_{i \in I}$ is indiscernible over A, but not totally indiscernible over A. By Lemma 7.1.1 we may assume $(I, <) = (\mathbb{Q}, <)$. Because any permutation on $\{1, \dots, n\}$ is a product of transpositions of neighbouring elements, there are some $L(A)$-formula $\varphi(x_1, \dots, x_n)$, rational numbers $r_1 < \dots < r_n$ and some $j \in \mathbb{Q}$ such that

$$\models \varphi(a_{r_1}, \dots, a_{r_j}, a_{r_{j+1}}, \dots, a_{r_n})$$

and

$$\models \neg\varphi(a_{r_1}, \dots, a_{r_{j+1}}, a_{r_j}, \dots, a_{r_n}).$$

The formula $\psi(x, y) = \varphi(a_{r_1}, \dots, x, y, \dots, a_{r_n})$ orders the elements (a_r), $(r_j < r < r_{j+1})$. By Exercise 8.2.1, this contradicts stability. ⊣

Let $p \in \mathrm{S}(A)$ be a stationary type. Recall that a Morley sequence in p is an A-indiscernible sequence of realisations of p independent over A. Morley sequences $(a_\alpha)_{\alpha<\lambda}$ are easy to construct as follows: choose a_α realising the unique non-forking extension of p to $A \cup \{a_\beta \mid \beta < \alpha\}$ (see Example 7.2.10). Since any well-ordered Morley sequence arises in this way, Morley sequences in p are uniquely determined by their order-type up to isomorphism over A.

THEOREM 9.1.2. *If T is stable, then any infinite sequence of indiscernibles over A is a Morley sequence for some stationary type defined over some extension of A.*

PROOF. Let $\mathcal{I} = (a_i)_{i \in I}$ be indiscernible over A. Notice that for all formulas $\varphi(x, \overline{b})$ the set

$$J_\varphi = \{i \in I \mid \models \varphi(a_i, \overline{b})\}$$

is finite or cofinite in I: otherwise for all $J \subseteq I$ the set of formulas

$$\{\varphi(a_i, \overline{y}) \mid i \in J\} \cup \{\neg\varphi(a_i, \overline{y}) \mid i \notin J\}$$

would be consistent. So there would be $2^{|I|}$ many φ-types over $\mathcal{I}$, contradicting stability of T.

This shows that for every φ either J_φ or $I \setminus J_\varphi$ is bounded by some k_φ (which depends only on φ).

The *average type* of $\mathcal{I}$ is a global type defined by

$$\mathrm{Av}(\mathcal{I}) = \{\varphi(x, \overline{b}) \mid \overline{b} \in \mathfrak{C}, \models \varphi(a_i, \overline{b}) \text{ for all but finitely many } i \in I\}.$$

By the preceding remarks, this is a complete type. Let I_0 be an infinite subset of I. Since $\varphi(x, \overline{b}) \in \mathrm{Av}(\mathcal{I})$ if and only if $\{i \in I_0 \mid \models \varphi(a_i, \overline{b})\}$ contains more than k_φ many (and hence infinitely many) elements, $\mathrm{Av}(\mathcal{I})$ is definable over $\mathcal{I}_0$. Hence $\mathrm{Av}(\mathcal{I})$ does not fork over $\mathcal{I}_0$ and its restriction to $\mathcal{I}_0$ is stationary (see Theorem 8.5.1 and Corollary 8.5.7.)

It is easy to see that all a_i, $i \in I \setminus I_0$, realise the type

$$p = \mathrm{Av}(\mathcal{I}) \upharpoonright A\mathcal{I}_0.$$

As this is also true for all $I_0' \supset I_0$, we see that a_i, $i \in I \setminus I_0$, forms a Morley sequence for p.

At the beginning of the proof we can now replace $\mathcal{I}$ by an infinite set of indiscernibles $\mathcal{I}'$ containing $\mathcal{I}$ as a coinfinite subset which shows $\mathcal{I}$ to be a Morley sequence for $p' = \mathrm{Av}(\mathcal{I}') \upharpoonright A(\mathcal{I}' \setminus \mathcal{I})$. ⊣

EXERCISE 9.1.1. If p is stationary and q a non-forking extension of p, then any Morley sequence of q is also a Morley sequence for p.

EXERCISE 9.1.2. Let $p \in \mathrm{S}(A)$ be stationary and $\mathcal{I}$ a Morley sequence of p.

a) Let $B \supset A$ and $\mathcal{I}_0 \subseteq \mathcal{I}$ such that $B \mathop{\smile\!\!\!\!\!\!\mid}_{A\mathcal{I}_0} \mathcal{I}$. Then $\mathcal{I} \setminus \mathcal{I}_0$ is a Morley sequence of the non-forking extension of p to B.
b) The type $\mathrm{Av}(\mathcal{I})$ is the non-forking global extension of p.

EXERCISE 9.1.3. We call indiscernibles $\mathcal{I}_0$ and $\mathcal{I}_1$ *parallel* if there is some infinite set $\mathcal{J}$ such that $\mathcal{I}_0\mathcal{J}$ and $\mathcal{I}_1\mathcal{J}$ are indiscernible. Show that $\mathcal{I}_0$ and $\mathcal{I}_1$ are parallel if and only if they have the same average type.

EXERCISE 9.1.4. Show that two types are parallel if and only if two (or all) of their infinite Morley sequences are parallel.

EXERCISE 9.1.5. Show the converse of Lemma 9.1.1: if all indiscernibles sequences are totally indiscernible, then T is stable.

9.2. Totally transcendental theories

Let T be a totally transcendental theory. We will prove the following theorem:

THEOREM 9.2.1 (Shelah [53]). *Let T be totally transcendental.*

1. *A model M is a prime extension of A if and only if M is atomic over A and does not contain an uncountable set of indiscernibles over A.*
2. *Prime extensions are unique.*

We first aim to show that a constructible set does not contain an uncountable set of indiscernibles.

LEMMA 9.2.2. *Let $\mathcal{I}$ be indiscernible over A and B a countable set. Then $\mathcal{I}$ contains a countable subset $\mathcal{I}_0$ such that $\mathcal{I} \setminus \mathcal{I}_0$ is indiscernible over $AB\mathcal{I}_0$.*

PROOF. By Theorem 9.1.2, $\mathcal{I}$ is a Morley sequence of some stationary types over some extension A' of A. Since T is totally transcendental, we only need to extend A by finitely many elements (this follows using Exercise 8.5.6 and Exercise 9.1.1). So we may assume $A' = A$. For every finite tuple $\overline{b}$ in B there is some finite set $\mathcal{I}_0$ such that $\overline{b} \mathop{\smile\!\!\!\!|}_{A\mathcal{I}_0} \mathcal{I}$. In this way we find a countable set $\mathcal{I}_0$ with

$$B\mathcal{I}_0 \mathop{\smile\!\!\!\!|}_{A\mathcal{I}_0} \mathcal{I}.$$

It now follows from Exercise 9.1.2 that $\mathcal{I} \setminus \mathcal{I}_0$ is a Morley sequence over $AB\mathcal{I}_0$. ⊣

We need a bit of set theory: a *club* $D \subseteq \omega_1$ is a closed and unbounded subset where closed means that $\sup(\alpha \cap D) \in D$ for all $\alpha \in \omega_1$.

THEOREM 9.2.3 (Fodor). *Let $D \subseteq \omega_1$ be a club and $f : D \to \omega_1$ a regressive function, i.e., $f(\alpha) < \alpha$ for all $\alpha \in D$. Then f is constant on an unbounded subset of ω_1.*

PROOF. Suppose not; then for each $\alpha \in \omega_1$, the fibre $D_\alpha = \{x \in D \mid f(x) = \alpha\}$ is bounded. Construct a sequence $\alpha_0 < \alpha_1 < \cdots$ of elements of D as follows. Let α_0 be arbitrary. If α_n is constructed, choose for α_{n+1} an upper bound of $\bigcup_{\beta<\alpha_n} D_\beta$. So we have for all $\gamma \in D$

$$\gamma \geq \alpha_{n+1} \Rightarrow f(\gamma) \geq \alpha_n.$$

For $\delta = \sup_{n<\omega} \alpha_n$, this implies $f(\delta) \geq \delta$. A contradiction. ⊣

Recall from Section 5.3 that a set $B = \{b_\alpha\}_{\alpha<\mu}$ is a *construction* over A if all $\text{tp}(b_\alpha/AB_\alpha)$ are isolated where as above $B_\alpha = \{b_\beta \mid \beta < \alpha\}$). A subset $C \subseteq B$ is called *construction closed* if for all $b_\alpha \in C$ the type $\text{tp}(b_\alpha/AB_\alpha)$ is isolated by some formula over $A \cup (B_\alpha \cap C)$.

The following lemma holds for arbitrary T.

LEMMA 9.2.4. *Let $B = \{b_\alpha\}_{\alpha<\mu}$ be a construction over A.*

1. *Any union of construction closed sets is construction closed.*
2. *Any $b \in B$ is contained in a finite construction closed subset of B.*
3. *If $C \subseteq B$ is construction closed, then B is constructible over AC.*

PROOF. The first part is clear. For the second part, let $b = b_\alpha \in B$. Since the type of $b = b_0$ is isolated over A, we can do induction on α. As $\{b_\beta\}_{\beta<\mu}$ is a construction over A, $\text{tp}(b_\alpha/AB_\alpha)$ is isolated by some formula $\varphi(x, \overline{c})$ where $\overline{c} = b_{\beta_1} \ldots b_{\beta_n}$ with $\beta_i < \alpha$. By induction, each b_{β_i} is contained in a finite construction closed set C_i. Thus $C_1 \cup \cdots \cup C_n \cup \{b_\alpha\}$ is finite and construction closed.

For Part 3 we assume $A = \emptyset$ to simplify notation. We will show that the type $\text{tp}(b_\alpha/CB_\alpha)$ is isolated for all α. This is clear if $b_\alpha \in C$. So assume $b_\alpha \notin C$. From the assumption it is easy to see that C is isolated over $B_{\alpha+1}$ where the isolating formulas only contain parameters from $B_{\alpha+1} \cap C \subseteq B_\alpha$.

We thus have

$$\text{tp}(C/B_\alpha) \vdash \text{tp}(C/B_\alpha b_\alpha).$$

If $\varphi(x)$ isolates the type $\text{tp}(b_\alpha/B_\alpha)$, then $\varphi(x)$ also isolates the type $\text{tp}(b_\alpha/CB_\alpha)$. ⊣

LEMMA 9.2.5. *If B is constructible over A, then B does not contain an uncountable set of indiscernibles over A.*

PROOF. We assume $A = \emptyset$. Let $\mathcal{I} = \{c_\alpha\}_{\alpha<\omega_1}$ be indiscernible. By Lemma 9.2.4 we can build a continuous sequence C_α of countable construction closed subsets of B such that $c_\alpha \in C_{\alpha+1}$. By Lemma 9.2.2 there is a club D consisting of limit ordinals such that for all $\delta \in D$ the set $\{c_\alpha \mid \alpha \geq \delta\}$ is indiscernible over C_δ. Each c_δ is isolated over C_δ by a formula with parameters from $C_{\delta'}$ with $\delta' < \delta$. By Fodor's Theorem there is some δ_0 such that for some cofinal set of δs from D the parameters can be chosen in C_{δ_0}. Assume that $\delta_1 < \delta_2$ are such elements. Then c_{δ_1} and c_{δ_2} have the same type over C_{δ_0}. But this is impossible since

$$\text{tp}(c_{\delta_2}/C_{\delta_0}) \vdash \text{tp}(c_{\delta_2}/C_{\delta_0} c_{\delta_1}).$$ ⊣

Let M be a model and $A \subseteq M$. We call a subset B of M *normal* in M over A if for every element $b \in B$ all realisations of $\text{tp}(b/A)$ in M are contained in B.

LEMMA 9.2.6. *Let T be a (not necessarily totally transcendental) theory which has prime extensions. If M is atomic over A and B is a normal subset of M containing A, then M is atomic over B.*

PROOF. Let c be a tuple from M, so $\mathrm{tp}(c/A)$ is isolated. Since the isolated types are dense over B by Theorem 4.5.7, there is some $d \in M$ atomic over B and realizing the type $\mathrm{tp}(c/A)$. Let $\mathrm{tp}(d/B)$ be isolated by $\psi(x, d_0)$ for some tuple $d_0 \in B$. Since c and d satisfy the same type over A there is some $c_0 \in M$ such that $\mathrm{tp}(cc_0/A) = \mathrm{tp}(dd_0/A)$. Then c satisfies $\psi(x, c_0)$ and as B is normal, we have $c_0 \in B$. It follows easily that $\psi(x, c_0)$ is complete over B as well. ⊣

PROOF OF THEOREM 9.2.1(1). If M is a prime extension of A, then M is atomic over A by Corollary 5.3.7 and since M can be embedded over A into some constructible prime extension, by Lemma 9.2.5 M does not contain an uncountable set of indiscernibles over A.

For the converse assume again $A = \emptyset$, i.e., suppose M is atomic over $\emptyset$ without uncountable set of indiscernibles. In order to prove that M can be embedded into any model N we enumerate all types over $\emptyset$ realised in M as $(p_\mu)_{\mu<\nu}$ and recursively extend the empty map to the normal sets $C_\mu = \bigcup_{\alpha<\mu} p_\alpha(M)$. That this is possible follows from the following.

CLAIM. *Let M be atomic over B, $p \in \mathrm{S}(B)$ and $B \subseteq C \subseteq M$ normal over B. Then any elementary map $C \longrightarrow N$ can be extended to $C \cup p(M)$.*

We proof the claim by induction on the Morley rank of p and note that the claim is clear if p is algebraic.

Assume inductively that the claim is proved for all types of Morley rank less than α (over arbitrary sets B). Then any given elementary map $f : C \longrightarrow N$ with $B \subseteq C \subseteq M$ and C normal over B can be extended to $C \cup \{a \in M \mid \mathrm{MR}(a/B) < \alpha\}$.

Let now $p \in \mathrm{S}(B)$ with $\mathrm{MR}(p) = \alpha$. Let $\{c_i\}_{i<\mu}$ be a maximal set of realisations of p in M independent over B. By Exercise 9.2.1, $\{c_i\}_{i<\mu}$ splits into a finite number of indiscernible sequences, which implies that μ is countable, and we can assume that $\mu \le \omega$. Let $B_i = B \cup \{c_0, \ldots, c_{i-1}\}$ and $C_i = C \cup \{a \in M \mid \mathrm{MR}(a/B_i) < \alpha\}$, so $B = B_0$. By maximality we have $p(M) \subseteq \bigcup_{i<\mu} C_i$. As M is atomic over B, it is also atomic over B_i and since C_i is normal over B_i even atomic over C_i. If f has been extended to C_i, we may extend f to $C_i B_{i+1}$ by the atomicity. By the induction hypothesis applied to B_{i+1} and $C_i B_{i+1}$ there is an extension to C_{i+1}.

The proof of 9.2.1(1) can easily be symmetrised to yield 9.2.1(2). ⊣

EXAMPLE (Shelah [55]). Let L contain a binary relation symbol E_α for every ordinal $\alpha < \omega_1$ and let T be the theory stating that each E_α is an equivalence relation such that E_0 consists of only one class and for any $\alpha < \beta < \omega_1$ each E_α-equivalence class is the union of infinitely many E_β-equivalence classes.

Then T is complete, stable (but not totally transcendental) and admits quantifier elimination. Every consistent $L(A)$-formula $\varphi(x)$ can be completed over A. Therefore there exists a prime model M which is constructible over the empty set.

In M any chain $(K_\alpha)_{\alpha<\omega_1}$ of E_α-equivalence classes has non-empty intersection: otherwise there would be a countable subset A of M and some limit ordinal $\eta < \omega_1$ such that:

(i) A is construction closed (with respect to a fixed construction);
(ii) $A \cap \bigcap_{\alpha<\eta} K_\alpha = \emptyset$;
(iii) $A \cap K_\alpha \neq \emptyset$ for all $\alpha < \eta$.

By (i) M/A is atomic. Let $c \in K_\eta$; then $\mathrm{tp}(c/A)$ is isolated by some formula $\varphi(x;\overline{a})$; by (ii) there exists some $\alpha < \eta$ such that $\overline{a} \cap K_\alpha = \emptyset$. By (iii) there is some $d \in A \cap K_\alpha$; since $\overline{a} \cap K_\alpha = \emptyset$, it would imply $\models \varphi(d;\overline{a})$, but this is impossible as φ isolates $\mathrm{tp}(c/A)$. Therefore we have $\bigcap(K_\alpha)_{\alpha<\omega_1} \neq \emptyset$.

Now let $a \in M$ and let N be the set of all b from M for which there is some ordinal $\alpha < \omega_1$ with $\models \neg aE_\alpha b$. Then N is also a prime model, but M and N are not isomorphic.

EXERCISE 9.2.1. Let T be totally transcendental, and $\mathcal{I}$ be an independent set of realisations of $p \in \mathrm{S}(A)$. Then $\mathcal{I}$ can be decomposed into a finite number $\mathcal{I}_1, \dots ,\mathcal{I}_n$ of indiscernible sets over A.

9.3. Countable stable theories

We assume throughout this section that T is countable and stable. For such T we will show that prime extensions, if they exist, are unique. The main point is Shelah's result that in this situation subsets of constructible sets are again constructible.

The proof of Theorem 9.2.1(1) showed that atomic extensions of A without uncountable sets of indiscernibles are constructible and hence prime. The uniqueness of such prime extensions then also follows directly from the following theorem which holds for arbitrary theories.

THEOREM 9.3.1 (Ressayre). *Constructible prime extensions are unique.*

PROOF. It suffices to prove the theorem for constructible prime extensions M and M' over the empty set. Let $f_0\colon E_0 \to E_0'$ be a maximal elementary map between construction closed subsets $E_0 \subseteq M$ and $E_0' \subseteq M'$. If $E_0 \neq M$, there is some proper construction closed finite extension E_1 of E_0. Since E_1 is atomic over E_0 by Lemma 9.2.4(3) and Corollary 5.3.6 there is an extension $f_1\colon E_1 \to E_1'$ of f_0. Then E_1' need not be construction closed, but there is a construction closed finite extension E_2' of E_1'. Similarly there exists a (not necessarily construction closed) set $E_2 \subseteq M$ and some extension

f_1 to an isomorphism $f_2\colon E_2 \to E_2'$. Continuing in this way we obtain an ascending chain of elementary isomorphisms $f_i\colon E_i \to E_i'$. Then $E_\infty := \bigcup E_i$ and $E_\infty' := \bigcup E_i'$ are construction closed and $f_\infty := \bigcup f_i$ is an elementary isomorphism from E_∞ to E_∞', contradicting the maximality of f_0. ⊣

THEOREM 9.3.2 (Shelah [55]). *If T is stable and countable, then any subset of a set constructible over A is again constructible over A.*

We immediately obtain the following corollary.

COROLLARY 9.3.3. *If T is a countable stable theory with prime extensions (see Exercise 5.3.2), then all prime extensions are unique.*

For the proof of the theorem we need the following lemma.

LEMMA 9.3.4. *Let T be countable and stable; if A and B are independent over C and B' is countable, then there is a countable subset C' of A with A and BB' independent over CC'.*

PROOF. Using the properties of forking, we find a countable subset $C' \subseteq A$ with $ABC \mathop{\smile\!\!\!\!\!|}_{BCC'} B'$; then $A \mathop{\smile\!\!\!\!\!|}_{BCC'} B'$, and since $A \mathop{\smile\!\!\!\!\!|}_{CC'} B$ we have $A \mathop{\smile\!\!\!\!\!|}_{CC'} BB'$. ⊣

PROOF OF THEOREM 9.3.2. Let B be constructible over A and D a subset of B. We may assume that B is infinite since finite sets are constructible. If E is an arbitrary construction closed subset of B and $E' \subseteq B$ a countable extension of E (i.e., $E' \setminus E$ is countable), then there is some countable construction closed extension E'' of E'. Similarly, for any E with $D \mathop{\smile\!\!\!\!\!|}_{A(D\cap E)} E$ and for every countable extension E' of E, there is a countable extension E'' with $D \mathop{\smile\!\!\!\!\!|}_{A(D\cap E'')} E''$ (Lemma 9.3.4).

Applying these closure procedures alternatingly countably many times, one sees that for any construction closed subset E of B with

$$D \mathop{\smile\!\!\!\!\!|}_{A(D\cap E)} E$$

and for any countable extension E' of E there exists a construction closed countable extension E'' of E' such that

$$D \mathop{\smile\!\!\!\!\!|}_{A(D\cap E'')} E''.$$

In this way we obtain a continuous chain $(C_\alpha)_{\alpha<\xi}$ of construction closed sets with $C_0 = \emptyset$, $\bigcup_{\alpha<\xi} C_\alpha = B$, countable differences $C_{\alpha+1} \setminus C_\alpha$ and

$$D \mathop{\smile\!\!\!\!\!|}_{A(D\cap C_\alpha)} C_\alpha.$$

For each α we can choose an ω-enumeration of $D \cap (C_{\alpha+1} \setminus C_\alpha)$. These enumerations can be composed to an enumeration of D. In order to show that this enumeration is a construction of D it suffices to show that every initial segment $\overline{d}$ of $D \cap (C_{\alpha+1} \setminus C_\alpha)$ is atomic over $A(D \cap C_\alpha)$. This follows from the Open Mapping Theorem (8.5.9) as $\overline{d}$ is atomic over AC_α. ⊣

Chapter 10

THE FINE STRUCTURE OF $\aleph_1$-CATEGORICAL THEORIES

10.1. Internal types

By the results in Sections 5.8 and 6.3 we know that models of $\aleph_1$-categorical theories are (minimal) prime extensions of strongly minimal sets. We will see in the next section that in this case the prime extensions M of $\varphi(M)$ are obtained in a particularly simple way, namely as an iterated fibration where each fibre is the epimorphic image of some $\varphi(M)^k$. *Unless stated otherwise, we assume in this section that T is totally transcendental.*

We need the concept of an *internal* type. We fix a 0-definable infinite subclass $\mathbb{F} = \varphi_0(\mathfrak{C})$ of $\mathfrak{C}$.

Definition 10.1.1. A partial type $\pi(x)$ over the empty set is called $\mathbb{F}$-*internal* if for some set B, the class $\pi(\mathfrak{C})$ is contained in $\operatorname{dcl}(\mathbb{F}B)$.

Lemma 10.1.2. *A partial type π is $\mathbb{F}$-internal if and only if there is some finite conjunction φ of formulas in π which is $\mathbb{F}$-internal.*

Proof. The complement of $\operatorname{dcl}(\mathbb{F}B)$ can be defined by a partial type $\sigma(x)$. So $\pi(\mathfrak{C}) \subseteq \operatorname{dcl}(\mathbb{F}B)$ if and only if $\pi \cup \sigma$ is inconsistent and this is witnessed by a finite part of π. ⊣

Example. Let G be a group. Let

$$M = (G, A)$$

be a two-sorted structure where A is a copy of G without the group structure. Instead, the structure M contains the map

$$\pi\colon G \times A \to A$$

defined as $\pi(g, a) = ga$. Then M is prime over G and clearly the type p of an element in A is G-internal, in fact, $p(M) \subseteq \operatorname{dcl}(G, a)$ for any $a \in A$. We will see in Corollary 10.1.7 that this is the typical picture for internal types.

Lemma 10.1.3. *A type $p \in S(A)$ is $\mathbb{F}$-internal if and only if there is some set of parameters B and some realisation e of p such that $e \in \operatorname{dcl}(\mathbb{F}B)$ and $e \mathop{\smile\!\!\!\!\!\!|}_A B$.*

PROOF. If $p(\mathfrak{C}) \subseteq \operatorname{dcl}(\mathbb{F}B)$, we just choose e as a realisation of p independent of B over A.

Conversely, given a realisation e of p and some $b \in B$ with $e \mathop{\smile\!\!\!\!\!\!\!\mid}_A b$ and $e \in \operatorname{dcl}(\mathbb{F}b)$, we choose a Morley sequence $b_0, b_1, \ldots$ of $\operatorname{tp}(b/\operatorname{acl}^{\mathrm{eq}}(A))$ of length $|T|^+$. By Exercise 7.2.1 for each $e' \in p(\mathfrak{C})$ there is some i such that $e' \mathop{\smile\!\!\!\!\!\!\!\mid}_A b_i$. If p is stationary, then eb and $e'b_i$ realise the same type over A and hence $e' \in \operatorname{dcl}(\mathbb{F}b_i)$. So $p(\mathfrak{C}) \subseteq \operatorname{dcl}(\mathbb{F}B)$ for $B = \{b_0, b_1, \ldots\}$. If p is not stationary, then by the previous argument applied to *each* extension of p to $\operatorname{acl}^{\mathrm{eq}}(A)$ these are all $\mathbb{F}$-internal. By taking unions we obtain a set B such that the realisations of all extensions of p to $\operatorname{acl}^{\mathrm{eq}}(A)$ are contained in $\operatorname{dcl}(\mathbb{F}B)$. $\dashv$

LEMMA 10.1.4. *A consistent formula φ is $\mathbb{F}$-internal if and only if there is a definable surjection from some $\mathbb{F}^n$ onto $\varphi(\mathfrak{C})$.*

PROOF. If there is a B-definable surjection $\mathbb{F}^n \to \varphi(\mathfrak{C})$, $\varphi(\mathfrak{C})$ is contained in $\operatorname{dcl}(\mathbb{F}B)$. For the converse assume that $\varphi(\mathfrak{C}) \subseteq \operatorname{dcl}(\mathbb{F}B)$. Then, by Exercise 6.1.12, for every $e \in \varphi(\mathfrak{C})$ there is a B-definable class $\mathbb{D}_e \subseteq \mathbb{F}^{n_e}$ and a B-definable map $f_e \colon \mathbb{D}_e \to \mathfrak{C}$ such that e is in the image of f_e. A compactness argument shows that there is a finite number of definable classes $\mathbb{D}_1, \ldots, \mathbb{D}_m \subseteq \mathbb{F}^n$ and definable maps $f_i \colon \mathbb{D}_i \to \mathfrak{C}$ such that $\varphi(\mathfrak{C})$ is contained in the union of the $f_i(\mathbb{D}_i)$. Fix a sequence of distinct elements $a_1, \ldots, a_m \in \mathbb{F}$ and an element b of $\varphi(\mathfrak{C})$. Define $f \colon \mathbb{F}^{n+1} \to \varphi(\mathfrak{C})$ by setting $f(\overline{x}, y) = f_i(\overline{x})$ if $y = a_i$ and $\overline{x} \in \mathbb{D}_i$ and $f(\overline{x}, y) = b$ for some $b \in \varphi(\mathfrak{C})$ otherwise. Then f is a surjection from $\mathbb{F}^{n+1}$ onto $\varphi(\mathfrak{C})$. $\dashv$

LEMMA 10.1.5. *Let T be an arbitrary theory and $\mathbb{F}$ a stably embedded 0-definable class. If a and b have the same type over $\mathbb{F}$, they are conjugate under an element of* $\operatorname{Aut}(\mathfrak{C}/\mathbb{F})$.

PROOF. We construct the automorphism as the union of a long ascending sequence of elementary maps $\alpha \colon A \cup \mathbb{F} \to B \cup \mathbb{F}$ which are the identity on $\mathbb{F}$. Assume that α is constructed and consider an element $c \in \mathfrak{C}$. Since $\mathbb{F}$ is stably embedded, the type of cA over $\mathbb{F}$ is definable over some subset C of $\mathbb{F}$. Choose some $d \in \mathfrak{C}$ with $\operatorname{tp}(dB/C) = \operatorname{tp}(cA/C)$. We can then extend α to an elementary map $\alpha' \colon \{c\} \cup A \cup \mathbb{F} \to \{d\} \cup B \cup \mathbb{F}$. To see this assume that $\models \varphi(c, a, f)$, where $a \in A$ and $f \in \mathbb{F}$. Then $\varphi(x, y, f)$ belongs to the type of cA over $\mathbb{F}$, so $\varphi(x, y, f)$ belongs also to the type of dB over $\mathbb{F}$ which shows $\models \varphi(d, \alpha(a), f)$. $\dashv$

A *groupoid* is a (small) category in which all morphisms are invertible, i.e., isomorphisms. A groupoid is connected if there are morphisms between any two objects. A *definable* groupoid $\mathcal{G}$ is one whose objects are given by a definable family $(O_i)_{i \in I}$ of classes and whose morphisms by a definable family $(M_{i,j})_{i,j \in I}$ of bijections $O_i \to O_j$. We use the notation $\operatorname{Hom}_{\mathcal{G}}(O_i, O_j)$ and $\operatorname{Aut}_{\mathcal{G}}(O_i)$ for $M_{i,j}$ and $M_{i,i}$.

We denote by $\mathbb{F}^{\text{eq}} \subseteq \mathfrak{C}^{\text{eq}}$ the collection of those 0-definable equivalence classes having representatives in $\mathbb{F}$.

THEOREM 10.1.6 (Hrushovski's Binding Groupoid). *Let T be totally transcendental, $\mathbb{E}$ and $\mathbb{F}$ be 0-definable and assume that $\mathbb{E}$ is $\mathbb{F}$-internal and non-empty. Then $\mathbb{E}$ is an object of a 0-definable connected groupoid $\mathcal{G}$ with the following properties*:

a) *$\mathbb{E}$ is not the only object in $\mathcal{G}$. The objects other than $\mathbb{E}$ are subsets of $\mathbb{F}^{\text{eq}}$.*
b) $\text{Aut}_{\mathcal{G}}(\mathbb{E}) = \text{Aut}(\mathbb{E}/\mathbb{F})$.

The group $\text{Aut}_{\mathcal{G}}(\mathbb{E})$ is Zilber's *binding group* (see also [46]).

PROOF. By Lemma 10.1.4 there is a definable surjection from some power $\mathbb{F}^n$ to $\mathbb{E}$. This induces a bijection $f : O \to \mathbb{E}$ for some definable class O of $\mathbb{F}^{\text{eq}}$. Let $f = f_c$ be defined from a parameter c and $O = O_d$ from a parameter d. By Lemma 8.3.3 we can find d in $\mathbb{F}(M_0)$ for an atomic model M_0 of T. Since, by Lemma 5.3.4, the isolated types are dense over any parameter set, we may assume that the type of c over $\mathbb{F} \cap M_0$ is isolated, say by a formula $\varphi(x, a)$. It is easy to see that $\varphi(x, a)$ isolates a type over $\mathbb{F}$. By extending d if necessary we may assume that $d = a$. Now let $\psi(y)$ isolate the type of d.

As objects of $\mathcal{G}$ we take $\mathbb{E}$ and the O_e where e realises $\psi(y)$ and we let $\text{Hom}_{\mathcal{G}}(O_e, \mathbb{E})$ be the set of all $f_{c'}$ with c' realising $\varphi(x, e)$. We claim that $\text{Hom}_{\mathcal{G}}(O_e, \mathbb{E})$ is a right coset of $\text{Aut}(\mathbb{E}/\mathbb{F})$, namely, if c' realises $\varphi(x, e)$, we have

$$\text{Hom}_{\mathcal{G}}(O_e, \mathbb{E}) = \text{Aut}(\mathbb{E}/\mathbb{F}) \circ f_{c'}.$$

This follows easily from the fact that, by Lemma 10.1.5, the elements of $\text{Hom}_{\mathcal{G}}(O_e, \mathbb{E})$ are of the form $f_{\alpha(c')}$ for some automorphism $\alpha \in \text{Aut}(\mathfrak{C}/\mathbb{F})$ and from the formula

$$f_{\alpha(c')} = \alpha \circ f_{c'}.$$

We can now set

$$\begin{aligned}
\text{Hom}_{\mathcal{G}}(\mathbb{E}, O_e) &= \{f^{-1} \mid f \in \text{Hom}_{\mathcal{G}}(O_e, \mathbb{E})\} \\
\text{Hom}_{\mathcal{G}}(\mathbb{E}, \mathbb{E}) &= \{f \circ g^{-1} \mid f, g \in \text{Hom}_{\mathcal{G}}(O_e, \mathbb{E})\} \\
\text{Hom}_{\mathcal{G}}(O_e, O_{e'}) &= \{f^{-1} \circ g \mid f \in \text{Hom}_{\mathcal{G}}(O_{e'}, \mathbb{E}), g \in \text{Hom}_{\mathcal{G}}(O_e, \mathbb{E})\}. \dashv
\end{aligned}$$

Recall that a group G acts *regularly* on a set A if for all $a, b \in A$ there exists a unique $g \in G$ with $ga = b$.

COROLLARY 10.1.7 (Binding group). *Let T be a totally transcendental theory, $\mathbb{E}$ and $\mathbb{F}$ be 0-definable and assume that $\mathbb{E}$ is $\mathbb{F}$-internal. Then the following holds.*

1. *There is a definable group $\mathbb{G} \subseteq \mathbb{F}^{\text{eq}}$, the binding group, and a definable class $\mathbb{A}$ on which $\mathbb{G}$ acts regularly and such that $\mathbb{E} \subseteq \text{dcl}(\mathbb{F}a)$ for all $a \in \mathbb{A}$.*

The group $\mathbb{G}$, the class $\mathbb{A}$, the group operation and the action of $\mathbb{G}$ on $\mathbb{A}$ are definable with parameters from $\mathbb{F}$.

2. $\mathrm{Aut}(\mathbb{E}/\mathbb{F})$ *is a 0-definable permutation group,* $\mathrm{Aut}(\mathbb{E}/\mathbb{F})$ *acts regularly on* $\mathbb{A}$ *and is definably isomorphic to* $\mathbb{G}$.

PROOF. Let $\mathcal{G}$ be the groupoid of Theorem 10.1.6. Fix any object O_e different from $\mathbb{E}$. Set $\mathbb{G} = \mathrm{Aut}_{\mathcal{G}}(O_e)$ and $\mathbb{A} = \mathrm{Hom}_{\mathcal{G}}(O_e, \mathbb{E})$. Now replace the definable bijections in $\mathbb{G}$ and $\mathbb{A}$ by their canonical parameters in order to obtain elements of $\mathbb{F}^{\mathrm{eq}}$ and $\mathfrak{C}^{\mathrm{eq}}$ respectively. $\dashv$

REMARK 10.1.8. Note that in the last corollary $\mathrm{Aut}(\mathbb{E}/\mathbb{F})$ is infinite if $\mathbb{E} \not\subseteq \mathrm{acl}^{\mathrm{eq}}(\mathbb{F})$.

EXERCISE 10.1.1. Let $\mathfrak{C}$ be $\mathbb{F}$-internal. Show that every model M is a minimal extension of $\mathbb{F}(M)$.

EXERCISE 10.1.2. Let T, $\mathbb{E}$ and $\mathbb{F}$ be as in Theorem 10.1.6.

1. Show that there is a finite subset A of $\mathbb{E}$ such that $\mathbb{E}$ is contained in $\mathrm{dcl}(\mathbb{F} \cup A)$.
2. Show that the converse of Remark 10.1.8 is also true: $\mathrm{Aut}(\mathbb{E}/\mathbb{F})$ is finite if $\mathbb{E}$ is contained in $\mathrm{acl}(\mathbb{F})$.

EXERCISE 10.1.3. Prove that every element of $\mathrm{dcl}^{\mathrm{eq}}(\mathbb{F})$ is interdefinable with an element of $\mathbb{F}^{\mathrm{eq}}$.

EXERCISE 10.1.4. Let T be arbitrary, $\mathbb{F}$ 0-definable and C a subset of $\mathbb{F}$. Show that $\mathrm{tp}(a/\mathbb{F})$ is definable over C if and only if $\mathrm{tp}(a/C) \vdash \mathrm{tp}(a/\mathbb{F})$.

EXERCISE 10.1.5 (Chatzidakis–Hrushovski, [17]). Let T be arbitrary and $\mathbb{F}$ 0-definable. Show that the following are equivalent:

a) $\mathbb{F}$ is stably embedded.
b) Every type $\mathrm{tp}(a/\mathbb{F})$ is definable over a subset C of $\mathbb{F}$.
c) For every a there is a subset C of $\mathbb{F}$ such that $\mathrm{tp}(a/C) \vdash \mathrm{tp}(a/\mathbb{F})$.
d) Every automorphism of $\mathbb{F}$ extends to an automorphism of $\mathfrak{C}$.

10.2. Analysable types

Throughout this section we assume that T is a stable theory eliminating imaginaries.

DEFINITION 10.2.1. Let $\mathbb{F}$ be a 0-definable class. A type $p \in S(\emptyset)$ is called $\mathbb{F}$-*analysable* if for every realisation a of p there is a sequence of tuples $a_0, \ldots, a_n = a$ in $\mathrm{dcl}(a)$ such that $\mathrm{tp}(a_i/a_0 \ldots a_{i-1})$ is $\mathbb{F}$-internal for $i = 0, \ldots, n$.

THEOREM 10.2.2 (Hrushovski [27]). *Let T be $\aleph_1$-categorical and $\mathbb{F}$ a 0-definable strongly minimal set. Then every type $p \in S(\emptyset)$ is $\mathbb{F}$-analysable.*

We need some preparation in order to prove this theorem.

THEOREM 10.2.3. *Let $p \in S(A)$ be a stationary type and $\mathcal{I}$ an infinite Morley sequence for p. Then* $\mathrm{Cb}(p) \subseteq \mathrm{dcl}(\mathcal{I})$.

PROOF. By Exercise 9.1.2(b) the average type $\mathrm{Av}(\mathcal{I})$ is the non-forking extension of p to the monster model. The proof of Theorem 9.1.2 and Lemma 8.5.8 imply that $\mathrm{Av}(\mathcal{I})$ is based on $\mathcal{I}$. ⊣

DEFINITION 10.2.4. 1. We call two types $p, q \in S(A)$ *almost orthogonal* if any realisation of p is independent over A from any realisation of q.
2. Two types over possibly different domains are *orthogonal* if all non-forking extensions to any common domain are orthogonal.
3. A theory T is called *unidimensional* if all stationary non-algebraic types are pairwise non-orthogonal.

Note that algebraic types are orthogonal to all types.

Let $\mathbb{F}$ be a strongly minimal set. We call a type $p \in S(A)$ orthogonal to $\mathbb{F}$ if for every realisation b of p, any $c \in \mathbb{F}$ and any extension B of A over which $\mathbb{F}$ is defined we have

$$b \mathop{\downarrow}_A B \Longrightarrow b \mathop{\downarrow}_B c.$$

If q is a type of Morley rank 1 containing $\mathbb{F}(x)$, this is equivalent to p being orthogonal to q.

LEMMA 10.2.5. *Let T be $\aleph_1$-categorical and $\mathbb{F}$ a strongly minimal set. Then no non-algebraic type is orthogonal to $\mathbb{F}$.*

PROOF. Let $\mathbb{F}$ be defined over A and suppose that $p \in S(A)$ is orthogonal to $\mathbb{F}$. Choose a model M containing A, a realisation b of p independent from M over A, and a model N prime over Mb. Then b is independent from $\mathbb{F}(N)$ over M. By Theorem 5.8.1, there is some $c \in \mathbb{F}(N) \setminus M$. Let $\varphi(x, y) \in L(M)$ so that $\varphi(x, b)$ isolates the type of c over Mb. Then $\varphi(x, b)$ cannot be realised in $\mathbb{F}(M)$ and hence must be algebraic. So $\mathrm{tp}(c/Mb)$ and by symmetry $\mathrm{tp}(b/Mc)$ fork over M, contradicting the choice of b. ⊣

PROOF OF THEOREM 10.2.2. By induction on $\alpha = \mathrm{MR}(p)$. If $\alpha = 0$, p is algebraic and hence trivially internal (in any definable class). If $\alpha > 0$, we apply Lemma 10.2.5 and find a realisation b of p, some $c \in \mathbb{F}$ and some set of parameters B such that $b \mathop{\downarrow} B$ and $b \not\mathop{\downarrow}_B c$. By finite character of forking we may assume that B is finite.

Let D be the canonical base of $\mathrm{tp}(cB/\mathrm{acl}(b))$. As $cB \mathop{\downarrow}_D b$, but $cB \not\mathop{\downarrow} b$, we have $b \not\mathop{\downarrow} D$. Let $(c_i B_i)$ be an infinite Morley sequence for $\mathrm{tp}(cB/D)$. By Theorem 10.2.3, we have

$$D \subseteq \mathrm{dcl}(c_0 B_0 c_1 B_1 \dots).$$

It follows from $b \mathop{\smile\!\!\!\!\!\!\mid}\, B$ (and $D \subseteq \operatorname{acl}(b)$) that we must have $D \mathop{\smile\!\!\!\!\!\!\mid}\, B$ and hence $D \mathop{\smile\!\!\!\!\!\!\mid}\, B_0 B_1 \ldots$. This shows that the type of any tuple $d \in D$ is $\mathbb{F}$-internal by Lemma 10.1.3. We choose a finite tuple $d \in D$ with $b \not\mathop{\smile\!\!\!\!\!\!\mid}\, d$. Then we have $\operatorname{MR}(b/d) < \alpha$. We may absorb the parameter d into the language and apply the induction hypothesis to $T(d) = T \cup \{\varphi(d) \mid \varphi \in \operatorname{tp}(d)\}$ to find a sequence $b_1, \ldots, b_n = b$ such that $b_i \in \operatorname{dcl}(db)$ and the types $\operatorname{tp}(b_i/db_1 \ldots b_{i-1})$ are $\mathbb{F}$-internal. Setting $b_0 = d$, we would be done if we knew that $d \in \operatorname{dcl}(b)$. For this we replace d by the canonical parameter d' of the finite set $\{d_1, \ldots, d_k\}$ of conjugates of d over b. We have thus achieved $d' \in \operatorname{dcl}(b)$. Because $d' \in \operatorname{dcl}(d_1 \ldots d_k)$ the type of d' is $\mathbb{F}$-internal and since $d \in \operatorname{acl}(d')$, we have $\operatorname{MR}(b/d') \leq \operatorname{MR}(b/d) < \alpha$. We can use this d' to finish the proof. ⊣

A complete theory is called *almost strongly minimal* if there is a strongly minimal formula φ (possibly containing parameters) such that $\mathfrak{C}$ is in the algebraic closure of $\varphi(\mathfrak{C}) \cup A$ for some set $A \subseteq \mathfrak{C}$.

COROLLARY 10.2.6 (Zilber). *Let T be an $\aleph_1$-categorical theory. If there is no infinite group definable in T^{eq}, then T is almost strongly minimal.*

PROOF. Let T be an $\aleph_1$-categorical theory and let $\mathbb{F}$ be a 0-definable strongly minimal set, possibly after adding parameters. If T is not almost strongly minimal, we can use Theorem 10.2.2 to find a definable class $\mathbb{E}$ which is $\mathbb{F}$-internal but not contained in $\operatorname{acl}^{\mathrm{eq}}(\mathbb{F})$. Then $\operatorname{Aut}(\mathbb{E}/\mathbb{F})$ is an infinite definable group by Corollary 10.1.7 and Remark 10.1.8. ⊣

THEOREM 10.2.7 (Baldwin [3]). *Any $\aleph_1$-categorical theory has finite Morley rank.*

To prove Theorem 10.2.7 we need the following definition which allows us to extend additivity of Morley rank beyond strongly minimal sets (see also Proposition 6.4.9 and Exercise 6.4.3).

DEFINITION 10.2.8. Let $f : \mathbb{B} \to \mathbb{A}$ be a definable surjection. We say that the *fibres of f have definable Morley rank* if there is a finite bound for the Morley rank of the fibres $f^{-1}(a)$ and if for every definable $\mathbb{B}' \subseteq \mathbb{B}$ and every k' the class $\{a \in \mathbb{A} \mid \operatorname{MR}(f^{-1}(a) \cap \mathbb{B}') = k'\}$ is definable.

REMARK 10.2.9. If $\mathbb{B}$ is a power of a strongly minimal set, the fibres of f have definable Morley rank by Corollary 6.4.4. ⊣

For the next statement remember that the Morley rank of the empty set is defined as $-\infty$.

LEMMA 10.2.10. *If the fibres of $f : \mathbb{B} \to \mathbb{A}$ have definable Morley rank and $\operatorname{MR}(\mathbb{A})$ is finite, we have*

$$\operatorname{MR}(\mathbb{B}) = \max_{k<\omega}(\operatorname{MR}(\mathbb{A}_k) + k),$$

where $\mathbb{A}_k = \{a \in \mathbb{A} \mid \operatorname{MR}(f^{-1}(a)) = k\}$.

PROOF. We leave it as an exercise (Exercise 10.2.1) to show that $\mathrm{MR}\,\mathbb{B} \geq \mathrm{MR}\,\mathbb{A}_k + k$ for all k. For the converse we may assume that all fibres have Morley rank k and that $\mathbb{A}$ has Morley degree 1 and Morley rank β. We show $\mathrm{MR}\,\mathbb{B} \leq \beta + k$ by induction on β.

Let $\mathbb{B}^i$ be an infinite family of disjoint definable subsets of $\mathbb{B}$. We want to show that one of the $\mathbb{B}^i$ has smaller Morley rank than $\beta + k$. Consider any $a \in \mathbb{A}$. Then for some i the fibre $f^{-1}(a) \cap \mathbb{B}^i$ must have rank smaller than k. So the intersection of all $\mathbb{A}_k^i = \{a \mid \mathrm{MR}(f^{-1}(a) \cap \mathbb{B}^i) = k\}$ is empty, which implies that one of the $\mathbb{A}_k^i$ has smaller rank that β. Induction yields $\mathrm{MR}\,\mathbb{B}^i < \beta + k$. ⊣

PROOF OF THEOREM 10.2.7. It is enough to prove that every element a has finite Morley rank over $\emptyset$. Each a has an *analysing* sequence $a_0, \ldots, a_n = a$ where all types $\mathrm{tp}(a_i/a_0 \ldots a_{i-1})$ are $\mathbb{F}$-internal. We prove by induction on n that the tuple $a_0 \ldots a_n$ has finite Morley rank. By Lemma 6.4.1 this implies that a_n has finite Morley rank.

By the induction hypothesis, $a_0 \ldots a_{n-1}$ is contained in a 0-definable set $\mathbb{A}$ of finite Morley rank. Since $\mathrm{tp}(a_n/a_0 \ldots a_{n-1})$ is $\mathbb{F}$-internal, a_n is contained in an $(a_0 \ldots a_{n-1})$-definable set which is an image of some power of $\mathbb{F}$ by a definable map. So we may assume that $a_0 \ldots a_n$ belongs to a 0-definable set $\mathbb{B}$ which projects onto $\mathbb{A}$ by the restriction map $\pi \colon \mathbb{B} \to \mathbb{A}$ and such that the fibres $\pi^{-1}(a)$ are definable images of some power of $\mathbb{F}$. By Corollary 10.2.9 the fibres of π have *definable Morley rank*. If the rank of the fibres is bounded by k, Lemma 10.2.10 bounds the rank of $\mathbb{B}$ by $\mathrm{MR}\,\mathbb{A} + k$. ⊣

We end this section with a different characterisation of $\aleph_1$-categorical theories due to Erimbetov [18].

THEOREM 10.2.11. *A countable theory T is $\aleph_1$-categorical if and only if it is ω-stable and unidimensional.*

PROOF. Assume first that T is $\aleph_1$-categorical. Let $\mathbb{F}$ be strongly minimal, defined over A, p and q be two stationary types over A. By Lemma 10.2.5 there is an extension $A \subseteq B$, realisations a, b, c_1, c_2 of p, q and $\mathbb{F}$ such that $a \mathop{\smile\!\!\!\!\mid}_A B, b \mathop{\smile\!\!\!\!\mid}_A B, a \not\mathop{\smile\!\!\!\!\mid}_B c_1, b \not\mathop{\smile\!\!\!\!\mid}_B c_2$. That means that $c_1 \in \mathrm{acl}(aB) \setminus \mathrm{acl}(B)$ and $c_2 \in \mathrm{acl}(bB) \setminus \mathrm{acl}(B)$. So c_1 and c_2 have the same type over B and we may assume that $c_1 = c_2$. But then $c_1 \not\mathop{\smile\!\!\!\!\mid}_B c_1$ implies $a \not\mathop{\smile\!\!\!\!\mid}_B b$ and p and q are not orthogonal.

For the converse assume that T is ω-stable and unidimensional. The proof of the Baldwin–Lachlan Theorem (5.8.1) shows that it is enough to prove that there are no Vaughtian pairs $M \precneqq N$ for strongly minimal formulas $\varphi(y)$ defined over M. Let a be any element in $N \setminus M$ and $p(x) = \mathrm{tp}(a/M)$. By assumption there is an extension M' of M and an element $c \in \varphi(\mathfrak{C}) \setminus M'$ such that $a \mathop{\smile\!\!\!\!\mid}_M M'$ and $c \in \mathrm{acl}(aM')$. Let $\delta(a, m', y)$ be a formula which isolates the type of c over aM'. Then the following sentences are true in M'

$$\mathrm{d}_p\, x \exists y \delta(x, m', y)$$
$$\mathrm{d}_p\, x \forall y(\delta(x, m', y) \to \varphi(y))$$
$$\forall y\, \mathrm{d}_p\, x \neg\delta(x, m', y).$$

So we find an $m \in M$ for which the corresponding sentences are true in M. This implies that there is a $b \in B$ such that $N \models \delta(a, m, b)$ and that all such b lie in $\varphi(N) \setminus M$. So $M \precneqq N$ is not a Vaughtian pair for φ. ⊣

EXERCISE 10.2.1. Let $f : \mathbb{B} \to \mathbb{A}$ definable and assume that all fibres have Morley rank at least α. Then $\mathrm{MR}\,\mathbb{B} \geq \alpha + \mathrm{MR}\,\mathbb{A}$.

EXERCISE 10.2.2. Two types $p \in \mathrm{S}_m(A)$ and $q \in \mathrm{S}_n(A)$ are *weakly orthogonal*, if $p(x) \cup q(y)$ axiomatises a complete type in $\mathrm{S}_{m+n}(A)$. Show

1. p and q are weakly orthogonal if and only if for any realisation a of p, q has a unique extension to Aa.
2. In a simple theory two stationary types p and q are weakly orthogonal if and only if they are almost orthogonal.

EXERCISE 10.2.3. Assume T simple and p and p' two stationary parallel types. Then p is orthogonal to a type q if and only if p' is.

EXERCISE 10.2.4. In a simple theory call a stationary, non-algebraic type $p \in \mathrm{S}(A)$ *regular* if it is orthogonal to every forking extension. Prove the following

1. If T is stable, then also every type parallel to p is regular.
2. $\mathrm{cl}(B) = \{c \in p(\mathfrak{C}) \mid B \not\perp_A c\}$ defines a pregeometry on $p(\mathfrak{C})$.
3. Dependence is transitive if the middle element realises a regular type: if $b \not\perp_A c$, $c \not\perp_A d$ and c realises p, then $b \not\perp_A d$. It follows that non-orthogonality is an equivalence relation on the class of regular types.

10.3. Locally modular strongly minimal sets

In this section, we let T denote a complete stable theory and $\varphi(x)$ a strongly minimal formula without parameters.

We call φ *modular* if its pregeometry is modular in the sense of Definition C.1.9, i.e., if for all relatively algebraically closed A, B in $\varphi(\mathfrak{C})$

$$\dim(A \cup B) + \dim(A \cap B) = \dim(A) + \dim(B). \tag{10.1}$$

We say T is *locally modular* if (10.1) holds whenever $A \cap B$ contains an element not in $\mathrm{acl}(\emptyset)$.

It is easy to see that φ is modular if and only if any two relatively algebraically closed subsets A and B of $\varphi(\mathfrak{C})$ are independent over their intersection:

$$A \underset{A\cap B}{\perp} B;$$

see Lemma C.1.10. In fact this holds for arbitrary sets B, not necessarily contained in $\varphi(\mathfrak{C})$.

LEMMA 10.3.1. *If φ is modular, then*

$$A \underset{A\cap B}{\perp} B$$

for algebraically closed B and any A which is relatively algebraically closed in $\varphi(\mathfrak{C})$.

PROOF. Let C be the intersection of B and $\varphi(\mathfrak{C})$. It is enough to show that A is independent from B over C. For this we may assume that B is the algebraic closure of a finite set and the elements of A are algebraically independent over C. We have to show that the elements of A remain algebraically independent over B. Choose a B-independent sequence $A_0, A_1, \ldots$ of sets realising the same type as A over B. For any i the intersection of $\mathrm{acl}(A_0 \ldots A_i)$ and $\mathrm{acl}(A_{i+1})$ is contained in B. So by local modularity $A_0 \ldots A_i$ and A_{i+1} are independent over C. This implies that the elements of $A_0 \cup A_1 \cup \cdots$ are algebraically independent over C. By Exercise 9.1.2 for some i the elements of $A_i \cup A_{i+1} \cup \cdots$ are algebraically independent over B. Hence also the elements of A are algebraically independent over B. ⊣

DEFINITION 10.3.2. A formula $\psi(x)$ without parameters is 1*-based* if

$$A \underset{\mathrm{acl}^{\mathrm{eq}}(A)\cap \mathrm{acl}^{\mathrm{eq}}(B)}{\perp} B$$

for all B and all subsets A of $\psi(\mathfrak{C})$.

For a set B and a tuple a, the strong type $\mathrm{stp}(a/B) = \mathrm{tp}(a/\mathrm{acl}^{\mathrm{eq}}(B))$ is stationary (see Exercises 8.4.9 and 8.5.4). We denote by $\mathrm{Cb}(a/B)$ the canonical basis of $\mathrm{stp}(a/B)$. Note that $\mathrm{Cb}(a/B)$ is a subset of $\mathfrak{C}^{\mathrm{eq}}$.

LEMMA 10.3.3. *A formula ψ is* 1*-based if and only if*

$$\mathrm{Cb}(a/B) \subseteq \mathrm{acl}^{\mathrm{eq}}(a)$$

for all sets B and finite tuples a in $\psi(\mathfrak{C})$.

PROOF. Let C be the intersection $\mathrm{acl}^{\mathrm{eq}}(a) \cap \mathrm{acl}^{\mathrm{eq}}(B)$. If T is 1-based, the strong type of a over B does not fork over C. So by Lemma 8.5.8 $\mathrm{Cb}(a/B)$ is contained in $C \subseteq \mathrm{acl}^{\mathrm{eq}}(a)$. If conversely $\mathrm{Cb}(a/B)$ is contained in $\mathrm{acl}^{\mathrm{eq}}(a)$, it is also contained in C and $a \perp_{\mathrm{Cb}(a/B)} B$ implies $a \perp_C B$. ⊣

COROLLARY 10.3.4. 1. *1-basedness is preserved under adding and removing parameters, i.e., ψ is* 1*-based if and only if ψ is* 1*-based in $\mathfrak{C}_A$ for any set A of parameters.*

2. *If ψ is* 1*-based and if every element of $\psi'(\mathfrak{C})$ is algebraic over $\psi(\mathfrak{C})$, then $\psi'(\mathfrak{C})$ is* 1*-based.*

PROOF. 1. If ψ is 1-based, then $\mathrm{Cb}_A(a/B) = \mathrm{Cb}(a/AB) \subseteq \mathrm{acl}^{\mathrm{eq}}(a) \subseteq \mathrm{acl}^{\mathrm{eq}}{}_A(a)$. If conversely ψ is 1-based in $\mathfrak{C}_A$ and $a \in \psi(\mathfrak{C})$ and B are given, we may assume that a, B are independent from A. We have then $\mathrm{Cb}(a/B) = \mathrm{Cb}(a/AB) = \mathrm{Cb}_A(a/B) \subseteq \mathrm{acl}^{\mathrm{eq}}{}_A(a)$. Since $\mathrm{Cb}(a/B)$ is also contained in $\mathrm{acl}^{\mathrm{eq}}(B)$, we conclude $\mathrm{Cb}(a/B) \subseteq \mathrm{acl}^{\mathrm{eq}}(a)$.

2. First note that if c is algebraic over a, then, for any set B, we have that c and B are independent over $\mathrm{Cb}(a/B)$, hence $\mathrm{Cb}(c/B) \subseteq \mathrm{acl}^{\mathrm{eq}}\,\mathrm{Cb}(a/B)$. Now let a' be a finite tuple from $\psi'(\mathfrak{C})$ and B any set. Choose a tuple a_1 from $\varphi(\mathfrak{C})$ over which a' is algebraic. If a_1 and a' are interalgebraic, we are done. Otherwise choose a_2 which realises the type of a_1 over a' and is independent from a_1 over a'. Then $\mathrm{Cb}(a'/B) \subseteq \mathrm{acl}^{\mathrm{eq}}\,\mathrm{Cb}(a_1/B) \cap \mathrm{acl}^{\mathrm{eq}}\,\mathrm{Cb}(a_2/B) \subseteq \mathrm{acl}^{\mathrm{eq}}(a_1) \cap \mathrm{acl}^{\mathrm{eq}}(a_2) \subseteq \mathrm{acl}^{\mathrm{eq}}(a')$. ⊣

THEOREM 10.3.5. *Let T be totally transcendental and φ a strongly minimal formula without parameters. Then the following are equivalent.*

a) *φ is locally modular.*

b) *φ is* 1*-based.*

c) *Every family of plane curves in φ has dimension at most* 1*. This means that for all B and elements a, b of $\varphi(\mathfrak{C})$, if* $\mathrm{tp}(ab/B)$ *has Morley rank* 1*, then* $\mathrm{Cb}(ab/B)$ *has Morley rank at most* 1 *over the empty set.*

PROOF. a) $\Rightarrow$ b): If φ is locally modular, φ becomes modular if we add a name for any element $x \in \varphi(\mathfrak{C}) \setminus \mathrm{acl}^{\mathrm{eq}}(\emptyset)$ to the language. If B and $a \in \varphi(\mathfrak{C})$ are given, we choose x independent from aB (over the empty set). It follows from Lemma 10.3.1 that a and Bx are independent over $\mathrm{acl}^{\mathrm{eq}}(ax) \cap \mathrm{acl}^{\mathrm{eq}}(Bx)$. This implies $\mathrm{Cb}(a/Bx) \subseteq \mathrm{acl}^{\mathrm{eq}}(ax)$. By the choice of x we have $\mathrm{Cb}(a/Bx) = \mathrm{Cb}(a/B) \subseteq \mathrm{acl}^{\mathrm{eq}}(B)$. Since x and B are independent over a, we have $\mathrm{acl}^{\mathrm{eq}}(ax) \cap \mathrm{acl}^{\mathrm{eq}}(B) \subseteq \mathrm{acl}^{\mathrm{eq}}(a)$. This implies $\mathrm{Cb}(a/B) \subseteq \mathrm{acl}^{\mathrm{eq}}(a)$.

b) $\Rightarrow$ c): Write $d = \mathrm{Cb}(ab/B)$. Then $\mathrm{MR}(ab/d) = 1$. If the Morley rank of d is not zero, ab and d are dependent over the empty set. By the definition of 1-basedness and Lemma 10.3.3, we have $d \in \mathrm{acl}^{\mathrm{eq}}(ab)$, and so $\mathrm{MR}(abd) \leq 2$. Since $\mathrm{MR}(ab/d) = 1$, we have $\mathrm{MR}(d) \leq 1$ using Proposition 6.4.9.

c) $\Rightarrow$ a): Let x be a non-algebraic element of $\varphi(\mathfrak{C})$. By Lemma C.1.11 we have to show the following: for all elements a, b and sets B in $\varphi(\mathfrak{C})$ with $\mathrm{MR}(ab/x) = 2$ and $\mathrm{MR}(ab/Bx) = 1$, there is some $c \in \mathrm{acl}(Bx)$ such that $\mathrm{MR}(ab/cx) = 1$. We may assume that $a \notin \mathrm{acl}(Bx)$. Consider the imaginary element $d = \mathrm{Cb}(ab/Bx)$. Since d is contained in $\mathrm{acl}^{\mathrm{eq}}(Bx)$, a is not algebraic over d. Also, since $x \notin \mathrm{acl}(ab)$ and d is algebraic over ab by assumption, x is not algebraic over d. So a and x have the same type over d and we can find an element c such that xc and ab have the same type over d. Since $b \in \mathrm{acl}(ad)$ and $d \in \mathrm{acl}^{\mathrm{eq}}(ab)$ this implies $c \in$

$\operatorname{acl}(xd) \subseteq \operatorname{acl}(Bx)$ and $d \in \operatorname{acl}^{\mathrm{eq}}(xc)$. So we have $\mathrm{MR}(ab/cx) = 1$ as required. ⊣

By a theorem of Zilber any ω-categorical strongly minimal theory is locally modular (see [64]). On the other hand (see Exercise 4.3.1) we have the following, which holds more generally for stable theories, see [43].

PROPOSITION 10.3.6. *A totally transcendental theory T which contains an infinite definable field is not* 1*-based.*

PROOF. Let K be an infinite definable set with a definable field structure. We may assume that everything is definable over the empty set. Let α be the Morley rank of K. We call an element x of K *generic* if $\mathrm{MR}(x) = \alpha$. We note first that if $p = (x, y)$ is an element of the line $g_{a,b} = \{(x, y) \mid ax + b = y\}$ which is not algebraic over a, b, then $a, b \in \operatorname{dcl} \operatorname{Cb}(p/a, b)$. This follows from the fact that two lines intersect in at most one point. So it is enough to find such p and $g_{a,b}$ with (a, b) not algebraic over p.

For this we choose four independent generic elements a, b, a', b' and let $p = (x, y)$ be the intersection of $g_{a,b}$ and $g_{a',b'}$. Since x and b' are interdefinable over a, b, a', we can conclude that x, a, b, a' are generic and independent. So p is not algebraic over a, b. Since y and b are interdefinable over x, a we have that x, y, a, a' are generic and independent. This implies that a, b is not algebraic over p. ⊣

The converse of the previous proposition for strongly minimal theories was known as *Zilber's Conjecture*. This Conjecture, namely that for any non-locally modular strongly minimal theory T an infinite field is definable in T^{eq}, was refuted by Hrushovski in [28]. A variant of his construction of a new strongly minimal set will be given in the next section. However, Hrushovski and Zilber proved in their fundamental work [30] that the Conjecture holds for so-called Zariski structures.

10.4. Hrushovski's examples

To end, we present a modification of Hrushovski's *ab initio* example of a new strongly minimal set [28]. This counterexample to Zilber's Conjecture has been the starting point of a whole new industry constructing new uncountably categorical groups [7], fields [6], [8], and geometries [5], [58]. The dimension function defined below also reappears in Zilber's work around Shanuel's Conjecture [65].

Following Baldwin [5] (see also [58]), we construct an almost strongly minimal *projective plane* as a modified Fraïssé limit: instead of considering structures with all their substructures we restrict the amalgamation to so-called *strong* substructures and embeddings. This will allow us to keep control over the algebraic closure of sets so that the resulting structure is uncountably categorical.

Recall that a projective plane is a point-line geometry such that any two lines meet in a unique point, any two points determine a unique line through them and there are four points no three of which are collinear. For convenience, we consider a projective plane as a *bipartite graph* whose vertices are the points and lines of the projective plane and in which the incidence between a point and a line is represented by an edge, making this graph naturally bipartite. In these terms a projective plane is a bipartite graph with the property that the distance between any two vertices is at most 3, the smallest cycles in the graph have length 6, and any element has at least 3 neighbours.

We fix a language $L = \{P, E\}$ for bipartite graphs where P is a predicate denoting the colouring and E denotes the edges. For a finite graph A, let $e(A)$ denote the number of edges in A and $|A|$ the number of vertices. We define $\delta(A) = 2|A| - e(A)$ and put $\delta(A/B) = \delta(AB) - \delta(B)$. Then δ satisfies the submodular law of dimension functions for pregeometries (see page 207):

$$\delta(AB) + \delta(A \cap B) \leq \delta(A) + \delta(B)$$

or equivalently

$$\delta(A/B) \leq \delta(A/A \cap B).$$

We call a finite set $B \subseteq M$ *strong* in M, $B \leq M$, if $\delta(A/B) \geq 0$ for all finite $A \subseteq M$. It follows easily from submodularity that this is a transitive relation, see Exercise 10.4.1.

Let $\mathcal{K}$ be the class of graphs M, bipartite with respect to P, not containing any 4-cycles, and such that $\delta(A) \geq 4$ for any finite subgraph A of M with $|A| \geq 3$. Note that this implies $\delta(A) \geq 2$ for all finite $A \in \mathcal{K}$. Any finite subset A of M is contained in a finite strong subset F of M: we can choose F to be any finite extension of A with $\delta(F)$ minimal.

Given graphs $A \subseteq M, N$ we denote by $M \otimes_A N$ the trivial amalgamation of M and N over A, obtained as the graph whose set of vertices is the disjoint union $(M \setminus A) \cup (N \setminus A) \cup A$ with incidence and predicate P induced by B and C.

LEMMA 10.4.1. *If $A \leq N$ and $\delta(F) \geq 4$ for all finite subgraphs F of M and N with $|F| \geq 3$, the same is true for $M \otimes_A N$. If M is finite, then M is strong in $M \otimes_A N$.*

PROOF. This follows from the fact that every finite $F \subseteq M \otimes_A N$ has the form $M' \otimes_{A'} N'$ where M, A', N' are the intersection of F with M, A, N respectively, and from the formula $\delta(F) = \delta(M') + \delta(N'/A')$. $\dashv$

DEFINITION 10.4.2. We call a proper strong extension F over A *minimal* if it cannot be split into two proper strong extensions $A \leq C \leq F$. We call a minimal extension i-minimal if $\delta(F/A) = i$. We use this terminology also for pairs (A, B) of disjoint sets, which we call i-minimal – or we say that B is i-minimal over A – if AB is an i-minimal extension of A. A 0-minimal pair (A, B) is called a *simple pair* if B is not 0-minimal over any proper subset of A.

The following is easy to see.

REMARK 10.4.3. Let B be 0-minimal over A and A_0 the set of elements of A which are connected to an element of B. Then (A_0, B) is simple and $A \cup B = A \otimes_{A_0} A_0B$.

LEMMA 10.4.4. *A proper strong extension $A \leq F$ is minimal if and only if $\delta(C/A) > \delta(F/A)$ for all C properly between A and F.*

PROOF. Let C be a set properly between A and F with $c = \delta(C/A) \leq \delta(F/A)$ minimal. Then F is a strong extension of C. ⊣

COROLLARY 10.4.5. *A minimal extension F of A is i-minimal for $i = 0, 1$ or 2. The 1-minimal and 2-minimal extensions are of the form $F = A \cup \{b\}$ where b is connected to at most one element of A.*

PROOF. Let F be an i-minimal extension of A. Assume that $B = F \setminus A$ has more than one element and $i > 0$. Since $\delta(b/A) \leq 2$ for any $b \in B$, by the previous lemma we must have $i = 1$ and no element of B is connected with A. Thus $1 = \delta(B)$, which is impossible for $B \in \mathcal{K}$. ⊣

We next fix a function μ from simple pairs (A, B) into the natural numbers satisfying the following properties:

1. $\mu(A, B)$ depends only on the isomorphism type of (A, B).
2. $\mu(A, B) \geq \delta(A)$.

We will be only interested in simple pairs (A, B) where AB belongs to $\mathcal{K}$. Note that this implies $A \neq \emptyset$ and hence $\mu(A, B) \geq 1$.

For any graph N and any simple pair (A, B) with $A \subseteq N$ we define $\chi^N(A, B)$ to be the maximal number of pairwise disjoint graphs $B' \subseteq N$ such that B and B' are isomorphic over A.

Let now $\mathcal{K}_\mu$ be the subclass of $\mathcal{K}$ consisting of those $N \in \mathcal{K}$ satisfying $\chi^N(A, B) \leq \mu(A, B)$ for every simple pair (A, B) with $A \subseteq N$. Clearly $\mathcal{K}_\mu$ depends only on the values $\mu(A, B)$, where AB belongs to $\mathcal{K}$.

LEMMA 10.4.6. *Let $N \in \mathcal{K}_\mu$ contain two finite subgraphs $A \leq F$. If $\delta(F/A) = 0$, then N contains only finitely many copies of F over A.*

PROOF. It suffices to consider the case that $F = A \cup B$ for B simple over A. Assume that B has infinitely many copies over A in N. Consider a finite extension C of A which is strong in N. There is a copy of B which is not contained in C. It follows from minimality that B is disjoint from C. So we can construct an infinite sequence of disjoint copies, contradicting $\chi^N(A, B) \leq \mu(A, B)$. ⊣

We need the following lemma:

LEMMA 10.4.7. *Let M be in $\mathcal{K}_\mu$, A a finite subgraph of M and (A, B) a simple pair. If $N = M \otimes_A AB \notin \mathcal{K}_\mu$ is witnessed by $\chi^N(A', B') > \mu(A', B')$, there are two possibilities for (A', B'):*

1. *$A' = A$ and B' is an isomorphic copy of B over A.*
2. a) *A' is contained in $A \cup B$, but not a subset of A.*
   b) *B contains an isomorphic copy of B' over A'.*

PROOF. First consider the case $A' \subseteq M$. Since $M \in \mathcal{K}_\mu$ there is some copy B'' of B' over A' which intersects B. If $B'' \not\subseteq B$, then $A' \leq A' \cup (M \cap B'') \leq A' \cup B''$ contradicting the minimality of B'' over A'. So $B'' \subseteq B$. Since (A', B'') is simple, every element of A' is connected with some element of B'' and so A' must be a subset of A. If B'' were a proper subset of B, 0-minimality of (A, B) would imply that $0 < \delta(B''/A) \leq \delta(B''/A')$ which is not possible. So $B'' = B$, which implies $A = A'$ by simplicity of (A, B).

Next consider the case $A' \not\subseteq M$, so $A' \cap B \neq \emptyset$. Then since B' is simple over A', no copy of B' over A' is contained in $M \setminus A$. Now suppose that there are k disjoint copies $B'_1, \ldots, B'_k$ of B' over A' contained in M and that the disjoint copies $B'_{k+1}, \ldots, B'_{k+l}$ intersect both M and B. Since each B'_i contains vertices which are connected to vertices of $A' \cap B$, it follows immediately that $\delta(A'/M) \leq \delta(A'/M \cap A') - k \leq \delta(A') - k$. Note that $\delta(M \cup A') \leq \delta(A')$ since $M \cup A'$ is strong in A'.

Since the B'_i are 0-minimal over A', we have for each $i = k+1, \ldots, k+l$:

$$\delta\Big(B'_i/M \cup A' \cup B'_{k+1} \cup \cdots \cup B'_{i-1}\Big) < 0.$$

This implies

$$\delta\Big(\bigcup_{i=k+1}^{k+l} B'_i/M \cup A'\Big) \leq -l.$$

Hence

$$0 \leq \delta\Big(\bigcup_{i=k+1}^{k+l} B'_i \cup A'/M\Big) \leq \delta(A'/M) - l \leq \delta(A') - (k+l).$$

Thus at most $\delta(A')$ many disjoint copies of B' over A' are *not* contained in B, leaving at least one copy of B' over A' inside B. Since each element of A' is connected to some element of this copy, we see that A must be contained in $A \cup B$, finishing the proof. ⊣

As in Section 4.4 we say that $M \in \mathcal{K}_\mu$ is $\mathcal{K}_\mu$-saturated if for all finite $A \leq M$ and strong extensions C of A with $C \in \mathcal{K}_\mu$ there is a strong embedding of C into M fixing A elementwise. Since the empty graph belongs to $\mathcal{K}_\mu$ and is strongly embedded in every $A \in \mathcal{K}_\mu$, this implies that every finite $A \in \mathcal{K}_\mu$ is strongly embeddable in M.

THEOREM 10.4.8. *The class $\mathcal{K}_\mu^{\text{fin}}$ of finite elements of $\mathcal{K}_\mu$ is closed under substructures, and has the joint embedding and the amalgamation property with respect to* strong embeddings. *There exists a countable $\mathcal{K}_\mu$-saturated structure M_μ, which is unique up to isomorphism. This structure M_μ is a projective plane with infinitely many points per line and infinitely lines through any point.*

PROOF. Clearly, $\mathcal{K}_\mu^{\text{fin}}$ is closed under substructures. We will show that $\mathcal{K}_\mu^{\text{fin}}$ has the joint embedding and the amalgamation property with respect to strong embeddings. Then exactly as in the proof of Theorem 4.4.4 we obtain a countable $\mathcal{K}_\mu$-saturated structure M_μ, which is unique up to isomorphism. In particular, in M_μ any partial isomorphism $f : A \to A'$ with $A, A' \leq M_\mu$ extends to an automorphism of M_μ. Since the empty graph is in $\mathcal{K}_\mu$ and strong in $A \in \mathcal{K}_\mu$, it suffices to prove the amalgamation property. Let $C_0, C_1, C_2 \in \mathcal{K}_\mu^{\text{fin}}$, $C_0 \leq C_1, C_2$. We have to find some $D \in \mathcal{K}_\mu^{\text{fin}}$ which contains C_1 and C_2 as strong subgraphs. We prove the amalgamation property by induction on the cardinality of $C_2 \setminus C_0$.

Case 1: C_2 is not a minimal extension of C_0. Then there is a set $C_2' \leq C_2$ which lies properly between C_0 and C_2. By the induction hypothesis we may amalgamate C_1 with C_2' over C_0 to get D', and then amalgamate D' with C_2 over C_2' to obtain D.

Case 2: C_2 is a minimal extension of C_0. We will show that either $D = C_1 \otimes_{C_0} C_1$ belongs to $\mathcal{K}_\mu$ or C_1 strongly contains a copy of C_2 over C_0.

By Corollary 10.4.5 there are three cases:

Case 2.i): C_2 is a 0-minimal extension of C_0. We assume $D = C_1 \otimes_{C_0} C_2 \notin \mathcal{K}_\mu$ and show that C_2 contains a copy of C_2' of C_2. Since $\delta(C_2'/C_0) = 0$ this then implies that C_2' is strong in C_1.

That $D \notin \mathcal{K}_\mu$ can have two reasons. First there might be a 4-cycle in D. This cycle must consist of a_0, a_0' in C_0, $b_1 \in C_1 \setminus C_0$ and $b_2 \in C_2 \setminus C_0$ such that b_1 and b_2 are connected with both a_0 and a_0'. But then minimality implies that $C_2 = C_0 \cup \{b_2\}$ and $C_0 \cup \{b_1\}$ is a copy of C_2 over C_0. Note that b_1 and b_2 must have the same colour.

The second reason might be that $\chi^D(A', B') > \mu(A', B')$ for a simple pair (A', B'). Let A be the set of elements in C_0 which are connected to a vertex in $B = C_2 \setminus C_0$. Then (A, B) is a simple pair and we have $D = C_1 \otimes_A AB$. We can now apply Lemma 10.4.7. The second case of the Lemma cannot occur since then every copy B'' of B' over A' in D must intersect C_2 and, since C_2 is strong in D_2, must be contained in C_2. This would imply that $\chi^{C_2}(A', B') = \chi^D(A', B')$. So the first case applies and we have $A' = A$ and B' is a copy of B over A. All other copies B'' of B over A are contained in C_1 by simplicity. Since B'' is minimal over A, either B'' must be a subset of C_0 or a subset of $C_1 \subseteq C_0$. Since C_2 is in $\mathcal{K}_\mu$, there is a B'' contained in $C_2 \setminus C_0$. Then $C_0 \cup B''$ is over C_0 isomorphic to $C_1 = C_0 \otimes_A AB$.

Case 2.ii): C_2 is a 1–minimal extension of C_0. Then $C_2 = C_0 \cup \{b\}$ and b is connected with a single $a \in C_0$. We show that $D = C_1 \otimes_{C_0} C_2$ is in $\mathcal{K}_\mu$. Clearly, D does not contain more cycles than C_1. So consider a simple pair (A, B) in D. Since $A \leq B$ and $\delta(B/A) = 0$, b cannot be contained in B. If $b \in A$, then $1 = \chi^D(A, B) \leq \mu(A, B)$ as b is connected to a unique element of D. So we have $D \in \mathcal{K}_\mu$.

Case 2.iii): C_2 is a 2-minimal extension of C_0. Then $C_2 = C_0 \cup \{b\}$ where b is not connected with C_0. The same argument as in the last case shows that $D = C_1 \otimes_{C_0} C_2$ is in $\mathcal{K}_\mu$.

Using the fact that any partial isomorphism $f : A \to A'$ with $A, A' \leq M_\mu$ extends to an automorphism of M_μ it is easy to see that M_μ is a projective plane: since any two vertices of the same colour form a strong substructure of M_μ, for any two pairs of such vertices with the same colouring there is an automorphism taking one pair to the other. Since there are pairs of vertices of the same partition at distance 2 in the graph, the same is true for any such pair. Thus any two points lie on a common line and any two lines intersect in a point. Uniqueness is immediate since there are no 4-cycles. Similarly, for any $n \in \omega$ the graph consisting of a vertex x_0 and neighbours $x_1, \dots x_n$ of x_0 (in either colouring) lies in $\mathcal{K}_\mu$ and is a strong extension of x_0. It follows that in M_μ every vertex has infinitely many neighbours. Translated into the language of point-line geometries this says that there are infinitely many points per line and infinitely many lines through any point. This of course already implies the existence of four points in M_μ no three of which are collinear. But this also follows from the fact that the corresponding graph, an 8-cycle of pairwise distinct elements is contained in $\mathcal{K}_\mu$. $\dashv$

We now turn to the model-theoretic properties of M_μ.

THEOREM 10.4.9. *Let T_μ be the theory (in the language of bipartite graphs) axiomatising the class of models M such that*:

1. *Every vertex of M has infinitely many neighbours.*
2. *$M \in \mathcal{K}_\mu$;*
3. *$M \otimes_A AB \notin \mathcal{K}_\mu$ for each simple pair (A, B) with $A \subseteq M$.*

Then $T_\mu = \mathrm{Th}(M_\mu)$.

PROOF. Note first that this forms an elementary class whose theory T_μ is contained in $\mathrm{Th}(M_\mu)$: clearly, Part 1 is a first-order property, which holds in M_μ by Theorem 10.4.8. For each simple pair (A, B) we can express that $\chi^M(A, B) \leq \mu(A, B)$, so Part 2 is first-order expressible and holds in M_μ by construction. For Part 3 notice that if $D = M \otimes_A AB \notin \mathcal{K}_\mu$, then by Lemma 10.4.7 to express the existence of a simple pair (A', B') with $\chi^D(A', B') > \mu(A', B')$ one can restrict to pairs which are contained in $A \cup B$. So this can be expressed in a first-order way. To see that this is true in M_μ we argue as follows. Assume $D \in \mathcal{K}_\mu$. Then for every finite $C \leq M_\mu$ which contains A, the graph $C \leq C \otimes_A AB$ belongs to $\mathcal{K}_\mu$ and so M_μ contains a copy of B over C. So we can construct in M_μ an infinite sequence of disjoint copies of B over A. This is not possible.

Let M be a model of T_μ. We have to show that M is elementarily equivalent to M_μ. Choose an ω-saturated $M' \equiv M$. By (one direction of) the next claim M' is $\mathcal{K}_\mu$-saturated. As in Exercise 4.4.1 M' and M_μ are partially isomorphic and therefore elementarily equivalent.

CLAIM. *The structure M is an ω-saturated model of T_μ if and only if it is $\mathcal{K}_\mu$-saturated.*

PROOF OF CLAIM. Let $M \models T_\mu$ be ω-saturated. To show that M is $\mathcal{K}_\mu$-saturated, let $A \leq M$ and $A \leq F \in \mathcal{K}_\mu^{\text{fin}}$. By induction we may assume that F is a minimal extension of A. There are three cases:

Case 1: F is a 0-minimal extension of A. By condition 3 $M \otimes_A F$ does not belong to $\mathcal{K}_\mu$. Then by Claim 2 of the proof of Theorem 10.4.8 M contains a strong copy of F over A.

Case 2: F is a 1-minimal extension of A. Then $F = A \cup \{b\}$ where b is connected to exactly one vertex $a \in A$. Since a is connected to infinitely many $b' \in M$, there is such a b' that is not algebraic over A. Then $F' = A \cup \{b'\}$ is isomorphic to F and strong in M, since by Lemma 10.4.6 F' is not contained in any $C \subseteq M$ with $\delta(C/A) = 0$.

Case 3: F is a 2-minimal extension of A. Then $F = A \cup \{b\}$ where b is not connected to A. By the previous case there are b' and b'' such that $A \leq A \cup \{b'\} \leq A \cup \{b', b''\} \leq M$, b' is connected with exactly one vertex from A and b'' is connected with b' but not with A. Then $F' = A \cup \{b''\}$ is isomorphic to F and $F \leq A \cup \{b', b''\}$ implies $F' \leq M$.

Conversely, suppose M is $\mathcal{K}_\mu$-saturated. Since M is partially isomorphic to M_μ, it is a model of T_μ. Choose an ω-saturated $M' \equiv M$. Then by the above M' is $\mathcal{K}_\mu$-saturated. So M' and M are partially isomorphic, which implies that M is ω-saturated by Exercise 4.3.13. $\dashv$

DEFINITION 10.4.10 (Coordinatisation). Let $\Pi = (\mathcal{P}, \mathcal{L}, \mathcal{I})$ be a projective plane, let $\ell \in \mathcal{L}$ be a line and let $a_1, a_2, a_3 \in \mathcal{P}$ be non-collinear points outside ℓ. Let D_ℓ denote the set of points on ℓ. Then every element of Π is in the definable closure of $D_\ell \cup \{a_1, a_2, a_3\}$: if the point $x \in \mathcal{P}$ lies for example, not on the line (a_1, a_2), let x_1 and x_2, respectively, denote the intersections of the lines (x, a_1) and (x, a_2) with ℓ. Then $x \in \operatorname{dcl}(a_1, a_2, x_1, x_2)$. A similar argument shows that also every line is definable from a_1, a_2, a_3 and two elements of ℓ. This process is called *coordinatisation.*

Let $\operatorname{cl}(B) = \operatorname{cl}_M(B)$ be the smallest strong subgraph of M containing B (see Exercise 10.4.1). We also define

$$d(A) = \min\{\delta(B) \mid A \subseteq B \subseteq M\} = \delta(\operatorname{cl}(A)).$$

Similarly, we put $d(A/B) = d(AB) - d(B)$.

We will show that for any vertex $a \in M_\mu$ the set D_a of neighbours of a is strongly minimal. To this end we start with the following easy lemma.

LEMMA 10.4.11. *Let M and M' be models of T_μ. Then tuples $a \in M$ and $a' \in M'$ have the same type if and only if the map $a \mapsto a'$ extends to an isomorphism of $\mathrm{cl}(a)$ to $\mathrm{cl}(a')$. In particular, $d(a)$ depends only on the type of a.* ⊣

LEMMA 10.4.12. *Let M be a model of T_μ and A a finite subset of M. Then a is algebraic over A if and only if $d(a/A) = 0$.*

PROOF. Clearly $\mathrm{cl}(A)$ is algebraic over A. If $d(a/A) = 0$, there is an extension B of $\mathrm{cl}(A)$ with $\delta(B/\mathrm{cl}(A)) = 0$. By Lemma 10.4.6 B is algebraic over $\mathrm{cl}(A)$.

For the converse we may assume that M is ω-saturated. If $d(a/A) > 0$, we decompose the extension $\mathrm{cl}(A) \leq \mathrm{cl}(Aa)$ into a series of minimal extensions $\mathrm{cl}(A) = F_0 \leq \cdots \leq F_n = \mathrm{cl}(Aa)$. One extension $F_k \leq F_{k+1}$ must be i-minimal for $i = 1, 2$. By the proof of Theorem 10.4.9, F_{k+1} has infinitely many conjugates over F_k. So $\mathrm{cl}(Aa)$ and therefore also a are not algebraic over A. ⊣

PROPOSITION 10.4.13. *For any model M of T_μ and any $a \in M$, the set D_a is strongly minimal.*

PROOF. Let A be a strong finite subset of M which contains a and let b be an element of D_a. Then $d(b/A)$ is 0 or 1. If $d(b/A) = 0$, then b is algebraic over A by the previous lemma. If $d(b/A) = 1$, then a is the only element of A connected with b. Thus Ab is also strong in M. So by Lemma 10.4.11 the type of b over A is uniquely determined. The claim now follows from Lemma 5.7.3. ⊣

Together with coordinatisation, the strong minimality of D_a now implies that T_μ is almost strongly minimal.

THEOREM 10.4.14. *The theory T_μ is almost strongly minimal, not 1-based and of Morley rank 2. For finite sets A, F we have $\mathrm{MR}(F/A) = d(F/A)$.*

PROOF. By coordinatisation there is a line a and finite set A of parameters such that every element is definable from A and two points of D_a. Since D_a is strongly minimal, this implies that every element has rank at most 2.

To see that T_μ is not 1-based, let ℓ be a line of M. Since $\{\ell, p\}$ is a strong subset, the type of (ℓ, p) is uniquely determined. It follows that p is not algebraic over ℓ, which implies $\ell \in \mathrm{Cb}(p/\ell)$. Also ℓ is not algebraic over p, which implies that T_μ is not 1-based.

To compute the Morley rank of F over A we may assume that $A \leq F \leq M_\mu$. By Proposition 6.4.9 (see also Exercise 6.4.2) we know that Morley rank is additive in T_μ. This shows that we may assume that F is a minimal extension of A. If $\delta(F/A) = 0$, F is algebraic over A by Lemma 10.4.12. If

$\delta(F/A) = 1$, we have $F = A \cup \{b\}$, where b is connected with some element of $a \in A$. Then $\mathrm{MR}(F/A) = 1$ by the proof of Proposition 10.4.13. If $\delta(F/A) = 2$, the proof of Theorem 10.4.9 shows that F has a 0-minimal extension F' which can be reached from A by two 1-minimal extensions. So $\mathrm{MR}(F/A) = \mathrm{MR}(F'/A) = 2$. ⊣

To show that this indeed yields a counterexample to Zilber's Conjecture, we finish by showing that

PROPOSITION 10.4.15. *There is no infinite group definable in T_μ^{eq}.*

PROOF. Assume that there is group G of Morley rank n is definable in T_μ^{eq}. To simplify notation we assume that G is definable without parameters. We apply now Exercise 8.5.9 to obtain a group configuration:

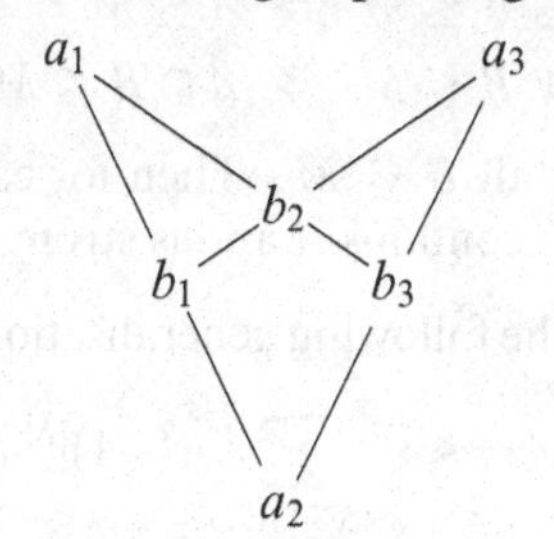

The a_i and b_i have Morley rank n; each element of a triple forming a line in the diagram is algebraic over the two other elements of this triple; any three non-collinear elements are independent. We will show that such a configuration cannot exist.

Choose finite closed sets A_i and B_i in M such that $a_i \in \mathrm{dcl}^{\mathrm{eq}}(A_i)$ and $b_i \in \mathrm{dcl}^{\mathrm{eq}}(B_i)$. We may assume that A_i is independent over a_i from the rest of the diagram and similarly for B_i. We may also assume that all ranks $\mathrm{MR}(A_i/a_i)$, $\mathrm{MR}(B_j/b_j)$ are the same, say k.

By additivity of Morley rank (Proposition 6.4.9; see also Exercise 6.4.2) we have then that the Morley rank of all six points $A_1 \dots B_3$ together is $3n + 6k$.

Consider the four "lines" $E_0 = \mathrm{cl}(A_1, B_1, A_2)$, $E_1 = \mathrm{cl}(A_2, B_3, A_3)$, $E_2 = \mathrm{cl}(A_1, B_2, B_3)$ and $E_3 = \mathrm{cl}(B_1, B_2, A_3)$ corresponding to the four lines of the configuration. Since the union E of the E_i has Morley rank $3n + 6k$, we have $\delta(E) \geq 3n + 6k$. On the other hand, we can bound $\delta(E)$ by the inequality of Exercise 10.4.2. First we note that $\delta(E_i) = 2n + 3k$. Then for different i, j we have $\delta(E_i \cap E_j) = n + k$, since we have e.g., $A_2 \subseteq E_1 \cap E_2 \subseteq \mathrm{acl}(A_2)$. Similarly the intersection of three of the E_i is contained in $\mathrm{acl}(\emptyset) = \emptyset$. We then have

$$\delta(E) \leq 4(2n + 3k) - 6(n + k) = 2n + 6k,$$

which is only possible if $n = 0$, so G is finite. ⊣

We now have the promised counterexample to Zilber's Conjecture.

COROLLARY 10.4.16. *Let T_a be the induced theory of T_μ on the strongly minimal set D_a (after adding the parameter a to the language). Then T_a is strongly minimal, not locally modular and does not interpret an infinite field.*

PROOF. By Corollary 10.3.4 since T_μ is almost strongly minimal over D_a and not 1-based, the strongly minimal set D_a itself cannot be locally modular. Hence the induced theory T_a is strongly minimal and not 1-based. If T_a did interpret an infinite field, then so would T_μ, which it doesn't. ⊣

EXERCISE 10.4.1. Let A and B be finite subgraphs of M and N an arbitrary subgraph. Prove:

$$A \leq M \Rightarrow A \cap N \leq N$$
$$A \leq B \leq M \Rightarrow A \leq M$$
$$A, B \leq M \Rightarrow A \cap B \leq M.$$

Assume that $\delta(B) \geq 0$ for all $B \subseteq M$. Then for each A there is a smallest finite subgraph $\mathrm{cl}(A)$ which contains A and is strong in M.

EXERCISE 10.4.2. Prove the following generalisation of the submodular law:

$$\delta(A_1 \cup \cdots \cup A_k) \leq \sum_{\emptyset \neq \Delta \subseteq \{1,\ldots,k\}} (-1)^{|\Delta|}\, \delta\Big(\bigcap_{i \in \Delta} A_i\Big).$$

EXERCISE 10.4.3. Show that for any $a \in \mathrm{Aut}(M_\mu)$ acts 2-transitively on D_a: for any two pairs of elements $x_1 \neq x_2$ and $y_1 \neq y_2$ in D_a, there is some $g \in \mathrm{Aut}(B_\mu)$ such that $g(x_1) = y_1$ and $g(x_2) = y_2$.

The following exercise shows directly that T_μ is not ω-categorical.

EXERCISE 10.4.4. Let $M = M_\mu$, and let $A = \{x_0, \ldots, x_{12}\}$ be a 12-cycle of pairwise different elements in M. (Such a set exists in every projective plane.) Show that $\mathrm{acl}(A)$ is infinite. (Hint: For any $k \geq 3$, a $2k$-cycle $(x_0, x_1, \ldots, x_{2k-1}, x_{2k} = x_0)$ is a 0-minimal extension of A if every x_i has a unique neighbour in A. Show that such an extension of A is in $\mathcal{K}_\mu$ by noting that for any simple pair (C, D) any element of D has at least 2 neighbours in D.)

Appendix A

SET THEORY

In this appendix we collect some facts from set theory presented from the naive point of view and refer the reader to [31] for more details. In order to talk about classes as well as sets we begin with a brief axiomatic treatment.

A.1. Sets and classes

In modern mathematics, the underlying axioms are mostly taken to be ZFC, i.e., the Zermelo–Fraenkel axioms (ZF) including the axiom of choice (AC) (see e.g., [31, p. 3]). For the monster model we may work in Bernays–Gödel set theory (BG) which is formulated in a two-sorted language, one type of objects being *sets* and the other type of objects being *classes*, with the element-relation defined between sets and sets and between sets and classes only. Since the axioms are less commonly known, we give them here following [31, p. 70]. We use lower case letters as variables for sets and capital letters for classes. BG has the following axioms.

1. (a) Extensionality: Sets containing the same elements are equal.
   (b) Empty set: The empty set exists.
   (c) Pairing: For any sets a and b, $\{a, b\}$ is a set. This means that there is a set which has exactly the elements a and b.
   (d) Union: For every set a, the union $\bigcup a = \{z \mid \exists y\ z \in y \in a\}$ is a set.
   (e) Power Set: For every set a, the power set $\mathfrak{P}(a) = \{y \mid y \subseteq a\}$ is a set.
   (f) Infinity: There is an infinite set. This can be expressed by saying that there is a set which contains the empty set and is closed under the successor operation $x \cup \{x\}$.
2. (a) Class extensionality: Classes containing the same elements are equal.
   (b) Comprehension: If $\varphi(x, y_1, \ldots, y_m, Y_1, \ldots, Y_n)$ is a formula in which only set-variables are quantified, and if $b_1, \ldots, b_m, B_1, \ldots, B_n$

are sets and classes, respectively, then

$$\{x|\varphi(x, b_1, \ldots, b_m, B_1, \ldots, B_n)\}$$

is a class.

(c) Replacement: If a class F is a function, i.e., if for every set b there is a unique set $c = F(b)$ such that $(b, c) = \{\{b\}, \{b, c\}\}$ belongs to F, then for every set a the image $\{F(z)|z \in a\}$ is a set.

3. Regularity: Every nonempty set has an $\in$-minimal element.

For BGC we add:

5. Global Choice: There is a function F such that $F(a) \in a$ for every nonempty set a.

The set-part of a model of BG is a model of ZF. Conversely, a model M of ZF becomes a model of BG by taking the definable subsets of M as classes. This shows that BG is a *conservative extension* of ZF: any set-theoretical statement provable in BG is also provable in ZF. Similarly, BGC is a conservative extension of ZFC, see [20]. For a historical discussion see also [9].

A.2. Ordinals

DEFINITION A.2.1. A *well-ordering* of a class X is a linear ordering of X such that any non-empty subclass of X contains a smallest element or, equivalently, such that X does not contain infinite properly descending chains. If X is a proper class rather than a set, we also ask that for all $x \in X$ the set $\{y|y < x\}$ of predecessors is a set.

The well-ordering of X is equivalent the following *principle of transfinite induction.*

Let $\mathcal{E}$ be a subclass of X. Assume that whenever all elements less than x are in $\mathcal{E}$, then x itself belongs to $\mathcal{E}$. Then $\mathcal{E} = X$.

Well-orderings can be used to define functions by recursion.

THEOREM A.2.2 (Recursion Theorem). *Let G be a function which takes functions defined on proper initial segments of a well-ordered set X as arguments. Then there is a unique function F defined on X satisfying the recursion formula*

$$F(x) = G(F \restriction \{y|y < x\})$$

PROOF. It is easy to see by induction that for all x there is a unique function f_x defined on $\{y|y \leq x\}$ and satisfying the recursion formula for all $x' \leq x$. Put $F = \bigcup_{x \in X} f_x$. ⊣

DEFINITION A.2.3 (v. Neumann). An *ordinal* is a well-ordered set in which every element equals its set of predecessors.

Note that the well-ordering of an ordinal is given by $\in$, so we can identify an ordinal with its set of elements. We denote the class of all ordinals by On. Elements of ordinals are again ordinals, so we have

$$\alpha = \{\beta \in \mathrm{On} | \beta < \alpha\}.$$

PROPOSITION A.2.4.
1. *Every well-ordered set* $(x, <)$ *is isomorphic to a unique ordinal* α.
2. On *is a proper class, well-ordered by* $\in$.

We call $\alpha = \mathrm{otp}(x, <)$ the *order type* of $(x, <)$. For ordinals we write $\alpha < \beta$ for $\alpha \in \beta$

PROOF. 1) Define F on x recursively by $F(y) = \{F(z) | z < y\}$. The image of F is an ordinal which is isomorphic to $(x, <)$ via F. Note that F is the only possible isomorphism between $(x, <)$ and an ordinal.

2) Consider two different ordinals α and β. We have to show that either $\alpha \in \beta$ or $\beta \in \alpha$. If not, $x = \alpha \cap \beta$ would be a proper initial segment of α and β and therefore itself an element of α and β, which is impossible. The class On is proper because otherwise it would itself be an ordinal. $\dashv$

The proof shows also that every well-ordered proper class is isomorphic to On.

For any ordinal α its *successor* is defined as $\alpha \cup \{\alpha\}$: it is the smallest ordinal greater than α. Starting from the smallest ordinal $0 = \emptyset$, its successor is $1 = \{0\}$; then $2 = \{0, 1\}$ and so on, yielding the natural numbers. The order type of the natural numbers is denoted by $\omega = \{0, 1, \ldots\}$; the next ordinal is $\omega + 1 = \{0, 1, \ldots, \omega\}$, et cetera.

By definition, a successor ordinal β contains a maximal element α (so β is the successor of α) and we write $\beta = \alpha + 1$. For natural numbers n, we put

$$\alpha + n = \alpha + \underbrace{1 + \cdots + 1}_{n \text{ times}}.$$

Ordinals greater than 0 which are not successor ordinals are called *limit ordinals*. Whenever $\{\alpha_i | i \in I\}$ is a non-empty set of ordinals without biggest element, $\sup_{i \in I} \alpha_i$ is a limit ordinal. Any ordinal can be uniquely written as

$$\lambda + n,$$

with $\lambda = 0$ or a limit ordinal.

We finish with a quick proof of the Well-ordering Theorem, which like Zorn's Lemma (see below) is equivalent to the Axiom of Choice, see [31, 5.1].

PROPOSITION A.2.5 (Well-ordering Theorem). *Every set has a well-ordering.*

PROOF. Let a be a set. Fix a set b which does not belong to a and define a function $F : \mathrm{On} \to a \cup \{b\}$ by the following recursion:

$$F(\alpha) = \begin{cases} \text{some element in } a \setminus \{F(\beta) | \beta < \alpha\} & \text{if this set is not empty.} \\ b & \text{otherwise.} \end{cases}$$

Then $\gamma = \{\alpha | F(\alpha) \neq b\}$ is an ordinal and F defines a bijection between γ and a. ⊣

Zorn's Lemma states that every partially ordered set in which every ordered subset has an upper bound contains a maximal element. We omit its proof.

A.3. Cardinals

Two sets are said to have the same cardinality if there is a bijection between them. By the well-ordering theorem any set x has the same cardinality as some ordinal. We call the smallest such ordinal the *cardinality* $|x|$ of x.

Ordinals occurring in this way are called *cardinals*. They are characterised by the property that all smaller ordinals have smaller cardinality. All natural numbers and ω are cardinals. The cardinality of a finite set is a natural number, a set of cardinality ω is called *countable*.

PROPOSITION A.3.1. *The class of all cardinals is a closed and unbounded subclass of* On.

PROOF. Being closed in On means that the supremum $\sup_{i \in I} \kappa_i$ of a set of cardinals is again a cardinal. This is easy to check. For the second part assume that there is a largest cardinal κ. Then every ordinal above κ would be the order type of a suitable well-ordering of κ. Since the well-orderings on κ form a set this would imply that On is a set. ⊣

The isomorphism between On and the class of all infinite cardinals is denoted by $\alpha \mapsto \aleph_\alpha$, which can be recursively defined by

$$\aleph_\alpha = \begin{cases} \omega & \text{if } \alpha = 0 \\ \aleph_\beta^+ & \text{if } \alpha = \beta + 1 \\ \sup_{\beta < \alpha} \aleph_\beta & \text{if } \alpha \text{ is a limit ordinal} \end{cases}$$

where κ^+ denotes the smallest cardinal greater than κ, the *successor cardinal* of κ. Positive cardinals which are not successor cardinals are *limit cardinals*.

Sums, products, and powers of cardinals are defined by disjoint union, Cartesian power and sets of maps, respectively. Thus

$$|x| + |y| = |x \cup y| \tag{A.1}$$

$$|x| \cdot |y| = |x \times y| \tag{A.2}$$

$$|x|^{|y|} = |{}^{y}x| \tag{A.3}$$

where we assume in (A.1) that x and y are disjoint. In (A.3), the set ${}^y x$ denotes the set of all functions from y to x.

It is easy to see that these operations satisfy the same rules as the corresponding operations on the natural numbers, e.g., $(\kappa^\lambda)^\mu = \kappa^{\lambda\cdot\mu}$. The following theorem shows that addition and multiplication are actually trivial for infinite cardinals.

THEOREM A.3.2 (Cantor's Theorem). 1. *If κ is infinite, then $\kappa \cdot \kappa = \kappa$.*
2. $2^\kappa > \kappa$. ⊣

PROOF. The proof of Part 2 is a generalisation of the well-known proof that there are uncountably many reals: if $(f_\alpha)_{\alpha<\kappa}$ is a sequence of functions from κ to 2, find a function $g : \kappa \to 2$ such that $g(\alpha) \neq f_\alpha(\alpha)$ for all $\alpha < \kappa$. We will prove Part 1 in Lemma A.3.7 below. ⊣

COROLLARY A.3.3. 1. *If λ is infinite, then $\kappa + \lambda = \max(\kappa, \lambda)$.*
2. *If $\kappa > 0$ and λ are infinite, then $\kappa \cdot \lambda = \max(\kappa, \lambda)$.*
3. *If κ is infinite, then $\kappa^\kappa = 2^\kappa$.*

PROOF. Let $\mu = \max(\kappa, \lambda)$. Then $\mu \leq \kappa + \lambda \leq \mu + \mu \leq 2 \cdot \mu \leq \mu \cdot \mu = \mu$, and if $\kappa > 0$, then $\mu \leq \kappa \cdot \lambda \leq \mu \cdot \mu = \mu$.

Finally,

$$2^\kappa \leq \kappa^\kappa \leq (2^\kappa)^\kappa = 2^{\kappa\cdot\kappa} = 2^\kappa. \qquad ⊣$$

COROLLARY A.3.4. *The set*

$${}^{<\omega}x = \bigcup_{n<\omega} {}^n x$$

of all finite sequences of elements of a non-empty set x has cardinality $\max(|x|, \aleph_0)$.

Note that with this notation ${}^{<\omega}2$ is the set of finite sequences in 0 and 1.

PROOF. Let κ be the cardinality of all finite sequences in x. Clearly, $|x| \leq \kappa$ and $\aleph_0 \leq \kappa$. On the other hand

$$\kappa = \sum_{n\in\mathbb{N}} |x|^n \leq \Big(\sup_{n\in\mathbb{N}} |x|^n\Big) \cdot \aleph_0 = \max(|x|, \aleph_0),$$

because

$$\sup_{n\in\mathbb{N}} |x|^n = \begin{cases} 1, & \text{if } |x| = 1 \\ \aleph_0, & \text{if } 2 \leq |x| \leq \aleph_0 \\ |x|, & \text{if } \aleph_0 \leq |x|. \end{cases} \qquad ⊣$$

The Continuum Hypothesis (CH) states that there is no cardinal strictly between ω and the cardinality of the continuum $\mathbb{R}$, i.e.,

$$\aleph_1 = 2^{\aleph_0}.$$

The Generalised Continuum Hypothesis (GCH) states more generally that

$$\kappa^+ = 2^\kappa \text{ for all infinite } \kappa.$$

Both CH and GCH are independent of ZFC (assuming these axioms are consistent, see e.g., [31] 14.32).

For every cardinal μ the *beth function* is defined as

$$\beth_\alpha(\mu) = \begin{cases} \mu, & \text{if } \alpha = 0, \\ 2^{\beth_\beta(\mu)}, & \text{if } \alpha = \beta + 1, \\ \sup_{\beta<\alpha} \beth_\beta(\mu), & \text{if } \alpha \text{ is a limit ordinal.} \end{cases}$$

For any linear order $(X, <)$ we can easily construct a well-ordered *cofinal subset*, i.e., a subset Y such that for any $x \in X$ there is some $y \in Y$ with $x \leq y$.

DEFINITION A.3.5. The *cofinality* $\mathrm{cf}(X)$ is the smallest order type of a well ordered cofinal subset of X.

It is easy to see that $\mathrm{cf}(X)$ is a *regular* cardinal where an infinite cardinal κ is regular if $\mathrm{cf}(\kappa) = \kappa$. Successor cardinals and ω are regular. The existence of *weakly inaccessible cardinals*, i.e., uncountable regular limit cardinals, cannot be proven in ZFC.

The following is a generalisation of Theorem A.3.2(2), and has a similar proof, see [31, 3.11].

THEOREM A.3.6. *If κ is an infinite cardinal, we have $\kappa^{\mathrm{cf}(\kappa)} > \kappa$.*

We conclude this section with a lemma which implies Theorem A.3.2(1).

LEMMA A.3.7 (The Gödel well-ordering, see [31, 3.5]). *There is a bijection* $\mathrm{On} \to \mathrm{On} \times \mathrm{On}$ *which induces a bijection $\kappa \to \kappa \times \kappa$ for all infinite cardinals κ.*

PROOF. Define

$$(\alpha, \beta) < (\alpha', \beta') \Leftrightarrow (\max(\alpha, \beta), \alpha, \beta) <_{\text{lex}} (\max(\alpha', \beta'), \alpha', \beta')$$

where $<_{\text{lex}}$ is the lexicographical ordering on triples. Since this is a well-ordering, there is a unique order-preserving bijection $\gamma\colon \mathrm{On} \times \mathrm{On} \to \mathrm{On}$. We show by induction that γ maps $\kappa \times \kappa$ to κ for every infinite cardinal κ, which in turn implies $\kappa \cdot \kappa = \kappa$. Since the image of $\kappa \times \kappa$ is an initial segment, it suffices to show that the set $X_{\alpha,\beta}$ of predecessors of (α, β) has smaller cardinality than κ for every $\alpha, \beta < \kappa$. We note first that $X_{\alpha,\beta}$ is contained in $\delta \times \delta$ with $\delta = \max(\alpha, \beta) + 1$. Since κ is infinite, we have that the cardinality of δ is smaller than κ. Hence by induction $|X_{\alpha,\beta}| \leq |\delta| \cdot |\delta| < \kappa$. $\dashv$

Appendix B

FIELDS

B.1. Ordered fields

Let R be an integral domain. A linear $<$ ordering on R is compatible with the ring structure if for all $x, y, z \in R$

$$x < y \rightarrow x + z < y + z$$
$$x < y \wedge 0 < z \rightarrow xz < yz.$$

A field $(K, <)$ together with a compatible ordering is an *ordered field*.

LEMMA B.1.1. *Let R be an integral domain and $<$ a compatible ordering of R. Then the ordering $<$ can be uniquely extended to an ordering of the quotient field of R.*

PROOF. Put $\frac{a}{b} > 0 \Leftrightarrow ab > 0$. ⊣

It is easy to see that in an ordered field sums of squares can never be negative. In particular, $1, 2, \ldots$ are always positive and so the characteristic of an ordered field is 0. A field K in which -1 is not a sum of squares is called *formally real.*

LEMMA B.1.2. *A field has an ordering if and only if it is formally real.*

PROOF. A field with an ordering is formally real by the previous remark. For the converse first notice that $\Sigma\Box$, the set of all sums of squares in K, is a *semi-positive cone,* i.e., a set P such that

$$\Sigma\Box \subseteq P \tag{B.1}$$
$$P + P \subseteq P \tag{B.2}$$
$$P \cdot P \subseteq P \tag{B.3}$$
$$-1 \notin P. \tag{B.4}$$

The first and third condition easily imply

$$x \in P \setminus 0 \Rightarrow \frac{1}{x} \in P.$$

Therefore, condition (B.4) is equivalent to $P \cap (-P) = 0$. It is also easy to see that for all b, the set $P + bP$ has all the properties of a semi-positive cone, except possibly (B.4). Condition (B.4) holds if and only if $b = 0$ or $-b \notin P$. We now choose P as a maximal semi-positive cone. Then

$$x \leq y \Leftrightarrow y - x \in P$$

and we obtain a compatible ordering of K. ⊣

COROLLARY B.1.3 (of the proof). *Let K be a field of characteristic different from 2 and let a be an element of K. There is an ordering of K making a negative if and only if a is not the sum of squares.*

PROOF. If $a \notin \Sigma\square$, then $\Sigma\square - a \cdot \Sigma\square$ is a semi-positive cone. ⊣

DEFINITION B.1.4. An *ordered field* $(R, <)$ is real closed if

a) every positive element is a square,
b) every polynomial of odd degree has a zero.

We call $(R, <)$ a *real closure* of the subfield $(K, <)$ if R is real closed and algebraic over K.

The field of real numbers is real closed. Similarly, the field of real algebraic numbers is a real closure of $\mathbb{Q}$. More generally, any field which is relatively algebraically closed in a real closed field is itself real closed.

THEOREM B.1.5. *Every ordered field $(K, <)$ has a real closure, and this is uniquely determined up to isomorphism over K.* ⊣

PROOF (SKETCH). Existence: Let $K_{\geq 0}$ be the semi-positive cone of $(K, <)$ and let L be a field extension of K. It is easy to see that the ordering of K can be extended to L if and only if

$$P = \{x_1 y_1^2 + \cdots + x_n y_n^2 \mid x_i \in K_{\geq 0},\ y_i \in L\}$$

is a semi-positive cone of L, i.e., if $-1 \notin P$. Therefore we may apply Zorn's Lemma to obtain a maximal algebraic extension R of K with an ordering extending the ordering of K. We claim that R is real closed.

Let r be a positive element of R and assume that r is not a square. Since the ordering cannot be extended to $L = R(\sqrt{r})$, there are $r_i \in R_{\geq 0}$ and $s_i, t_i \in R$ such that

$$-1 = \sum r_i (s_i \sqrt{r} + t_i)^2.$$

Then $-1 = \sum r_i (s_i^2 r + t_i^2)$, contradicting the fact that the right hand side is positive. Thus every element of R is a square (and the ordering of R is unique).

Let $f \in R[X]$ be a polynomial of minimal odd degree n without zero (in R). Clearly, f is irreducible. Let α be a zero of f (in the algebraic closure of R) and put $L = R(\alpha)$. Since L cannot be ordered, -1 is a sum of squares in L. So there are polynomials $g_i \in R[X]$ of degree less than n such that f divides $h = 1 + \sum g_i^2$. The leading coefficients of the g_i^2 are squares in R and so cannot cancel out. Hence the degree of h is even and less than $2n$. But then the polynomial hf^{-1} has odd degree less than n and no zero in R since h does not have a zero. A contradiction.

Uniqueness: Let R and S be real closures of $(K, <)$. It suffices to show that R and S are isomorphic over K as fields. By (the easy characteristic 0 case of) Lemma B.3.13 below it is enough to show that an irreducible polynomial $f \in K[X]$ has a zero in R if and only if it has one in S. This follows from the next lemma which we prove at the end of the section. ⊣

LEMMA B.1.6 (J. J. Sylvester). *Let f be irreducible*[1] *in $K[X]$ and $(R, <)$ a real closed extension of $(K, <)$. The number of zeros of f in R equals the signature of the trace form of the K-algebra $K[X]/(f)$.*

Note that a formally real field may have different orderings leading to non-isomorphic real closures. However, in a real closed field the ordering is uniquely determined by the field structure. Hence it makes sense to say that a field is real closed without specifying its ordering.

The Fundamental Theorem of Algebra holds for arbitrary real closed fields:

THEOREM B.1.7. *Let R be real closed. Then $C = R(\sqrt{-1})$ is algebraically closed.*

PROOF. Notice that all elements of C are squares: one square root of $a + b\sqrt{-1}$ is given by

$$\sqrt{\frac{\sqrt{a^2+b^2}+a}{2}} \pm \sqrt{\frac{\sqrt{a^2+b^2}-a}{2}}\sqrt{-1},$$

where we choose $\pm$ according to the sign of b.

Let F be a finite extension of C. We claim that $F = C$. We may assume that F is a Galois extension of R. Let G be a 2-Sylow subgroup of $\mathrm{Aut}(F/R)$ and L the fixed field of G. Then the degree L/R is odd. The minimal polynomial of a generating element of this extension has the same degree and is irreducible. But since all irreducible polynomials over R of odd degree are linear, we have $L = R$. Therefore $G = \mathrm{Aut}(F/R)$ and hence also $H = \mathrm{Aut}(F/C)$ are 2-groups. Now 2-groups are soluble (even nilpotent). So if H is non-trivial, it has a subgroup of index 2 and thus C has a field extension of degree 2; but this is impossible since every element is a square. Hence $H = 1$ and $F = C$. ⊣

COROLLARY B.1.8. *If R is real closed, the only monic irreducible polynomials are:*

- *Linear polynomials*

$$X - a,$$

  $(a \in R)$.
- *Quadratic polynomials*

$$(X - b)^2 + c,$$

  $(b, c \in R, c > 0)$.

---

[1] It suffices that f is non-constant and all zeros in $\mathrm{acl}(K)$ have multiplicity 1.

PROOF. Since all non-constant polynomials $f \in R[X]$ have a zero in $R(\sqrt{-1})$, all irreducible polynomials must be linear or quadratic. Any monic polynomial of degree 2 is of the form $(X-b)^2+c$. It is reducible if and only if it has a zero x in R if and only if $c \leq 0$ (namely $x = b \pm \sqrt{-c}$). ⊣

Finally we prove *Sylvester's Lemma* (B.1.6). Let K be an ordered field, R a real closure of K and $f \in K[X]$ irreducible. Consider the finite dimensional K-algebra $A = K[X]/(f)$. The *trace* $\mathrm{Tr}_K(a)$ of $a \in A$ is the trace of left multiplication by a, considered as a vector space endomorphism. The trace form is a symmetric bilinear form given by

$$(a,b)_K = \mathrm{Tr}_K(ab).$$

Let $a_1, \ldots, a_n$ be a basis of A diagonalising $(\ ,\ )_K$, i.e., $(a_i, a_j)_K = \lambda_i \delta_{ij}$. The *signature* of such a form is defined as the number of positive λ_i minus the number of negative λ_i. Sylvester's Theorem (from linear algebra) states that the signature is independent of the diagonalising basis. By tensoring A with R, we obtain the R-algebra

$$A_R = A \otimes R \cong R[X]/(f).$$

The basis $a_1, \ldots, a_n$ is also a diagonalising R-basis for the trace form of A_R with the same λ_i and hence the same signature as the trace form of A. We now split f in $R[X]$ into irreducible polynomials $g_1, \ldots, g_m$. Since f does not have multiple zeros, the g_i are pairwise distinct. By the Chinese Remainder Theorem we have

$$R[X]/(f) \cong R[X]/(g_1) \times \cdots \times R[X]/(g_m).$$

This shows that the trace form of $R[X]/(f)$ is the direct sum of the trace forms of the $R[X]/(g_i)$, and hence its signature is the sum of the corresponding signatures. Sylvester's Lemma follows once we show that the signature of the trace form is equal to 1 for a linear polynomial, while the signature is 0 for an irreducible polynomial of degree 2.

If g is linear, then $R[X]/(g) = R$. The trace form $(x,y)_R = xy$ has signature 1. If g is irreducible of degree 2, then $R[X]/(g) = R(\sqrt{-1})$ and $\mathrm{Tr}_R(x + y\sqrt{-1}) = 2x$. The trace form is diagonalised by the basis $1, \sqrt{-1}$. With respect to this basis we have $\lambda_1 = 2$ and $\lambda_2 = -2$ and so its signature is zero.

B.2. Differential fields

In this section, all rings considered have characteristic 0. Let R be a commutative ring and S an R-module. (We mainly consider the case that S is a ring containing R.) An additive map $d : R \to S$ is called a *derivation* if

$$d(rs) = (dr)s + r(ds).$$

For any ring S the *ring of dual numbers* is defined as

$$S[\varepsilon] = \{a + b\varepsilon \mid a, b \in S\},$$

where $\varepsilon^2 = 0$. The following is an easy observation.

LEMMA B.2.1. *Let S be a ring containing R and let $\pi \colon S[\varepsilon] \to S$; $a+b\varepsilon \mapsto a$. Any derivation $d \colon R \to S$ defines a homomorphism*

$$\mathrm{t}_d \colon R \to S[\varepsilon];\ r \mapsto r + (dr)\varepsilon$$

inverting π. Conversely, any such homomorphism arises from a derivation.

LEMMA B.2.2. *Let S be a ring containing the polynomial ring $R' = R[x_1, \ldots, x_n]$ and let $d \colon R \to S$ be a derivation. For any sequence $s_1, \ldots, s_n$ of elements of S there is a unique extension of d to a derivation $d' \colon R' \to S$ taking x_i to s_i for $i = 1, \ldots, n$.*

PROOF. Extend $\mathrm{t}_d \colon R \to S[\varepsilon]$ via $\mathrm{t}_{d'}(x_i) = x_i + s_i\varepsilon$ to a homomorphism $\mathrm{t}_{d'} \colon R' \to S[\varepsilon]$. ⊣

LEMMA B.2.3. *Let R be a subring of the field K and let R' be an intermediate field algebraic over R. Then every derivation $d \colon R \to K$ can be uniquely extended to R'.*

PROOF. It suffices to consider the following two cases:

1. R' is the quotient field of R. If a is a unit, then $a + b\varepsilon$ is a unit in $K[e]$ (with inverse $a^{-1} - ba^{-2}\varepsilon$.) Therefore, t_d takes non-zero elements of R to units in $K[\varepsilon]$ and thus can be uniquely extended to R'.

2. R is a field and $R' = R[a]$ a simple algebraic extension. Let $f(x)$ be the minimal polynomial of a over R. Then t_d can be extended to R' if and only if $(\mathrm{t}_d f)(x)$ has a zero of the form $a + b\varepsilon$ in $K[\varepsilon]$. But $(\mathrm{t}_d f)(x) = f(x) + (f^d\varepsilon)(x)$, hence

$$(\mathrm{t}_d f)(a + b\varepsilon) = f(a + b\varepsilon) + f^d(a)\varepsilon = \left(\frac{\partial f}{\partial x}(a)b + f^d(a)\right)\varepsilon. \tag{B.5}$$

Since $\mathrm{char}(K) = 0$, f is separable, so $\frac{\partial f}{\partial x}(a) \neq 0$ and d can be extended to R' by

$$d'(a) = a - f^d(a)\left(\frac{\partial f}{\partial x}(a)\right)^{-1}.$$

⊣

Let $C = \{c \in K \mid dc = 0\}$ denote the set of *constants* of K. Clearly, C is a subring containing 1. The previous lemma implies that C is a subfield which is relatively algebraically closed in K.

COROLLARY B.2.4. *Let R be a subring of the field K. Any derivation $d \colon R \to K$ can be extended to a derivation of K.* ⊣

REMARK B.2.5. Let (K, d) be a differential field and F a field extension of K, $a, b \in F^n$. Assume that the ideal of all f in $K[x]$ with $f(a) = 0$ is generated by I_0. Then the following are equivalent:

a) There is an extension of d to F with $da = b$.
b) For all $f \in I_0$ we have

$$\frac{\partial f}{\partial x}(a)b + f^d(a) = 0.$$

PROOF. It is clear that a) implies b) because

$$d(f(a)) = \frac{\partial f}{\partial x}(a)b + f^d(a).$$

In order to show the converse we have to extend the homomorphism $t_d : K \to F[\varepsilon]$ to $K[a]$ in such a way that a is mapped to $a + b\varepsilon$. This is possible if and only if $(t_d f)(a + b\varepsilon) = 0$ for all $f \in I_0$. By (B.5) this implies $\frac{\partial f}{\partial x}(a)b + f^d(a) = 0$. ⊣

Let (F, d) be a saturated model of DCF_0 and $V \subseteq F^n$ an irreducible affine variety (see [51, Chapter 1] for basic algebraic geometry). The *torsor* $\mathcal{T}(V) \subseteq F^{2n}$ is defined by the equations

$$f(x) = 0 \quad \text{and} \quad \frac{\partial f}{\partial x}(x)y + f^d(x) = 0$$

for all f in the vanishing ideal of V (see p. 199). Clearly $(a, da) \in \mathcal{T}(V)$ for all $a \in V$.

Remark B.2.5 states that for any small subfield K over which V is defined, and any $(a_0, b_0) \in \mathcal{T}(V)$ such that $a_0 \in V$ is generic over K there is a pair $(a, da) \in \mathcal{T}(V)$ satisfying the same field-type over K as (a, b). This already proves one direction of the following equivalence:

REMARK B.2.6. [41] An algebraically closed differential field (K, d) is a model of DCF_0 if and only if for every irreducible affine variety V defined over K and every regular section[2] $s : V \to \mathcal{T}(V)$ of the projection $\mathcal{T}(V) \to V$ there is some $a \in V(K)$ with $da = s(a)$.

PROOF. We show that an algebraically closed differential field (K, d) with this property is a model of the axioms of DCF_0.

Let f and g be given with f irreducible. As in the proof of Theorem 3.3.22 we find a field extension $F = K(\alpha, \ldots, d^n\alpha)$ in which $\alpha, \ldots, d^{n-1}\alpha$ are algebraically independent over K and $f(\alpha, \ldots, d^n\alpha) = 0$. Let β be the inverse of $g(\alpha, \ldots, d^{n-1}\alpha)\frac{\partial f}{\partial x_n}(\alpha, \ldots, d^n\alpha)$. Note that $K[\alpha, \ldots, d^n\alpha, \beta]$ is closed under d. Putting $c = (\alpha, \ldots, d^n\alpha, \beta)$, let $(c, dc) = s(c)$ for some tuple s of polynomials in $K[x_0, \ldots, x_{n+1}]$. Then s defines a section $V \to \mathcal{T}(V)$ where $V \subseteq K^{n+1}$ is the variety defined by $f = 0$. By assumption there is some $(a, b) \in V(K)$ with $(a, b, da, db) = s(a, b)$. Hence $f(a, \ldots, d^n a) = 0$ and $g(a, \ldots, d^{n-1}a) \neq 0$. ⊣

---

[2]A *section* of a surjection $f : A \to B$ is a map $g : B \to A$ with $fg = \mathrm{id}_B$. A map between affine varieties is called *regular* if it is given by polynomials.

REMARK B.2.7 (Linear differential equations). Let K be a model of DCF_0 and A an $n \times n$-matrix with coefficients from K. The solution set of the system of differential equations

$$dy = Ay \tag{B.6}$$

is an n-dimensional C-vector space.

PROOF. Choose an $n \times n$-matrix Y over an extension of K whose coefficients are algebraically independent over K. Lemma B.2.2 shows that d can be extended to $K(Y)$ by $dY = AY$. Since K is existentially closed, there is a regular matrix over K, which we again denote by Y, such that $dY = AY$. The columns of Y are n linearly independent solutions of (B.6). Such a matrix is called a *fundamental system* of the differential equation. We show that any solution y is a C-linear combination of the columns of Y. Let z be a column over K with $y = Yz$. Then

$$AYz = d(Yz) = (dY)z + Ydz = AYz + Ydz.$$

Hence $Ydz = 0$, so $dz = 0$, i.e., the elements of z are constants. ⊣

REMARK B.2.8. Let K be a differential field and K' an extension of K with fields of constants C and C', respectively. Then K and C' are linearly disjoint over C, i.e., any set of C-linearly independent elements of K remains linearly independent over C' (see Section B.3).

PROOF. Let $a_0, \ldots, a_n$ be in K and linearly dependent over C'. Then the columns of

$$B = \begin{pmatrix} a_0 & \ldots & a_n \\ da_0 & \ldots & da_n \\ \vdots & & \vdots \\ d^n a_0 & \ldots & d^n a_n \end{pmatrix}$$

are linearly dependent. Let $m < n$ be maximal with

$$Y = \begin{pmatrix} a_0 & \ldots & a_m \\ da_0 & \ldots & da_m \\ \vdots & & \vdots \\ d^m a_0 & \ldots & d^m a_m \end{pmatrix}$$

regular. We want to conclude that $a_0, \ldots, a_{m+1}$ is linearly dependent over C. So we may assume $n = m+1$. Then the last row of B is a K-linear combination of the first $m+1$ rows. Thus for some $m \times m$ matrix A all columns of Y and

$$z = \begin{pmatrix} a_n \\ \vdots \\ d^m a_n \end{pmatrix}$$

are solutions of the differential equation $dy = Ay$. The proof of Remark B.2.7 shows that z is a C-linear combination of the columns of Y. ⊣

B.3. Separable and regular field extensions

DEFINITION B.3.1. Two rings R, S contained in a common field extension are said to be *linearly disjoint* over a common subfield k if any set of k-linearly independent elements of R remains linearly independent over S. Note that then R is also linearly disjoint over K from the quotient field of S.

REMARK B.3.2. It is easy to see that this is equivalent to saying that the ring generated by R and S is canonically isomorphic to the tensor product $R \otimes_k S$. So linear disjointness is a symmetric notion.

If S is a *normal* algebraic extension of K, so invariant under automorphisms of K^{alg}/K, then the linear disjointness of R and S over K does not depend on the choice of a common extension of R and S (see Exercise 8.4.4 for a similar phenomenon).

Recall that an algebraic field extension L over K is *separable* if every element of L is a zero of a separable polynomial $f \in K[X]$ and *Galois* if L is normal and separable over K.

LEMMA B.3.3. *If F and L are field extensions of K with L Galois over K, then F and L are linearly disjoint over K if and only if $F \cap L = K$.*

PROOF. One direction is clear (see also the remark below). For the other direction assume $F \cap L = K$. We may assume that L/K is finitely generated. Since L/K is separable, there is a primitive element, say $L = K(a)$. Let f be the minimal polynomial of a over F. Since all roots of f belong to L, f is in $L[X]$ and therefore in $K[X]$. It follows that $[K(a) : K] = [F(a) : F]$ and F and L are linearly disjoint over K. ⊣

That $h_1, \ldots, h_n$ are algebraically independent over L means that the monomials in $h_1, \ldots, h_n$ are linearly independent over L. This observation implies that L and H are algebraically independent over K if they are linearly disjoint over K. The converse holds if L/K is *regular*:

DEFINITION B.3.4. A field extension L over K is *regular* if L and K^{alg} are linearly disjoint over K in some common extension.

LEMMA B.3.5. *If L is a regular extension of K and H/K is algebraically independent from L/K, then L and H are linearly disjoint over K.*

The reader may note that in the special case where $H = K^{\mathrm{alg}}$ this is just the definition of regularity.

PROOF. Let $l_i \in L$, $h_i \in H$, $\sum_{i<n} l_i h_i = 0$, but not all $h_i = 0$. Since L and H are independent over K^{alg}, the type $\mathrm{tp}(L/K^{\mathrm{alg}}H)$ is an heir of $\mathrm{tp}(L/K^{\mathrm{alg}})$

by Corollaries 6.4.5 or 8.5.13. Thus, there is a non-trivial n-tuple $\bar{h}' \in K^{\mathrm{alg}}$ such that $\bar{l} \cdot \bar{h}' = 0$. Since L/K is regular, there is a non-trivial $\bar{h}'' \in K$ such that $\bar{l} \cdot \bar{h}'' = 0$. This proves the claim. $\dashv$

LEMMA B.3.6. *Let K be a field. There is a natural bijection between isomorphism types of field extensions $K(a_1, \ldots, a_n)$ of K with n generators*[3] *and prime ideals P in $K[X_1, \ldots, X_n]$. Regular extensions correspond to* absolutely prime *ideals, i.e., ideals which generate a prime ideal in $K^{\mathrm{alg}}[X_1, \ldots, X_n]$.*

PROOF. We associate with $L = K(a_1, \ldots, a_n)$ the *vanishing ideal* $P = \{f \in K[X_1, \ldots, X_n] \mid f(a_1, \ldots, a_n) = 0\}$. Conversely if P is a prime ideal the quotient field of $K[X_1, \ldots, X_n]/P$ is an extension of K generated by the cosets of $X_1, \ldots, X_n$.

Let I be the ideal generated by P in $K^{\mathrm{alg}}[X_1, \ldots, X_n]$. Then $K[a_1, \ldots, a_n] \otimes_K K^{\mathrm{alg}}$ and $K^{\mathrm{alg}}[X_1, \ldots, X_n]/I$ are isomorphic as K-algebras. Now L/K is regular if and only if $K[a_1, \ldots, a_n] \otimes_K K^{\mathrm{alg}}$ is an integral domain, which means that I is prime. $\dashv$

REMARK B.3.7. It is easy to see that the case $\operatorname{tr.deg}(a_1, \ldots, a_n) = n - 1$ corresponds exactly to the case where P is a principal ideal, generated by an irreducible polynomial $f \in K[X_1, \ldots, X_n]$: this polynomial is uniquely determined up to a constant factor. ([35, VII, Ex.26])

The *perfect hull* of a field K is the smallest perfect field containing it, so in characteristic 0 equals K, and in characteristic p is the union of all $K^{p^{-n}} = \{a \in K^{\mathrm{alg}} \mid a^{p^n} \in K\}$.

DEFINITION B.3.8. A field extension L over K is *separable* if, in some common extension, both L and the perfect hull of K are linearly disjoint over K. This extends the definition in the algebraic case.

Again this does not depend on the choice of the common extension L and the perfect hull of K. Note that regular extensions are separable.

Let K be a field of characteristic $p > 0$. For any subset A we call the field $K^p(A)$ the *p-closure* of A. This defines a pregeometry on K (see Section C.1) whose associated dimension is the *degree of imperfection*. A basis of K in the sense of this pregeometry is called a p-basis. If $b = (b_i \mid i \in I)$ is a p-basis of K, then the products $b^\nu = \prod_{i \in I} b_i^{\nu_i}$ defined for multi-indices $\nu = (\nu_i \mid i \in I)$ where $0 \leq \nu_i < p$ and almost all ν_i equal to zero, form a linear basis of K over K^p. If the degree of imperfection of K is a finite number e, it follows that $[K : K^p] = p^e$.

REMARK B.3.9. A field L is a separable extension of K if and only if L and $K^{p^{-1}}$ are linearly disjoint over K. It follows that L/K is separable if and only if a p-basis of K stays p-independent in L.

---

[3] By quantifier elimination this is just $\mathrm{S}_n(K)$ in the theory of K^{alg}.

PROOF. Consider a sequence (a_i) of elements of L which is linearly independent over K. Then (a_i) is independent over $K^{p^{-1}}$ by assumption which implies that (a_i^p) is independent over K. Then (a_i^p) is independent over $K^{p^{-1}}$ implying that (a_i) is independent over $K^{p^{-2}}$. In this way one can show inductively that (a_i) is independent over $K^{p^{-n}}$ for all n. ⊣

LEMMA B.3.10. *If $b_1, \dots, b_n$ are p-independent in K, they are algebraically independent over $\mathbb{F}_p$.*

PROOF. The $b_1, \dots, b_{n-1}$ form a p-basis of $F = \mathbb{F}_p(b_1, \dots, b_{n-1})$. So K/F is separable by Remark B.3.9. If an element $c \in K$ is algebraic over F, it is separably algebraic over F. So c is also separably algebraic over $K^p F$, which is only possible if c belongs to $K^p F$. Since b_n does not belong to $K^p F$, it follows that b_n is not algebraic over F. ⊣

LEMMA B.3.11. 1. *Let R be an integral domain of characteristic p and let b be a p-basis of R in the sense that every element of R is a unique R^p-linear combination of the b^ν. Then b is also a p-basis of the quotient field of R.*
2. *Let b be p-basis of K and L is a separable algebraic extension of K, then b is also a p-basis of L.*

PROOF. 1): Let K be the quotient field of R. Clearly b is p-independent in K. Since the elements of b are algebraic over K^p, it follows that $K^p[b]$ is a subfield of K which contains $R = R^p[b]$. So $K = K^p[b]$.

2): We may assume that L/K is finite. Since L^p and K are linearly disjoint over K^p, we have $[L^p K : K] = [L^p : K^p] = [L : K]$. It follows that $L^p K = L$. ⊣

LEMMA B.3.12. *The field $L = K(a_1, \dots, a_n)$ is separable over K if and only if there is a transcendence basis $\overline{a}' \subseteq \{a_1, \dots, a_n\}$ of L/K such that L is separably algebraic over $K(\overline{a}')$.*

PROOF. This follows from the fact that every irreducible polynomial in $K[X]$ is of the form $f(X^{p^k})$ for some $k \geq 0$ and some separable polynomial $f \in K[X]$. ⊣

LEMMA B.3.13 (See [16, Théorème 5.11]). *Algebraic field extensions of K are isomorphic over K if and only if the same polynomials in $K[X]$ have a zero in these extensions.*

PROOF. Consider two algebraic extensions L and L' of K. A compactness argument shows that L and L' are isomorphic over K if and only if they contain, up to isomorphism over K, the same finitely-generated subextensions. The condition that the same polynomials in $K[X]$ have zeros in L and L' is equivalent to L and L' having the same simple subextensions. So the lemma is clear if L and L' are separable over K which means that in the general case we may assume that K is infinite.

By symmetry it is enough to show that every finite subextension F/K of L/K has an isomorphic copy in L'/K, or in other words that $F \subseteq \alpha(L')$ for some $\alpha \in \operatorname{Aut}(K^{\mathrm{alg}}/K)$. By our assumption every $a \in F$ is contained in some $\alpha(L')$, so F is the union of all $F_\alpha = \alpha(L') \cap F$. There are only finitely many different F_α. To see this choose a finite normal extension N/K which contains F. Then F_α depends only on the restriction of α to N. Since F and the F_α are vector spaces over the infinite field K, we can apply Neumann's Lemma 3.3.9 and conclude that F equals one of the F_α. ⊣

EXERCISE B.3.1. Show that elementary extensions are regular.

B.4. Pseudo-finite fields and profinite groups

DEFINITION B.4.1. A field K is *pseudo algebraically closed* (PAC) if every absolutely irreducible affine variety defined over K has a K-rational point.

See [51, Chapter 1] for basic algebraic geometry.

LEMMA B.4.2. *Let K be a field. The following are equivalent:*

a) *K is PAC,*
b) *Let $f(X_1, \ldots, X_n, T) \in K[X_1, \ldots, X_n, T]$ be absolutely irreducible and of degree greater than 1 in T and let $g \in K[X_1, \ldots, X_n]$. Then there exists $(a, b) \in K^n \times K$ such that $f(a, b) = 0$ and $g(a) \neq 0$.*
c) *K is existentially closed in every regular field extension.*

PROOF. a) ⇒ b): Let $\mathfrak{C}$ be an algebraically closed field containing K. The zero set of f defines an absolutely irreducible variety $X \subseteq \mathfrak{C}^n \times \mathfrak{C}$ over K. Then the set $V = \{(a, b, c) \mid f(a, b) = 0,\ g(a)c = 1\}$ is absolutely irreducible as well and defined over K.

b) ⇒ c): Let L/K be a finitely generated regular extension. Since regular extensions are separable, by Lemma B.3.12 we find a transcendence basis $\overline{a}$ of L/K such that L is separable algebraic over $K(\overline{a})$. Then there is a b separable algebraic over $K(\overline{a})$ such that $L = K(a_1, \ldots, a_n, b)$. Let now $c_1, \ldots, c_m$ be elements of L satisfying certain equations over K.[4] We need $c_i' \in K$ satisfying the same equations. Write $c_i = \frac{h_i(\overline{a}, b)}{g(\overline{a})}$ for polynomials $h_i(\overline{x}, y)$ and $g(\overline{x})$ over K. By Remark B.3.7 the vanishing ideal of $(\overline{a}, b)$ over K is generated by some $f(\overline{x}, y)$, which is absolutely irreducible since L/K is regular. There are $\overline{a}', b' \in K$ such that $f(\overline{a}', b') = 0$ and $g(\overline{a}') \neq 0$. Hence the $c_i' = \frac{h_i(\overline{a}', b')}{g(\overline{a}')}$ satisfy all equations satisfied by the c_i.

c) ⇒ a): Let V be an absolutely irreducible variety over K. If $\overline{c} \in V$ is an element of $\mathfrak{C}$ generic over K, then $K(\overline{c})/K$ is regular by Lemma B.3.6. ⊣

COROLLARY B.4.3. *"PAC" is an elementary property.*

[4] As we are working in an infinite field, we don't need inequalities.

PROOF. Being absolutely irreducible is a quantifier-free property of the coefficients of $f(X_1,\dots,X_n,T)$ because the theory of algebraically closed fields has quantifier elimination. ⊣

DEFINITION B.4.4. A *profinite group* is a compact[5] topological group with a neighbourhood basis for 1 consisting of subgroups.

Note that by compactness open subgroups are of finite index.

Profinite groups can be presented as inverse limits

$$\varprojlim_{i\in I} A_i$$

of finite groups, with neighbourhood bases for the identity given by the inverse images of the A_i, see [62] for more details.

DEFINITION B.4.5. A profinite group is called *procyclic* if all finite (continuous) quotients are cyclic.

LEMMA B.4.6 (see [62]). *Let G be a profinite group. The following are equivalent*:

a) *G is procyclic*;
b) *G is an inverse limit of finite cyclic groups*;
c) *G is (topologically) generated by a single element a, i.e., the group abstractly generated by a is dense in G.* ⊣

By a generator for a profinite group we always mean a *topological* generator, i.e., an element generating a dense subgroup.

DEFINITION B.4.7. We denote by $\widehat{\mathbb{Z}}$ the inverse limit of all $\mathbb{Z}/n\mathbb{Z}$, with respect to the canonical projections.

LEMMA B.4.8. *A profinite group G is procyclic if and only if for every n there is at most one closed subgroup of index n. If G has exactly one closed subgroup of index n for each $n > 0$, then G is isomorphic to $\widehat{\mathbb{Z}}$.*

PROOF. It suffices to prove the lemma for finite G. So if G has order n, we have to show that G is cyclic if and only if for each k dividing n there is a unique subgroup of order k (hence of index n/k). If a is a generator for G and H a subgroup of G of order k, then $a^{n/k}$ is a generator for H, proving uniqueness. For the other direction, note that by assumption any two elements having the same order generate the same subgroup. Thus, if G has an element of order k, then G contains exactly $\varphi(k)$ such elements. The claim now follows from the equality

$$n = \sum_{k|n} \varphi(k).$$

⊣

[5]Note that for us compact spaces are Hausdorff.

LEMMA B.4.9. *Let $G \to H$ be an epimorphism of procyclic groups. Then any generator of H lifts to a generator of G.*

PROOF. First we consider the case where G is finite. We may then assume that G and H are p-groups. If $H = 0$, we choose any generator as the preimage. If $H \neq 0$, any preimage of a generator of H is a generator of G.

If G is infinite, let h be a generator of H. The finite case implies that for every open normal subgroup A of G we find an inverse image of h which generates G/A. The claim follows now by compactness. ⊣

For a field K we write $\mathrm{G}(K)$ for its *absolute Galois group* $\mathrm{Aut}(K^{\mathrm{alg}}/K)$.

DEFINITION B.4.10. A perfect field K is called *procyclic* if $\mathrm{G}(K)$ is procyclic, and 1-*free* if $\mathrm{G}(K) \cong \widehat{\mathbb{Z}}$.

By Lemma B.4.8 a perfect field is procyclic (1-free, respectively) if and only if it has at most one (exactly one, respectively) extension of degree n in K^{alg} for every n. Thus, being procyclic or 1-free is an elementary property of K.

DEFINITION B.4.11. A perfect 1-free PAC-field is called *pseudo-finite*.

REMARK B.4.12. Being pseudo-finite is an elementary property. A theorem of Lang and Weil (see [36]) on the number of points on varieties over finite fields shows that infinite ultraproducts of finite fields are PAC and hence pseudo-finite. Conversely, any pseudo-finite field is elementarily equivalent to an ultraproduct of finite fields (see [1] and Exercise 7.5.1).

If L/K is regular, then clearly K is relatively algebraically closed in L. If K is perfect, the converse holds as well:

PROPOSITION B.4.13. *Let K be perfect and L/K a field extension. Then the following are equivalent:*

a) *L/K regular;*
b) *K is relatively algebraically closed in L;*
c) *the natural map $\mathrm{G}(L) \to \mathrm{G}(K)$ is surjective.*

PROOF. Using Remark B.3.2 and the functoriality of tensor products, it is easy to see that a) implies c) (even without K being perfect). Clearly, c) implies b). To see that b) implies a) assume that K is relatively algebraically closed in L. Since every finite extension between K and K^{alg} is generated by a single element, it suffices to show for $a \in K^{\mathrm{alg}}$ that the minimal polynomial f of a over L has coefficients in K. But this follows as the coefficients of f can be expressed by conjugates of $a \in K^{\mathrm{alg}} \cap L$ over K. ⊣

If L is perfect and N/L and L/K are regular extensions, then also N/K is regular.

COROLLARY B.4.14. *Let L/K be regular, L procyclic and K 1-free. Then L is 1-free. If N/L is an extension of L such that N/K is regular, then N/L is regular.*

PROOF. If $\mathrm{G}(L)$ is procyclic and $\pi\colon \mathrm{G}(L) \to \widehat{\mathbb{Z}} = \mathrm{G}(K)$ is surjective, then π is an isomorphism. Let $\rho\colon \mathrm{G}(N) \to \mathrm{G}(L)$ be the natural map. If $\rho\pi$ is surjective, then so is ρ. ⊣

Any profinite group G acts faithfully and with finite orbits on the set S consisting of all left cosets of finite index subgroups. That point stabilisers are open subgroups of G just says that the action is continuous with respect to the discrete topology on S. We use this easy observation to prove the following:

LEMMA B.4.15. *Any procyclic field has a regular pseudo-finite field extension.*

PROOF. Let K be a procyclic field. Let S be as in the previous remark for $G = \widehat{\mathbb{Z}}$, let $L' = K(X_s)_{s\in S}$ with the action of G on L' given by the action of G on S and L be the fixed field of G. Then L' is a Galois extension of L with Galois group G and L' (and L) are regular extensions of K by Proposition B.4.13. Let $\sigma \in \mathrm{Aut}(K^{\mathrm{alg}}L'/L)$ be a common extension of a generator of $\mathrm{G}(K)$ and a generator of $\widehat{\mathbb{Z}} = \mathrm{Aut}(L'/L)$. Extend σ to some $\sigma' \in \mathrm{Aut}(L^{\mathrm{alg}}/L)$ and consider the fixed field L'' of σ'. Then $\mathrm{G}(L'')$ is generated by σ' and has $\widehat{\mathbb{Z}}$ as a homomorphic image, so L'' is 1-free. Also L'' is a regular extension of K by Proposition B.4.13. It is now easy to construct, by a long chain, a regular procyclic extension N of L'' which is existentially closed in all regular procyclic extensions. This extension N is again regular over K and 1-free.

We use the criterion given in Lemma B.4.2 to show that N is PAC. Let N' be a regular extension of N. Let $\pi \in \mathrm{G}(N')$ be a lift of a generator of $\mathrm{G}(N)$ and N'' the fixed field of π in N'^{alg}. Then N'' is a regular procyclic extension of N. By construction, N is existentially closed in N'' and hence also in N'. ⊣

Procyclic fields have the amalgamation property for regular extensions:

LEMMA B.4.16. *If L_1, L_2 are regular procyclic extensions of a field K, there is a common procyclic regular extension H of L_1 and L_2 in which L_1 and L_2 are linearly disjoint over K.*

PROOF. Let L_1 and L_2 be regular procyclic extensions of a field K (so K is also procyclic). We may assume that the L_i are algebraically independent over K in some common field extension. Let σ be a (topological) generator of $\mathrm{G}(K)$. By Lemma B.4.9, we can lift σ to generators σ_i of $\mathrm{G}(L_i)$. By Lemma B.3.5, L_1 and L_2 are linearly disjoint. As $L_1^{\mathrm{alg}}L_2^{\mathrm{alg}}$ is the quotient field of the tensor product of L_1^{alg} and L_2^{alg} over K^{alg}, the σ_i generate an automorphism of $L_1^{\mathrm{alg}}L_2^{\mathrm{alg}}$ which can be extended to an automorphism τ of $(L_1L_2)^{\mathrm{alg}}$. The fixed field of τ is procyclic with Galois group generated by τ and a regular extension of L_1 and L_2 by Proposition B.4.13. ⊣

By Lemma B.4.15 this also implies that pseudo-finite fields have the amalgamation property for regular extensions.

Appendix C

COMBINATORICS

C.1. Pregeometries

In this section, we collect the necessary facts and notions about pregeometries, existence of bases and hence a well-defined dimension, modularity laws etc. First recall Definition 5.6.5.

DEFINITION. A *pregeometry* (X, cl) is a set X with a closure operator $\mathrm{cl}\colon \mathfrak{P}(X) \to \mathfrak{P}(X)$ such that for all $A \subseteq X$ and $a, b \in X$

a) (REFLEXIVITY) $A \subseteq \mathrm{cl}(A)$
b) (FINITE CHARACTER) $\mathrm{cl}(A)$ is the union of all $\mathrm{cl}(A')$, where the A' range over all finite subsets of A.
c) (TRANSITIVITY) $\mathrm{cl}(\mathrm{cl}(A)) = \mathrm{cl}(A)$
d) (EXCHANGE) $a \in \mathrm{cl}(Ab) \setminus \mathrm{cl}(A) \Rightarrow b \in \mathrm{cl}(Aa)$

REMARK C.1.1. The following structures are pregeometries:

1. A vector space V with the linear closure operator.
2. For a field K with prime field F, the relative algebraic closure $\mathrm{cl}(A) = F(A)^{\mathrm{alg}}$
3. The p-closure in a field K of characteristic $p > 0$, i.e., $\mathrm{cl}(A) = K^p(A)$.

PROOF. We prove EXCHANGE for algebraic dependence in a field K as an example.[1] We may assume that A is a subfield of K. Let a, b be such that $a \in A(b)^{\mathrm{alg}}$. So there is a non-zero polynomial $F \in A[X, Y]$ with $F(a, b) = 0$. If we also assume that $b \notin A(a)^{\mathrm{alg}}$, it follows that $F(a, Y) = 0$. This implies $a \in A^{\mathrm{alg}}$. ⊣

A pregeometry in which points and the empty set are closed, i.e., in which

$$\mathrm{cl}'(\emptyset) = \emptyset \text{ and } \mathrm{cl}'(x) = \{x\} \text{ for all } x \in X,$$

is called *geometry*. For any pregeometry (X, cl), there is an associated geometry (X', cl') obtained by setting $X' = X^\bullet/\sim$, and $\mathrm{cl}'(A/\sim) = \mathrm{cl}(A)^\bullet/\sim$ where $\sim$ is the equivalence relation on $X^\bullet = X \setminus \mathrm{cl}(\emptyset)$ defined by $\mathrm{cl}(x) = \mathrm{cl}(y)$.

---

[1] For the p-closure see [10, § 13].

Starting from a vector space V, the geometry obtained in this way is the associated projective space $\mathrm{P}(V)$. The important properties of a pregeometry are in fact mostly properties of the associated geometry.

DEFINITION C.1.2. Let (X, cl) be pregeometry. A subset A of X is called

1. *independent* if $a \notin \mathrm{cl}(A \setminus \{a\})$ for all $a \in A$;
2. a *generating set* if $X = \mathrm{cl}(A)$;
3. a *basis* if A is an independent generating set.

LEMMA C.1.3. *Let (X, cl) be a pregeometry with generating set E. Any independent subset of E can be extended to a basis contained in E. In particular every pregeometry has a basis.*

PROOF. Let B be an independent set. If x is any element in $X \setminus \mathrm{cl}(B)$, $B \cup \{x\}$ is again independent. To see this consider an arbitrary $b \in B$. Then $b \notin \mathrm{cl}(B \setminus \{b\})$, whence $b \notin \mathrm{cl}(B \setminus \{b\} \cup \{x\})$ by exchange.

This implies that for a maximal independent subset B of E, we have $E \subseteq \mathrm{cl}(B)$ and therefore $X = \mathrm{cl}(B)$. ⊣

DEFINITION C.1.4. Let (X, cl) be a pregeometry. Any subset S gives rise to two new pregeometries, the *restriction* (S, cl^S) and the *relativisation* (X, cl_S), where

$$\mathrm{cl}^S(A) = \mathrm{cl}(A) \cap S,$$
$$\mathrm{cl}_S(A) = \mathrm{cl}(A \cup S).$$

REMARK C.1.5. Let A be a basis of (S, cl^S) and B a basis of (X, cl_S). Then the (disjoint) union $A \cup B$ is a basis of (X, cl).

PROOF. Clearly $A \cup B$ is a generating set. Since B is independent over S, we have $b \notin \mathrm{cl}_S(B \setminus \{b\}) = \mathrm{cl}(A \cup B \setminus \{b\})$ for all $b \in B$. Consider an $a \in A$. We have to show that $a \notin \mathrm{cl}(A' \cup B)$, where $A' = A \setminus \{a\}$. As $a \notin \mathrm{cl}(A')$, we let B' be a maximal subset of B with $a \notin \mathrm{cl}(A' \cup B')$. If $B' \neq B$ this would imply that $a \in \mathrm{cl}(A' \cup B' \cup \{b\})$ for any $b \in B \setminus B'$ which would in turn imply $b \in \mathrm{cl}(A \cup B')$, a contradiction. ⊣

We say A is a basis *of* S and B a basis *over* or *relative to* S.

LEMMA C.1.6. *All bases of a pregeometry have the same cardinality.*

PROOF. Let A be independent and B a generating subset of X. We show that

$$|A| \leq |B|.$$

Assume first that A is infinite. Then we extend A to a basis A'. Choose for every $b \in B$ a finite subset A_b of A' with $b \in \mathrm{cl}(A_b)$. Since the union of the A_b is a generating set, we have $A' = \bigcup_{b \in B} A_b$. This implies that B is infinite and $|A| \leq |A'| \leq |B|$.

Now assume that A is finite. That $|A| \leq |B|$ follows immediately from the following exchange principle: *Given any* $a \in A \setminus B$ *there is some* $b \in B \setminus A$ *such that* $A' = \{b\} \cup A \setminus \{a\}$ *is independent*. For, since $a \in \mathrm{cl}(B)$, B cannot be contained in $\mathrm{cl}(A \setminus \{a\})$. Choose b in B but not in $\mathrm{cl}(A \setminus \{a\})$. It follows from the exchange property that A' is independent. ⊣

DEFINITION C.1.7. The *dimension* $\dim(X)$ of a pregeometry (X, cl) is the cardinality of a basis. For a subset S of X let $\dim(S)$ be the dimension of (S, cl^S) and $\dim(X/S)$ the dimension of (X, cl_S).

By Remark C.1.5 we have

LEMMA C.1.8. $\quad \dim(X) = \dim(S) + \dim(X/S)$.

The dimensions in our three standard examples are:

- The dimension of a vector space.
- The transcendence degree of a field.
- The *degree of imperfection* of a field of finite characteristic (see [10, § 13, Ex.1]).

Let (X, cl) be a pregeometry. For arbitrary subsets A and B one sees easily that the *submodular law* holds:

$$\dim(A \cup B) + \dim(A \cap B) \leq \dim(A) + \dim(B).$$

One may hope for equality to hold if A and B are closed.

DEFINITION C.1.9. We call a pregeometry (X, cl) *modular*, if

$$\dim(A \cup B) + \dim(A \cap B) = \dim(A) + \dim(B) \tag{C.1}$$

for all cl-closed A and B.

The main examples are

- *trivial* pregeometries where $\mathrm{cl}(A \cup B) = \mathrm{cl}(A) \cup \mathrm{cl}(B)$ for all A, B.
- vector spaces with the linear closure operator.

Let K be a field of transcendence degree at least four over its prime field F. The following argument shows that the pregeometry of algebraic dependence on K is not modular. Choose $x, y, x', y' \in K$ algebraically independent over F. From these elements we can compute $a, b \in K$ such that $ax + b = y$ and $ax' + b = y'$. Since the elements x, x', a, b generate the same subfield as x, y, x', y', they are also algebraically independent. This implies that $F(x)$ and $F(x')$ are isomorphic over $F(a,b)^{\mathrm{cl}}$, where the superscript cl denotes the relative algebraic closure in K. This isomorphism maps y to y' and therefore we have

$$F(x,y)^{\mathrm{cl}} \cap F(a,b)^{\mathrm{cl}} \subseteq F(x,y)^{\mathrm{cl}} \cap F(x',y')^{\mathrm{cl}} = F^{\mathrm{cl}}.$$

So K is not modular since

$$\begin{aligned}\operatorname{tr.deg} F(x,y,a,b) + \operatorname{tr.deg} F &= 3+0\\ &< 2+2 = \operatorname{tr.deg} F(x,y) + \operatorname{tr.deg} F(a,b).\end{aligned}$$

Let us call two sets A and B (geometrically) *independent* over C if all subsets $A_0 \subseteq A$ and $B_0 \subseteq B$ which are both independent over C are disjoint and their union is again independent over C. The following is then easy to see.

LEMMA C.1.10. *For a pregeometry* (X, cl) *the following are equivalent*:

1. (X, cl) *is modular.*
2. *Any two closed* A *and* B *are independent over their intersection.*
3. *For any two closed sets* A *and* B *we have* $\dim(A/B) = \dim(A/A \cap B)$. ⊣

Considering the closed 1- and 2-dimensional subsets of a modular pregeometry (X, cl) as points and lines, respectively, these satisfy the Veblen–Young Axioms of Projective Geometry provided any line contains at least three points: namely, any two distinct points a, b lie on a unique line $\overline{ab}$; and for four distinct points a, b, c, d, if the lines $\overline{ab}$ and $\overline{cd}$ intersect, then so do $\overline{ac}$ and $\overline{bd}$. If the dimension of X is at least 4, then by the fundamental theorem of projective geometry, this is indeed isomorphic to the projective geometry of a vector space over some skew field (see e.g., [13], Thm. 1). The projective planes in Section 10.4 are examples of projective geometries of dimension 3 which do not arise from vector spaces over skew fields. Note that by Exercise C.1.4 a subset A of a modular geometry is closed if and only if for any distinct $a, b \in A$ the line containing them is also contained in A.

LEMMA C.1.11. *A pregeometry* (X, cl) *is modular if and only if for all* a, b, B *with* $\dim(ab) = 2$, $\dim(ab/B) = 1$, *there is* $c \in \mathrm{cl}(B)$ *such that* $\dim(ab/c) = 1$.

PROOF. If (X, cl) is modular and a, b, B are as in the lemma, then ab and B are dependent, but independent over the intersection of $\mathrm{cl}(ab)$ and $\mathrm{cl}(B)$. Let c be an element of the intersection which is not in $\mathrm{cl}(\emptyset)$. Then $\dim(ab/c) = 1$.

Assume that the property of the lemma holds. We show that the third condition of Lemma C.1.10 is satisfied. For this we may assume that $n = \dim(A/A \cap B)$ is finite and proceed by induction on n. The cases $n = 0, 1$ are trivial. So assume $n \geq 2$. Let $a_1, \ldots, a_n$ be a basis of A over $A \cap B$. We have to show that $\dim(a_1, \ldots, a_n/B) = n$. By induction we know that $\dim(a_1 \ldots a_{n-2}/B) = n-2$. So it is enough to show that $\dim(a_{n-1}a_n/B') = 2$ where B' be the closure of $\{a_1, \ldots, a_{n-2}\} \cup B$. If not, by our assumption there is $c \in B'$ such that $\dim(a_{n-1}a_n/c) = 1$. ⊣

DEFINITION C.1.12. A pregeometry (X, cl) is *locally modular* if (C.1) holds for all closed sets A, B with $\dim(A \cap B) > 0$.

REMARK C.1.13. Clearly, (X, cl) is locally modular if and only if for all $x \in X \setminus \mathrm{cl}(\emptyset)$ the relativised pregeometry (X, cl_x) is modular. ⊣

An *affine subspace* of a vector space V is a coset of a subvector space. The affine subspaces of V define a locally modular geometry, which is not modular, because of the existence of parallel lines. Note that in this example, points have dimension 1, lines dimension 2, etc.

The arguments on page 207 show also that a field of transcendence degree at least 5 is not locally modular. Just replace F by a subfield of transcendence degree 1.

EXERCISE C.1.1. Consider a pregeometry (X, cl). Let $A \mathop{\smile\!\!\!\!\!\!|}^{\mathrm{cl}}_C B$ be the relation of A and B being independent over C and $A \mathop{\smile\!\!\!\!\!\!|}^{0}_C B$ the relation $\mathrm{cl}(AC) \cap \mathrm{cl}(BC) = \mathrm{cl}(C)$. Show the following:

1. $\mathop{\smile\!\!\!\!\!\!|}^{\mathrm{cl}}$ has the following properties as listed in Theorem 7.3.13: MONOTONICITY, TRANSITIVITY, SYMMETRY, FINITE CHARACTER and LOCAL CHARACTER.
2. $\mathop{\smile\!\!\!\!\!\!|}^{0}$ has WEAK MONOTONICITY (see Theorem 8.5.10), SYMMETRY, FINITE CHARACTER and LOCAL CHARACTER. MONOTONICITY holds only if $\mathop{\smile\!\!\!\!\!\!|}^{0} = \mathop{\smile\!\!\!\!\!\!|}^{\mathrm{cl}}$, i.e., if X is modular.

EXERCISE C.1.2. Prove that trivial pregeometries are modular.

EXERCISE C.1.3 (P. Kowalski). Let K be a field of characteristic $p > 0$ with degree of imperfection at least 4. Prove that K with p-dependence is not locally modular.

EXERCISE C.1.4. (X, cl) is modular if and only if for all $c \in \mathrm{cl}(A \cup B)$ there are $a \in \mathrm{cl}(A)$ and $b \in \mathrm{cl}(B)$ such that $c \in \mathrm{cl}(a, b)$.

EXERCISE C.1.5. The set of all closed subsets of a pregeometry forms a lattice, where the infimum is intersection and the supremum of X and Y is $X \sqcup Y = \mathrm{cl}(X \cup Y)$. Show that a pregeometry is modular if and only if the lattice of closed sets is modular, i.e., if for all closed A, B, C

$$A \subseteq C \Rightarrow A \sqcup (B \cap C) = (A \sqcup B) \cap C.$$

EXERCISE C.1.6. Let (X, cl) be a pregeometry of uncountable dimension. Suppose that for all closed B of countable dimension the automorphism group $\mathrm{Aut}(X/B)$ acts transitively on $X \setminus B$. Then (X, cl) is locally modular if and only if for all closed B of countable dimension and all a, b with $\dim(a, b) = 2$ and $\dim(a, b/B) = 1$, and for every $\sigma \in \mathrm{Aut}(X/B)$ the two pairs (a, b) and $(\sigma(a), \sigma(b))$ are not independent over $\emptyset$. Conclude that for any finite A, X is locally modular if and only if X_A is locally modular.

C.2. The Erdős–Makkai Theorem

THEOREM C.2.1 (Erdős–Makkai). *Let B be an infinite set and $\mathcal{S}$ a set of subsets of B with $|B| < |\mathcal{S}|$. Then there are sequences $(b_i \mid i < \omega)$ of elements of B and $(S_i \mid i < \omega)$ of elements of $\mathcal{S}$ such that either*

$$b_i \in S_j \Leftrightarrow j < i \tag{C.2}$$

or

$$b_i \in S_j \Leftrightarrow i < j \tag{C.3}$$

for all $i, j \in \omega$.

PROOF. Choose a subset $\mathcal{S}'$ be a subset of $\mathcal{S}$ of the same cardinality as B such that any two finite subsets of B which can be separated by an element of $\mathcal{S}$ can be separated by an element of $\mathcal{S}'$. The hypothesis implies that there must be an element S^* of $\mathcal{S}$ which is not a Boolean[2] combination of elements of $\mathcal{S}'$.

Assume that for some n, three sequences $(b'_i \mid i < n)$ in S^*, $(b''_i \mid i < n)$ in $B \setminus S^*$ and $(S_i \mid i < n)$ in $\mathcal{S}'$ have already been constructed. Since S^* is not a Boolean combination of $S_0, \ldots, S_{n-1}$, there are $b'_n \in S^*$ and $b''_n \in B \setminus S^*$ such that for all $i < n$

$$b'_n \in S_i \;\Leftrightarrow\; b''_n \in S_i.$$

Choose S_n as any set in $\mathcal{S}'$ separating $\{b'_0, \ldots, b'_n\}$ and $\{b''_0, \ldots, b''_n\}$.

Now, an application of Ramsey's theorem shows that we may assume that either $b'_n \in S_i$ or $b'_n \notin S_i$ for all $i < n$. In the first case we set $b_i = b''_i$ and get (C.2), in the second case we set $b_i = b'_{i+1}$ and get (C.3). ⊣

C.3. The Erdős–Rado Theorem

DEFINITION C.3.1. For cardinals κ, λ, μ we write $\kappa \to (\lambda)^n_\mu$ (read as κ *arrows* λ) to express the fact that for any function $f : [\kappa]^n \to \mu$ there is some $A \subseteq \kappa$ with $|A| = \lambda$, such that f is constant on $[A]^n$. In other words, every partition of $[\kappa]^n$ into μ pieces has a homogeneous set of size λ.

With this notation Ramsey's Theorem 5.1.5 states that

$$\omega \to (\omega)^n_k \quad \text{for all } n, k < \omega.$$

In an analogous manner one can define a cardinal κ to be a *Ramsey cardinal* if $\kappa \to (\kappa)^{<\omega}_2$. In other words, if for any n a partition of $[\kappa]^n$ into two classes is given, there is a set of size κ simultaneously homogeneous for all partitions. A Ramsey cardinal κ satisfies $\kappa \to (\kappa)^{<\omega}_\gamma$ for all $\gamma < \kappa$ (see [33] 7.14). More

[2] It actually suffices to consider *positive* Boolean combinations.

generally, an uncountable cardinal κ is called *weakly compact* if it satisfies $\kappa \to (\kappa)^2_2$. Such cardinals are weakly inaccessible and their existence cannot be proven from ZFC.

THEOREM C.3.2 (Erdős–Rado). $\beth^+_n(\mu) \to (\mu^+)^{n+1}_\mu$.

PROOF. This follows from $\mu^+ \to (\mu^+)^1_\mu$ and the following lemma.

LEMMA C.3.3. *If* $\kappa^+ \to (\mu^+)^n_\mu$, *then* $(2^\kappa)^+ \to (\mu^+)^{n+1}_\mu$.

PROOF. We note first that the hypothesis implies $\mu \leq \kappa$. Now let B be a set of cardinality $(2^\kappa)^+$ and $f : [B]^{n+1} \to \mu$ be a colouring. If A is a subset of B, we call a function $p : [A]^n \to \mu$ a *type* over A. If $b \in B \setminus A$, the type $\text{tp}(b/A)$ is the function which maps each n-element subset s of A to $f(s \cup \{b\})$. If $|A| \leq \kappa$, there are at most 2^κ many types over A. Thus an argument as in the proof of Lemma 6.1.2 shows that there is some $B_0 \subseteq B$ of cardinality 2^κ such that, for every $A \subseteq B_0$ of cardinality at most κ, every type over A which is realised in B is already realised in B_0.

Fix an element $b \in B \setminus B_0$. We can easily construct a sequence $(a_\alpha)_{\alpha<\kappa^+}$ in B_0 such that every a_α has the same type over $\{a_\beta \mid \beta < \alpha\}$ as b. By assumption $\{a_\alpha \mid \alpha < \kappa^+\}$ contains a subset A of cardinality μ^+ such that $\text{tp}(b/A)$ is constant on $[A]^n$. Then f is constant on $[A]^{n+1}$. ⊣

⊣

Appendix D

SOLUTIONS TO EXERCISES

Exercises whose results are used in the book have their solutions marked with an asterisk.

Chapter 1. The basics

EXERCISE 1.2.3. We consider formulas which are built using $\neg, \wedge, \exists, \forall$ and we move the quantifiers outside using

$$\begin{aligned}
\neg\exists x\varphi &\sim \forall x\neg\varphi \\
\neg\forall x\varphi &\sim \exists x\neg\varphi \\
(\varphi \wedge \exists x\psi(x,\bar{y})) &\sim \exists z\, (\varphi \wedge \psi(z,\bar{y})) \\
(\varphi \wedge \forall x\psi(x,\bar{y})) &\sim \forall z\, (\varphi \wedge \psi(z,\bar{y})).
\end{aligned}$$

In the last equivalence we replaced the bounded variable x by a variable z which does not occur freely in φ.

EXERCISE 1.2.4. That $\Pi_{i\in I}\mathfrak{A}_i/\mathcal{F}$ is well defined is easy to see. Łos's Theorem is proved by induction on the complexity of φ. The case of atomic formulas is clear by construction. If φ is a conjunction, we use the fact that $X \in \mathcal{F}$ and $Y \in \mathcal{F} \;\Leftrightarrow\; X \cap Y \in \mathcal{F}$. If φ is a negation, we use $X \notin \mathcal{F} \;\Leftrightarrow\; I \setminus X \in \mathcal{F}$. If $X = \{i \in I \mid \mathfrak{A}_i \models \exists y\, \psi(\bar{a}_i, y)\} \in \mathcal{F}$, choose for every $i \in I$ some $b_i \in A_i$ such that $i \in X \Rightarrow \mathfrak{A}_i \models \psi(\bar{a}_i, b_i)$. Then by induction $\Pi_{i\in I}\mathfrak{A}_i/\mathcal{F} \models \psi((\bar{a}_i)_{\mathcal{F}}, (b_i)_{\mathcal{F}})$ and we have $\Pi_{i\in I}\mathfrak{A}_i/\mathcal{F} \models \exists y\, \psi((\bar{a}_i)_{\mathcal{F}}, y)$. The converse is also easy.

EXERCISE 1.3.1. The theory DLO of dense linear orders without endpoints is axiomatised in $\mathrm{L}_{\mathrm{Order}}$ by

- $\forall x\ \neg x < x$
- $\forall x, y, z\ (x < y \wedge y < z \rightarrow x < z)$
- $\forall x, y\ (x < y \vee x \doteq y \vee y < x)$
- $\forall x, z\ (x < z \rightarrow \exists y\, (x < y \wedge y < z))$
- $\forall x \exists y\ x < y$

- $\forall y \exists x\ x < y$.

The class of all algebraically closed fields can be axiomatised by the theory ACF:

- Field (field axioms)
- For all $n > 0$: $\forall x_0 \ldots x_{n-1}\ \exists y\ \ x_0 + x_1 y + \cdots + x_{n-1} y^{n-1} + y^n \doteq 0$.

EXERCISE 1.3.3. Let $\mathfrak{A}$ and $\mathfrak{B}$ be two elementarily equivalent L-structures. It is easy to see that $\mathfrak{A}$ and $\mathfrak{B}$ are isomorphic if L is finite. Let L be arbitrary and assume towards a contradiction that $\mathfrak{A}$ and $\mathfrak{B}$ are not isomorphic. Then for every bijection $f : A \to B$ there is a $Z_f \in L$ which is not respected by f. If L_0 is the set of all the Z_f, then $\mathfrak{A} \restriction L_0$ and $\mathfrak{B} \restriction L_0$ are not isomorphic, contradicting our first observation. One should note that isomorphism follows also from Exercise 6.1.1 and Lemma 4.3.3.

*EXERCISE 1.3.5. Show by induction on the complexity of φ that for all $f \in \mathcal{I}$ and all $\bar{a}$ in the domain of f we have $\mathfrak{A} \models \varphi(\bar{a}) \Leftrightarrow \mathfrak{B} \models \varphi(f(\bar{a}))$.

Chapter 2. Elementary extensions and compactness

EXERCISE 2.1.2. Hint for Part 1: We may assume that $\mathcal{C} = \{\mathfrak{A}_i \mid i \in I\}$ is a set. If $\mathfrak{M}$ is a model of T, choose an ultrafilter $\mathcal{F}$ on I which contains $\mathcal{F}_\varphi = \{i \in I \mid \mathfrak{A}_i \models \varphi\}$ for all $\varphi \in \mathrm{Th}(\mathfrak{M})$.

EXERCISE 2.2.1. Hint: Let T be a finitely satisfiable theory. Consider the set I of all finite subsets of T. For every $\Delta \in I$ choose a model $\mathfrak{A}_\Delta$. Find a suitable ultrafilter $\mathcal{F}$ on I such that $\prod_{\Delta \in I} \mathfrak{A}_\Delta / \mathcal{F}$ is a model of T.

EXERCISE 2.3.1. Part 2. If e is a new element of an elementary extension and if f and g are almost disjoint, then $f(e)$ and $g(e)$ are different.

3. Let $\mathfrak{Q}$ be a proper elementary extension. Show first that $\mathfrak{Q}$ contains a positive infinitesimal element e. Then show that for every $r \in \mathbb{R}$ there is an element q_r such that $P_r(q_r)$ and $Q_r(q_r + e)$ are true in $\mathfrak{Q}$.

EXERCISE 2.3.3. For every prime p, ACF_p has a model which is the union of a chain of finite fields.

Chapter 3. Quantifier elimination

*EXERCISE 3.1.1. We imitate the proof of Lemma 3.1.1. That a) implies b) is clear. For the converse consider an element y_1 of Y_1 and $\mathcal{H}_{y_1}$, the set of all elements of $\mathcal{H}$ containing y_1. Part b) implies that the intersection of the sets in $\mathcal{H}_{y_1}$ is disjoint from Y_2. So a finite intersection h_{y_1} of elements of $\mathcal{H}_{y_1}$ is disjoint from Y_2. The h_{y_i}, $y_1 \in Y_1$, cover Y_1. So Y_1 is contained in the union H of finitely many of the h_{y_i}. Hence H separates Y_1 from Y_2.

EXERCISE 3.2.3. Hint: For any simple existential formula write down the equivalence to a quantifier-free formula. Show that this is equivalent to an $\forall\exists$-sentence and that T is axiomatised by these.

EXERCISE 3.3.1. Like the proof of Theorem 3.3.2 with order-preserving automorphisms replaced by edge-preserving ones.

*EXERCISE 3.3.2. The cases ACF, RCF are easy. So we concentrate on DCF_0. Clearly, the algebraic closure of the differential field generated by A is contained in the model-theoretic algebraic closure. For the converse, let K_0 be an algebraically closed differential field and a_0 an element not in K_0. If $\dim(a_0/K_0)$ is infinite, then a_0 and all its derivatives have the same type over K_0, so a_0 is not not model-theoretically algebraic over K_0. If $\dim(a_0/K_0) = n > 0$, consider the minimal polynomial f of a_0 over K_0. Let K_1 be a d-closed extension of K_0 containing a_0. Then f remains irreducible over K_1 and there is some a_1 whose minimal polynomial over K_1 is f. Now extend K_1 to some field K_2 containing a_1 etc.. In this way we obtain an infinite sequence of distinct elements having the same type over K_0 as a_0, showing that a_0 is not algebraic over K_0 in the sense of model theory.

EXERCISE 3.3.3. Let K be algebraically closed, $X = \{a \mid \mathcal{K} \models \gamma(a, b)\}$ a definable subset of K^n and suppose that $f : X \to X$ is given by an n-tuple of polynomials $f(x, b)$. We may assume that $\gamma(x, z)$ is quantifier-free. We want to show that K satisfies

$$\begin{aligned} &\forall y \Big(\forall x (\varphi(x, y) \to \varphi(f(x, y), y)) \land \\ &\quad \forall x, x' (\varphi(x, y) \land \varphi(x', y) \land f(x, y) \doteq f(x', y) \to x \doteq x') \to \\ &\qquad\qquad \forall x' (\varphi(x', y) \to \exists x (\varphi(x, y) \land f(x, y) \doteq x')) \Big). \end{aligned}$$

This is obviously true in finite fields (even in all finite L_{Ring}-structures) and logically equivalent to an $\forall\exists$-sentence. So the claim follows from Exercise 2.3.3.

For the second part use Exercise 6.1.14 and proceed as before.

Chapter 4. Countable models

*EXERCISE 4.2.1. The set $[\varphi]$ is a singleton if and only if $[\varphi]$ is non-empty and cannot be divided into two non-empty clopen subsets $[\varphi \land \psi]$ and $[\varphi \land \neg\psi]$. This means that for all ψ either ψ or $\neg\psi$ follows from φ modulo T. So $[\varphi]$ is a singleton if and only if φ *generates* the type

$$\langle\varphi\rangle = \{\psi(\overline{x}) \mid T \vdash \forall\overline{x}\,(\varphi(\overline{x}) \to \psi(\overline{x}))\},$$

which of course must be the only element of $[\varphi]$.

This shows that $[\varphi] = \{p\}$ implies that φ isolates p. If, conversely, φ isolates p, this means that $\langle\varphi\rangle$ is consistent with T and contains p. Since p is a type, we have $p = \langle\varphi\rangle$.

EXERCISE 4.2.2. a): The sets $[\varphi]$ are a basis for the closed subsets of $\mathrm{S}_n(T)$. So the closed sets of $\mathrm{S}_n(T)$ are exactly the intersections $\bigcap_{\varphi\in\Sigma}[\varphi] = \{p \in \mathrm{S}_n(T) \mid \Sigma \subseteq p\}$.

b): The set X is the union of a sequence of countable nowhere dense sets X_i. We may assume that the X_i are closed, i.e., of the form $\{p \in \mathrm{S}_n(T) \mid \Sigma_i \subseteq p\}$. That X_i has no interior means that Σ_i is not isolated. The claim follows now from Corollary 4.1.3.

EXERCISE 4.2.3. Let $X = \{\mathrm{tp}(a_0, a_2, \dots) \mid \text{the } a_i \text{ enumerate a model of } T\}$. Consider for every formula $\varphi(\bar{v}, y)$ the set $X_\varphi = \{p \in \mathrm{S}_\omega \mid (\exists y\varphi(\bar{v}, y) \to \varphi(\bar{v}, v_i)) \in p \text{ for some } i\}$. The X_φ are open and dense and X is the intersection of the X_φ.

*EXERCISE 4.2.5. The homeomorphism from $\mathrm{S}_m(aB)$ to the fibre above $\mathrm{tp}(a/B)$ is given by $\mathrm{tp}(c/aB) \mapsto \mathrm{tp}(ca/B)$.

*EXERCISE 4.3.9. Assume that $\mathfrak{A}$ is κ-saturated, B a subset of A of smaller cardinality than κ and $p(x, \bar{y})$ a $(n+1)$-type over B. Let $\bar{b} \in A$ be a realisation of $q(\bar{y}) = p \restriction \bar{y}$ and $a \in A$ a realisation of $p(x, \bar{b})$. Then $(a, \bar{b})$ realises p.

*EXERCISE 4.3.13. If $\mathfrak{B}$ is ω-saturated and elementarily equivalent to $\mathfrak{A}$, then the set of all isomorphisms between finitely-generated substructures that are elementary partial maps in the sense of $\mathfrak{A}$ and $\mathfrak{B}$ is non-empty, and has the back-and-forth property.

Now assume that $\mathfrak{A}$ and $\mathfrak{B}$ are partially isomorphic via $\mathcal{I}$; they are elementarily equivalent by Exercise 1.3.5. Consider a finite subset B_0 of $\mathfrak{B}$ and a type $p \in \mathrm{S}(B_0)$. There is an $f \in \mathcal{I}$ which contains B_0 in its image. Choose a realisation a of $f^{-1}(p)$ in $\mathfrak{A}$ and an extension $g \in \mathcal{I}$ of f which is defined on a. Then $g(a)$ realises p.

*EXERCISE 4.4.1. If $\mathfrak{M}$ and $\mathfrak{M}'$ are $\mathcal{K}$-saturated, consider the set $\mathcal{I}$ of all isomorphisms between finitely-generated substructures of $\mathfrak{M}$ and $\mathfrak{M}'$.

EXERCISE 4.5.1. Let n be such that $\mathrm{S}_n(T)$ uncountable. Prove that there is a consistent formula φ such that both $[\varphi]$ and $[\neg\varphi]$ are uncountable. Inductively we obtain a binary tree of consistent formulas; see proof of Theorem 5.2.6(2).

Chapter 5. $\aleph_1$-categorical theories

EXERCISE 5.1.2. To ease notation we replace the partition by a function $\gamma : [A]^n \to \{1, \dots, k\}$. Fix a non-principal ultrafilter $\mathcal{U}$ on A. For each $s \in [A]^{n-1}$ choose $c(s)$ such that $\{a \in A \mid \gamma(s \cup \{a\}) = c(s)\}$ belongs to $\mathcal{U}$. Construct a sequence $a_0, a_1, \dots$ of distinct elements such that $\gamma(s \cup \{a_n\}) = c(s)$ for all $s \in [\{a_0, \dots, a_{n-1}\}]^{n-1}$. Apply induction to c restricted to $[\{a_0, a_1, \dots\}]^{n-1}$.

EXERCISE 5.2.1. Let T define a linear ordering of the universe. By Exercise 8.2.8 there is a linear ordering J of bigger cardinality than κ which has a dense subset of cardinality κ. A compactness argument shows that T has a model with a subset B which is order-isomorphic to J. Let A be a dense subset of B of cardinality κ. Then all elements of B have different types over A and so $|\,\mathrm{S}(A)| \geq |B| > \kappa$.

EXERCISE 5.2.3. As an example consider formulas $\varphi^0(x), \varphi^1(x), \dots$ without parameters such that $\varphi^n(x)$ implies $\varphi^{n+1}(x)$. We have to show that $\Phi = \{\varphi^n \mid n < \omega\}$ is realised in $\mathfrak{M} = \prod_{i<\omega} \mathfrak{A}_i/\mathcal{F}$ if each φ^n is realised in $\mathfrak{M}$. For each n there is a $B^n \in \mathcal{F}$ such that, for all $i \in B^n$, φ^n is realised in $\mathfrak{A}_i$. We may assume that the B^n are descending and have empty intersection since $\mathbb{F}$ is non-principal. Now for every $i \in B^0$ let n be maximal with $i \in B^n$. Choose an element a_i which realises φ^n in $\mathfrak{A}_i$. For i outside B^0 chose $a_i \in \mathfrak{A}_i$ arbitrary. Then the class of $(a_i)_{i<\omega}$ realises Φ in $\mathfrak{M}$.

*EXERCISE 5.2.5. If T is totally transcendental, each reduct is also totally transcendental. The converse follows from the observation that a binary tree contains only countably many formulas.

*EXERCISE 5.2.6. If T is κ-stable, then $|\,\mathrm{S}_n(\emptyset)| \leq \kappa$. Choose for any two n-types over the empty set a separating formula. Then any formula is logically equivalent to a finite Boolean combination of these κ-many formulas.

EXERCISE 5.3.2. a) $\Rightarrow$ b): Let A be a countable subset of the model $\mathfrak{M}$. A prime extension of A is just a prime model of $T_A = \mathrm{Th}(\mathfrak{M}_A)$. So the claim follows from Theorem 4.5.7.

b) $\Rightarrow$ c): Assume that A is contained in the model $\mathfrak{M}$ and that the isolated types are not dense over A. So there is a consistent $L(A)$-formula φ, which does not contain a complete $L(A)$-formula. Add a predicate P for the set A and consider the $L \cup \{P\}$-structure $(\mathfrak{M}, A)$. Choose a countable elementary substructure $(\mathfrak{M}_0, A_0)$ which contains the parameters of φ. Then φ does not contain a complete $L(A_0)$-formula.

c) $\Rightarrow$ a): Like the proof of 5.3.3.

EXERCISE 5.5.4. Clearly, $T(q)$ is complete λ-stable if and only if T is. It is also clear that if $T(q)$ has a Vaughtian pair then so does T. For the converse use a construction as in Theorem 5.5.2 to find a Vaughtian pair $\mathfrak{M} \prec \mathfrak{N}$ such that q is realised in $\mathfrak{N}$.

EXERCISE 5.5.7. Hint: Use Exercise 5.5.6.

EXERCISE 5.6.1. If $p \in \mathrm{S}(A)$ is not algebraic, then all n-types

$$q(x_1, \dots, x_n) = \{x_i, i = 1, \dots, n, \text{ satisfies } p \text{ and the } x_i \text{ are pairwise distinct}\}$$

are consistent and hence realised in $\mathfrak{M}$.

*EXERCISE 5.6.2. If $c_i \in \operatorname{acl}(Ac_n)$ for some $i < n$, let $a_0 \dots a_n$ realise $\operatorname{tp}(c_0 \dots c_n/A)$ with $a_0 \dots a_{n-1} \notin \operatorname{acl}(AB)$. Then $a_n \notin B$. If $c_i \notin \operatorname{acl}(Ac_n)$ for all $i < n$, realise $\operatorname{tp}(c_n/A)$ by some $a_n \notin B$ and then $\operatorname{tp}(c_0 \dots c_{n-1}/Aa_n)$ outside B.

EXERCISE 5.7.2. Prove that the theory eliminates quantifiers.

Chapter 6. Morley rank

*EXERCISE 6.1.2. Fix any model M which contains A. If b is not algebraic over A, then b has a conjugate over A which does not belong to M. This implies that M has a conjugate M' over A which does not contain b.

Note that Exercise 5.6.2 implies that if C is any set without elements algebraic over A, there is a conjugate M' of M which is disjoint from C.

EXERCISE 6.1.3. Choose special models $\mathfrak{A}_i$ of T_i of the same cardinality and observe that a reduct of a special model is again special.

EXERCISE 6.1.4. If θ does not exist, the set T' of all L-sentences θ such that $\vdash \varphi_1 \to \theta$ or $\vdash \neg\varphi_2 \to \theta$ is consistent. Choose a complete L-theory T which contains T' and apply Exercise 6.1.3 to $T_1 = \{\varphi_1\} \cup T$ and $T_2 = \{\neg\varphi_2\} \cup T'$.

EXERCISE 6.1.5. We use the criterion of Exercise 2.1.2(2). Let C be the class of all reducts of models of T' to L. It is easy to see that C is closed under ultraproducts. If $\mathfrak{A}$ belongs to C and $\mathfrak{B}$ is elementarily equivalent to $\mathfrak{A}$, consider an expansion $\mathfrak{A}'$ of $\mathfrak{A}$ to a model of T'. Now choose two special elementary extensions $\mathfrak{A}' \prec \mathfrak{D}'$, $\mathfrak{B} \prec \mathfrak{E}$ of the same cardinality. Then $\mathfrak{D} \cong \mathfrak{E}$ belongs to C.

EXERCISE 6.1.6. Let $|A| < \kappa$ and $p = p(x_i)_{i<\kappa}$ be a κ-type over p. Denote by p_α the restriction of p to the variables $(x_i)_{i<\alpha}$. Construct a realisation $(a_i)_{i<\kappa}$ of p inductively: if $(a_i)_{i<\alpha}$ realises p_α, choose a_α as a realisation of $p_{\alpha+1}((a_i)_{i<\alpha}, x_\alpha)$.

EXERCISE 6.1.8. 1) Construct an elementary chain $(\mathfrak{M}_\alpha)_{\alpha<\kappa}$ of structures of cardinality $2^{<\kappa}$ such that all types over subsets of $\mathfrak{M}_\alpha$ of cardinality less than κ are realised in $\mathfrak{M}_{\alpha+1}$. This is possible because for regular κ a set of cardinality $2^{<\kappa}$ has at most $2^{<\kappa}$-many subsets of cardinality less than κ.

Part 2) follows from 1) since $2^{<\kappa} = \kappa$.

*EXERCISE 6.1.11. If $B \subseteq \operatorname{dcl}(A)$, every formula with parameters in B is equivalent to a formula with parameters in A. So every type over A axiomatises a type over B. This proves 1) $\Rightarrow$ 2). For the converse show that $b \in \operatorname{dcl}(A)$ if $\operatorname{tp}(b/A)$ has a unique extension to Ab.

*EXERCISE 6.1.12. If $\varphi(a,b)$ is a formula witnessing $b \in \operatorname{dcl}(a)$, let $\mathbb{D}$ equal the class of elements x for which there is a unique y with $\varphi(x,y)$ and let $f\colon \mathbb{D} \to \mathbb{E}$ denote the corresponding map, which we may assume to be surjective. If furthermore $a \in \operatorname{dcl}(b)$ is witnessed by $\psi(y,x)$ and a function $g\colon \mathbb{E}_1 \to \mathbb{D}_1$, we get a 0-definable bijection $\{(x,y) \in \mathbb{E} \times \mathbb{D}_1 \mid f(x) = y \text{ and } g(y) = x\}$.

EXERCISE 6.1.13. Use Exercise 3.3.2.

EXERCISE 6.1.14. Use Exercise 6.1.13 and compactness.

EXERCISE 6.1.16. By induction on n. We distinguish two cases. First assume that for some $i < 0$, $H_n \cap H_i$ has finite index in H_n. We can then cover every coset of H_n by finitely many cosets of H_i. Since G is not a finite union of cosets of $H_0,\ldots,H_{n-1}$, we are done. Now assume that all $H_i' = H_n \cap H_i$ have infinite index in H_n. Assume towards a contradiction that G is a finite union of cosets of the $H_0,\ldots,H_n$. Since H_n has infinite index in G, there must be a coset of H_n which is covered by a finite number of cosets of $H_0,\ldots,H_{n-1}$. This implies that H_n is a finite union of cosets of the H_i', $i < n$, which is impossible by induction.

*EXERCISE 6.1.17. To see that Exercise 6.1.16 implies Exercise 6.1.15 consider the subgroups $H_i := G_{c_i}, i \leq n$, each of infinite index in G. The finitely many cosets $a_j H_i$ with $a_j(c_i) \in B, i \leq n$, do not cover G, so there is some $g \in G$ such that for all $i \leq n$ we have $g(c_i) \notin B$.

For the converse, let $H_1,\ldots,H_n \leq G$ be subgroups of infinite index, and consider the action of G on the disjoint union of the G/H_i by left translation. By Exercise 6.1.15, for any $a_1,\ldots,a_n \in G$ there is some $g \in G$ such that for all $i \leq n$ we have $g(1 \cdot H_i) \notin a_i H_i$, proving Exercise 6.1.16.

*EXERCISE 6.2.1. Let $\varphi(x,a)$ be defined from the parameter tuple $a \in M$. There is an infinite family $(\varphi_i(x,b_i))$ of pairwise inconsistent formulas of Morley rank $\geq \alpha$ which imply $\varphi(x,a)$. Since M is ω–saturated, there is a sequence (a_i) in M such that $\operatorname{tp}(a,a_0,a_1,\ldots) = \operatorname{tp}(a,b_0,b_1,\ldots)$.

EXERCISE 6.2.2. Let I be the set of all i for which $\varphi \wedge \psi_i$ has rank α. The hypothesis implies that all k-element subsets of I contain two indices i,j such that $\varphi \wedge \psi_i \not\sim_\alpha \varphi \wedge \psi_j$. So $|I| \leq (k-1)\operatorname{MD}\varphi$.

EXERCISE 6.2.3. Let G^0 be the intersection of all definable subgroups of finite index; it is definable by Remark 6.2.8. If N is a finite subgroup which is normalised by G, the centraliser of N in G is a definable group of finite index in G^0.

EXERCISE 6.2.4. Let M be an ω-saturated model and let p be a type over M of Morley rank α and degree n, witnessed by $\varphi(x,\overline{m}) \in p$. If $n > 1$, there is a formula $\psi(x,\overline{b})$ such that $\varphi(x,\overline{m}) \wedge \psi(x,\overline{b})$ and $\varphi(x,\overline{m}) \wedge \neg\psi(x,\overline{b})$ both have Morley rank α. Choose $\overline{a} \in M$ with $\mathrm{tp}(\overline{a}/\overline{m}) = \mathrm{tp}(\overline{b}/\overline{m})$. Then both formulas $\varphi(x,\overline{m}) \wedge \psi(x,\overline{a})$ and $\varphi(x,\overline{m}) \wedge \neg\psi(x,\overline{a})$ have rank α and degree less than n and one of these formulas belongs to p, contradicting the choice of $\varphi(x,\overline{m})$.

EXERCISE 6.2.5. Assume that there is a formula $\varphi(x,\overline{b})$ of rank $\geq |T|^+$. Construct a binary tree of formulas $(\varphi_s(x,\overline{y}_s) \mid s \in {}^{<\omega}2)$ below $\varphi(x,\overline{b})$ so that for all k and all $\alpha < |T|^+$ there are parameters $\overline{a}_s$ such that $\mathrm{MR}\,\varphi_s(x,\overline{a}_s) \geq \alpha$ for all s with $|s| = k$. Conclude that $\mathrm{MR}\,\varphi(x,\overline{b}) = \infty$,

*EXERCISE 6.2.8. Let a be in $\mathrm{acl}(A)$ and $a_1,\ldots,a_n$ the conjugates of a over A. Then $\varphi(x,a)$ and $\varphi(x,a_1) \wedge \cdots \wedge \varphi(x,a_n)$ have the same Morley rank.

EXERCISE 6.4.1. In a pregeometry a finite set A is independent from B over C if and only if $\dim(A/BC) = \dim(A/C)$. Now use Theorem 6.4.2.

*EXERCISE 6.4.2. Assume that abC is independent from B and apply Proposition 6.4.9.

EXERCISE 6.4.4. The first two inequalities follow easily from Lemma 6.4.1. For the third inequality we may assume that $\mathbb{A}$ has Morley degree 1. We distinguish two cases:

a) $\beta_{\mathrm{gen}} > 0$. Let $\mathbb{D}_i$ be an infinite family of disjoint definable subclasses of $\mathbb{B}$ defined over $C' \supset C$. Choose $a \in \mathbb{A}$ which has rank α over C'. For some i the rank of $f^{-1}(a) \cap \mathbb{D}_i$ is bounded by some $\beta'_{\mathrm{gen}} < \beta_{\mathrm{gen}}$. By induction we have $\mathrm{MR}(\mathbb{D}_i) \leq \beta \cdot \alpha + \beta'_{\mathrm{gen}} < \beta \cdot \alpha + \beta_{\mathrm{gen}}$.

b) $\beta_{\mathrm{gen}} = 0$. Then $\mathbb{A}$ contains a definable subclass $\mathbb{A}'$ of rank α over which all fibres are finite. By the second inequality we then have $\mathrm{MR}(f^{-1}(\mathbb{A}')) \leq \alpha \leq \beta \cdot \alpha$. If $\mathbb{A}' \neq \mathbb{A}$ we have $\mathrm{MR}(\mathbb{A} \setminus \mathbb{A}') = \alpha' < \alpha$ and by induction $\mathrm{MR}(f^{-1}(\mathbb{A} \setminus \mathbb{A}')) \leq \beta \cdot \alpha' + \beta \leq \beta \cdot \alpha$.

EXERCISE 6.4.5. The language L contains a binary relation symbol E and unary predicate P_n^i for all n and $i \leq n$, and T says that E is an equivalence relation with infinite classes, the P_n^i are infinite and disjoint and for each n the union of $P_n^0 \cup \c{d}ots \cup P_n^n$ is an E-equivalence class.

Chapter 7. Simple theories

EXERCISE 7.1.3. If p forks over A, there is some $\varphi(x, m) \in p$ which implies a conjunction $\bigvee_{\ell<d} \varphi_\ell(x, b)$ of formulas which fork over A. Choose a tuple b' in M which realises the type of b over Am. The formulas $\varphi_\ell(x, b')$ fork over A and one of them belongs to p.

*EXERCISE 7.1.4. Let q be A-invariant. Use Lemma 7.1.4 to show that q does not divide over A: if $\pi(x, b)$ belongs to q, then all the $\pi(x, b_i)$ also belong to q. That q does not fork over A follows from Exercise 7.1.3.

*EXERCISE 7.1.5. p contains a formula φ which divides over A. So there are κ many $\alpha_i \in \operatorname{Aut}(M/A)$ such that the system of all $\alpha_i(\varphi)$ is k-inconsistent. This implies that κ many of the $\alpha_i(p)$ must be distinct.

EXERCISE 7.1.6. The type $p(x)$ forks over the empty set since it implies the disjunction of $\mathrm{cyc}(0, x, 3)$ and $\mathrm{cyc}(2, x, 1)$.

*EXERCISE 7.1.7. This is just a variant of Proposition 7.1.6. Let $\mathcal{I} = (b_i \mid i < \omega)$ be a sequence of A-indiscernibles containing b such that $(\varphi(xb_i) \mid i < \omega)$ is k-inconsistent. Since $\mathrm{tp}(a/Ab)$ does not divide over A, we can assume that $\mathcal{I}$ is indiscernible over Aa.

EXERCISE 7.2.1. Let $(p_\alpha)_{\alpha\in|T|^+}$ be a chain of types with $p_\alpha \in S(A_\alpha)$. Their union $\bigcup_{\alpha\in|T|^+} p_\alpha$ does not fork over a subset A' of $\bigcup_{\alpha\in|T|^+} A_\alpha$ of cardinality at most $|T|$ by Proposition 7.2.5 (Local Character) and Proposition 7.2.15. Since $A' \subseteq A_\alpha$ for some sufficiently large $\alpha \in |T|^+$, from that value of the index onwards the chain no longer forks. (Note that this property is just a reformulation of Local Character for $\kappa = |T|^+$.)

For the last sentence note that otherwise the types $p_\alpha = \mathrm{tp}(c/A\{b_\beta \mid \beta < \alpha\})$ would contradict the first part.

*EXERCISE 7.2.2. Let $q = \mathrm{tp}(b/B)$ and $r = \mathrm{tp}(c/C)$. Find an A-automorphism which maps c to b and C to C' such that $B \mathop{\smile\!\!\!\!\!\!\,|}_{Ab} C'$. Then $r' = \mathrm{tp}(b/C')$ and $s = \mathrm{tp}(b/BC')$ are as required: we have $B \mathop{\smile\!\!\!\!\!\!\,|}_{A} b$, which together with $B \mathop{\smile\!\!\!\!\!\!\,|}_{Ab} C'$ yields $B \mathop{\smile\!\!\!\!\!\!\,|}_{A} C'b$ from which $b \mathop{\smile\!\!\!\!\!\!\,|}_{C'} B$.

EXERCISE 7.2.3. By monotonicity and symmetry it suffices to show that a_X and $a_{Y\setminus X}$ are independent over A. So we can assume that X and Y are disjoint and, by finite character, that X and Y are finite. We proceed by induction on $|X \cup Y|$. Let $z \in X \cup Y$ be maximal. By symmetry we may assume that $z \in Y$. Then a_z is independent from $a_{X\cup Y\setminus\{z\}}$. The claim now follows from the induction hypothesis and transitivity.

*EXERCISE 7.2.4. This follows from the fact that if $(b_i \mid i < \omega)$ is indiscernible over A, there is an A-conjugate B' of B such that $(b_i \mid i < \omega)$ is indiscernible over B'.

EXERCISE 7.2.5. Up to symmetry this is only a reformulation of Monotonicity and Transitivity.

EXERCISE 7.2.6. For ease of notation we restrict to the following special case: *If $b_1 b_2$ is independent from C, we have*

$$b_1 \mathop{\smile\!\!\!\!\!\!|}\, b_2 \Longleftrightarrow b_1 \mathop{\smile\!\!\!\!\!\!|}_{C} b_2$$

since by Corollary 7.2.18 both sides of this equivalence are equivalent to the triple (b_1, b_2, C) being independent. (For the direction from right to left it suffices in fact to assume that b_1 and b_2 are independent from C individually.)

*EXERCISE 7.2.7. The hypothesis implies $a \mathop{\smile\!\!\!\!\!\!|}_{AB} C$. Now the claim follows from Remark 7.1.3.

EXERCISE 7.3.1. Let $\mathcal{I}$ be the sequence $b = b_0, b_1, \ldots$. Use Proposition 7.3.6 and induction on n to show that $\bigcup\{\pi(x, b_i) \mid i < 2^n\}$ does not fork over A. So we find a realisation of $\bigcup\{\pi(x, b_i) \mid i < \omega\}$ which is independent from $\mathcal{I}$ over A. That one can choose c in such a way that $\mathcal{I}$ is indiscernible over Ac follows from Lemma 5.1.3 and FINITE CHARACTER.

EXERCISE 7.3.2. 1. Let $(c_i \mid i < \omega)$ be an antichain for $\theta_1 \wedge \theta_2$. Then by Ramsey's theorem there is an infinite $A \subseteq \omega$ such that $(c_i \mid i \in A)$ is an antichain for θ_1 or for θ_2.

2. If $\theta^{\sim}(x, y)$ is not thick, it has, by compactness, antichains $(c_i \mid i \in I)$ indexed by arbitrary linear orders I. If $I^{\sim}$ is the inverse order, $(c_i \mid i \in I^{\sim})$ is an antichain for θ.

3. Consider $\theta \wedge \theta^{\sim}$.

EXERCISE 7.3.3. Choose an A-conjugate c of b, different from b, and set $B = b$ and $C = c$.

EXERCISE 7.3.5. Show that $a \mathop{\smile\!\!\!\!\!\!|}^{0}_{A} A$ with Existence and use Monotonicity and Transitivity.

EXERCISE 7.3.6. Let $p \in \mathrm{S}(M)$ with two different non-forking extensions to $B \supset M$. Let $B_0, B_1, \ldots$ be an M-independent family of conjugates of B. Then on each B_i there are two different extensions q_i^0, q_i of p. Now by the independence theorem for any function $\varepsilon\colon \omega \to 2$ there is a non-forking extension q^{ε} of p to $\bigcup_{i<\omega} B_i$, which extends every each $q_i^{\varepsilon(i)}$.

EXERCISE 7.4.1. If $\alpha \in \mathrm{Aut}(\mathfrak{C})$ has the given property, choose a model M of size $|T|$ and $\beta \in \mathrm{Aut}_f(\mathfrak{C})$ with $\beta \restriction M = \alpha \restriction M$. Then $\alpha\beta^{-1} \in \mathrm{Aut}(\mathfrak{C}/M) \subseteq \mathrm{Aut}_f(\mathfrak{C})$.

EXERCISE 7.4.2. If θ is thick and defined over A, the conjunction of the A-conjugates of θ is thick and defined over A.

EXERCISE 7.4.3. Extend B and C to models M_B and M_C such that $M_B \mathop{\smile\!\!\!\!\!\!|}_A M_C$, $b \mathop{\smile\!\!\!\!\!\!|}_A M_B$ and $c \mathop{\smile\!\!\!\!\!\!|}_A M_C$. Now it suffices to find some d such that $d \mathop{\smile\!\!\!\!\!\!|}_A M_B M_C$, $\text{tp}(d/M_B) = \text{tp}(b/M_B)$ and $\text{tp}(d/M_C) = \text{tp}(c/M_C)$.

EXERCISE 7.4.4. We have to show that for every thick $\theta(x, y)$ the formula $\theta(x, a)$ does does not divide over A. So let $a = a_0, a_1, \ldots$ be indiscernible over A. Then $\models \theta(a_i, a_j)$ for all i, j. This shows that $\{\theta(x, a_i) \mid i < \omega\}$ is finitely satisfiable. If T is simple and B is any set, choose a'' independent from B over A such that $\text{nc}_A(a'', a)$. Finally choose a' such that $\text{tp}(a'/Aa) = \text{tp}(a''/Aa)$ and $a' \mathop{\smile\!\!\!\!\!\!|}_{Aa} B$.

EXERCISE 7.4.5. If $\mathcal{I}$ is an infinite sequence of indiscernibles over A, then $\mathcal{I}$ is indiscernible over some model which contains A.

EXERCISE 7.4.6. Use the Erdős–Rado Theorem C.3.2.

EXERCISE 7.4.7. Let R be bounded and A-invariant and $a_0, a_1, \ldots$ indiscernible over A. Show that $R(a_i, a_j)$ for all $i < j$.

EXERCISE 7.4.8. This follows from Exercise 7.4.7(c).

EXERCISE 7.5.1. If K is a subfield of $\mathbb{F}_p^{\text{alg}}$, consider the set $I = \{i \mid \mathbb{F}_{p^i} \subseteq K\}$. If $\mathcal{F}$ is an ultrafilter on I which contains $\{j \in I \mid i|j\}$ for all $i \in I$, then K is the absolute part of $\prod_{i \in I} \mathbb{F}_{p^i}/\mathcal{F}$.

EXERCISE 7.5.2. Use Corollary 7.5.3 and Remark B.4.12.

Chapter 8. Stable theories

EXERCISE 8.1.1. If q is a coheir of p, the sets $\varphi(M)$, $\varphi \in q$ are non-empty and closed under finite intersections. So there is an ultrafilter $\mathcal{F}$ on M which contains all $\varphi(M)$.

EXERCISE 8.1.2. Consider a Morley sequence of a global coheir extension of $\text{tp}(a/M) = \text{tp}(b/M)$ over M.

EXERCISE 8.1.3. Use Exercise 7.1.1.

EXERCISE 8.1.4. We may assume that ψ has Morley degree 1 and do induction on $\text{MR}\,\psi = \alpha$. If $\alpha = \beta + 1$, choose an infinite disjoint family of M-definable classes $\psi_i(\mathfrak{C}) \subseteq \psi(\mathfrak{C})$ of rank $\beta = \text{MR}(\psi \wedge \neg\varphi) < \text{MR}\,\psi$. Then $\text{MR}(\psi_i \wedge \varphi) = \beta$ for some i and $\psi_i \wedge \varphi$ is realised in M by induction. If α is a limit ordinal, choose some definable $\psi'(\mathfrak{C}) \subseteq \psi(\mathfrak{C})$ with $\text{MR}(\psi \wedge \neg\varphi) < \text{MR}(\psi') < \alpha$ and apply the induction hypothesis to ψ' and $\psi' \wedge \varphi$.

The second part follows from the first.

EXERCISE 8.1.5. If b' is another realisation of $q(y) \upharpoonright B$ and a' realises $p(x) \upharpoonright Bb'$ there is a B-automorphism α taking b' to b. If $p(x)$ is A-invariant, $\alpha(a')$ realises $\mathrm{tp}(a/Bb)$ and so $p(x) \otimes q(y)$ is well defined. The same proof shows $p(x) \otimes q(y)$ to be A-invariant if p, q both are A-invariant.

*EXERCISE 8.2.1. If $\varphi(\overline{x}, \overline{y})$ has the order property witnessed by $\overline{a}_0, \overline{a}_1, \ldots$ and $\overline{b}_0, \overline{b}_1, \ldots$, then the sequence $\overline{a}_0\overline{b}_0, \overline{a}_1\overline{b}_1, \ldots$ is ordered by the formula $\varphi'(\overline{x}\overline{y}, \overline{x}'\overline{y}') = \varphi(\overline{x}, \overline{y}')$. The converse is obvious.

EXERCISE 8.2.3. The formula xRy has the binary tree property.

EXERCISE 8.2.5. If $\varphi(x, y)$ has SOP, the formula $\psi(x, y_1, y_2) = \varphi(y_1, x) \wedge \neg\varphi(y_2, x)$ has the tree property with respect to $k = 2$.

EXERCISE 8.2.6. Hint: if T is unstable, there are a formula $\varphi(x, y)$ and indiscernibles $(a_i b_i \mid i \in \mathbb{Q})$ with $\models \varphi(a_i, b_j) \Leftrightarrow i < j$. If φ does not have the independence property, there are finite disjoint subsets J, K of $\mathbb{Q}$ such that $\Phi_{J,K}(y) = \{\varphi(a_i, y) \mid i \in J\} \cup \{\neg\varphi(a_i, y) \mid i \in K\}$ is inconsistent. Not all of J can be less than all elements of K. Choose J and K minimising the number of inversions $F = \{(j, k) \in J \times K \mid k < j\}$. Choose $(j, k) \in F$ so that the interval (k, j) does not contain any elements of $J \cup K$. Write $J = J_0 \cup \{j\}$ and $K = K_0 \cup \{k\}$. Then $\Phi_{J_0 \cup \{k\}, K_0 \cup \{j\}}(y)$ is consistent and the formula (with parameters)

$$\bigwedge \Phi_{J_0,K_0}(y) \wedge \neg\varphi(x, y)$$

has the strict order property.

*EXERCISE 8.2.7. a) $\Rightarrow$ b): A type $p \in \mathrm{S}(A)$ is determined by the family of all p_φ, the φ-parts of p. Hence

$$|\mathrm{S}(A)| \leq \prod_{\varphi} |\mathrm{S}_\varphi(A)| \leq \prod_{\varphi} |A| = |A|^{|T|}.$$

So if $|A| = \lambda$ and $\lambda^{|T|} = \lambda$, then $|\mathrm{S}(A)| \leq \lambda$.

b) $\Rightarrow$ c): Clear.

c) $\Rightarrow$ a): Follows directly from Theorem 8.2.3, a) $\Rightarrow$b).

The last assertion follows from Lemma 5.2.2.

EXERCISE 8.2.8. See the proof of 8.2.3, a) $\Rightarrow$ d). Let $A = \{0, \frac{1}{2}, 1\}$. We embed I into ${}^{\mu}A$ by extending sequences from I_0 to a sequence of length μ with constant value $\frac{1}{2}$. We order I by the ordering induced from the lexicographic ordering of ${}^{\mu}A$.

EXERCISE 8.2.9. If $\psi(x, y)$ is a Boolean combination of $\varphi_0(x, y)$ and $\varphi_1(x, y)$, show that the ψ-type of a tuple over B is determined by its $\varphi_0(x, y)$-type and its $\varphi_1(x, y)$-type. So we have $|\mathrm{S}_\psi(B)| \leq |\mathrm{S}_{\varphi_0}(B)| \cdot |\mathrm{S}_{\varphi_1}(B)|$.

EXERCISE 8.2.10. 1. Since $\mathrm{R}_{\varphi}(\psi) = \infty$, as in Theorem 6.2.7, there is a binary tree of consistent formulas of the form $\psi \wedge \delta$, $\delta \in \Phi$. Now we follow the proof of Theorem 5.2.6 and conclude first that over some countable A there are uncountable many φ-types which contain ψ. This implies then that $\varphi(x, y)$ has the binary tree property. Again by the proof of 6.2.7 this implies that $\mathrm{R}_{\varphi}(\psi) = \infty$.

2. If $\beta < \mathrm{R}_{\varphi}(\psi)$, there is a formula $\varphi(x, a)$ such that $\psi(x) \wedge \varphi(x, a)$ and $\psi(x) \wedge \neg\varphi(x, a)$ have rank at least β. So if $\omega \leq \mathrm{R}_{\varphi}(\psi)$, then φ would have binary trees of arbitrary finite height and so would have the binary tree property.

*EXERCISE 8.2.11. Assume that the tree property of φ is witnessed by the parameters $A = \{a_s \mid s \in {}^{<\omega}\omega\}$. If φ has the tree property with respect to $k = 2$, it is easy to see that φ has the binary tree property.

For the general case we make use of Exercise 8.2.10: if φ is stable, all φ-ranks are less than ω. It follows that there is a sequence $\sigma \in {}^{\omega}\omega$ such that the φ-ranks of the formulas $\bigwedge_{1 \leq i < n} \varphi(x, a_{\sigma \upharpoonright i})$ are strictly decreasing, which is impossible.

EXERCISE 8.3.1. Let T be the theory of an equivalence relations with three infinite classes. There is only one 1-type over the empty set, and this does not have a good definition.

EXERCISE 8.3.2. Six of the eight possible cases are realised by 1-types. For the cases ($\neg$D, C, I, $\neg$H), ($\neg$D, $\neg$C, I, H) use 2-types and for the case ($\neg$D, $\neg$C, I, $\neg$H) a 3-type.

EXERCISE 8.3.3. Let $\pi_n : \mathrm{S}_n(B) \to \mathrm{S}_{n+1}(B)$ be a continuous section. For any n-tuple c if π_n maps $\mathrm{tp}(c/B)$ to $p(x, y)$, then $p_c = p(x, c)$ is a type over cB. Continuity implies that for every $\varphi(x, y)$ there is a B–formula $\psi(y)$ such that for all c we have $\varphi(x, c) \in p_c \iff \models \psi(c)$. The following coherence condition ensures that $p = \bigcup_{c \in \mathfrak{C}} p_c$ is consistent. For any map $s : \{1, \ldots, m\} \to \{1, \ldots, n\}$ let $s^{\#} : \mathrm{S}_n(B) \to \mathrm{S}_m(B)$ and $s^{*} : \mathrm{S}_{n+1}(B) \to \mathrm{S}_{m+1}(B)$ be the associated natural restriction maps. Then coherence means $\pi_m \circ s^{\#} = s^{*} \circ \pi_n$.

EXERCISE 8.3.4. Consider the Boolean algebra of all M-definable subsets of $\varphi(M)^n$ and the subalgebra of $\varphi(M)$-definable subsets. The two algebras coincide if and only if they have the same Stone spaces (see p. 49). For the second part note that – if $\varphi(\mathfrak{C})$ has a least two elements – then for every $\psi(x, y)$ there is a formula $\chi(x, z)$ such that every class $\psi(x, b)$ which is a subclass of $\varphi(\mathfrak{C})^n$ has the form $\chi(\mathfrak{C}, c)$ for some $c \in \varphi(\mathfrak{C})$.

EXERCISE 8.3.5. 1. Prove that formulas with Morley rank are stable. The proof that totally transcendental theories are stable on page 134 is similar.

2. It is enough to find $\mathrm{d}_p x\, \varphi(x, y)$ for stable $\varphi(x, y)$. Use the proof of Theorem 8.3.1.

3. Use the proof of Corollary 8.3.3 and Remark 8.2.2.

EXERCISE 8.3.6. The proof follows the pattern of the proof of the Erdős–Makkai Theorem C.2.1. Assume that $B^* = \{b \in B \mid \varphi(x, b) \in p\}$ is not a positive Boolean combination of sets of the form $\{b \in B \mid\models \varphi(c, b)\}, c \in C$. Construct three sequences $(b'_i \mid i < n)$ in B^*, $(b''_i \mid i < n)$ in $B \setminus B^*$ and $(c_i \mid i < n)$ in C such that for all $i < n$

$$\models \varphi(c_i, b'_n) \Rightarrow \models \varphi(c_i, b''_n)$$

and for all $i \leq n$

$$\models \varphi(c_n, b'_i) \text{ and } \models \neg\varphi(c_n, b''_i).$$

EXERCISE 8.3.7. Show first that there is a finite sequence $\Delta_1, \ldots, \Delta_n$ such that every φ-type is definable by an instance of one of the Δ_i. (Otherwise the $L \cup \{P, c\}$-theory stating that in a model of T the φ-part of the type of c over P is not definable would be consistent.)

EXERCISE 8.3.8. Hint: If q is a weak heir of p, then $\mathrm{D}_\varphi(q) = \mathrm{D}_\varphi(p)$ where $\mathrm{D}_\varphi(p)$ is defined as the minimum of $\mathrm{D}_\varphi(\theta)$ for $\theta \in p$. The argument in Theorem 8.3.1 now shows that q is definable over M.

EXERCISE 8.3.9. Let p be the global extension of $\mathrm{tp}(a/M)$ which is definable over M. By Lemma 8.1.5 $\mathrm{tp}(a/Mb)$ is an heir of $\mathrm{tp}(a/M)$ if and only if $\mathrm{tp}(a/Mb) \subseteq q$, i.e., if and only if $\varphi(x, b) \in \mathrm{tp}(a/Mb) \Leftrightarrow \models \mathrm{d}_p x\varphi(x, b)$. So if q is the global M-definable extension of $\mathrm{tp}(b/M)$, we have that $\mathrm{tp}(a/Mb)$ is an heir of $\mathrm{tp}(a/M)$ if and only if $\varphi(x, b) \in \mathrm{tp}(a/Mb) \Leftrightarrow \models \mathrm{d}_p x\varphi(x, y) \in q(y)$. Lemma 8.3.4 implies now that $\mathrm{tp}(a/Mb)$ is an heir of $\mathrm{tp}(a/M)$ if and only if $\mathrm{tp}(b/Ma)$ is an heir of $\mathrm{tp}(b/M)$.

EXERCISE 8.3.10. One direction follows from Exercise 8.1.3. For the other direction assume that φ does not fork over A. Then φ is contained in a global type p which does not fork over A. Apply Corollary 8.3.7.

EXERCISE 8.4.1. $\mathbb{D}$ is definable from some some tuple $d \in D$. Any such d is a canonical parameter for $\mathbb{D}$.

EXERCISE 8.4.2. Let e be an imaginary and A the smallest algebraically closed set in the home sort over which e is definable. Then e is definable from a finite tuple $a \in A$. Since every automorphism which fixes e leaves A invariant, all elements of A are algebraic over e.

For the converse let $a \in \mathrm{acl}(e)$ be a real tuple over which e is definable. Then $A = \mathrm{acl}(a)$ is the smallest algebraically closed set over which e is definable.

EXERCISE 8.4.3. Infset and DLO have the following property: if A, B are finite sets and if the tuples a and b have the same type over $A \cap B$, then there is a sequence $a = a_0, b_0, \ldots, a_n, b_n = b$ such that a_i and b_i have the same type over A and b_i and a_{i+1} have the same type over B. This implies that for every definable class $\mathbb{D}$ there is a smallest set over which $\mathbb{D}$ is definable.

Infset does not eliminate imaginaries since no finite set with at least two elements has a canonical parameter.

*EXERCISE 8.4.4. Let q_1 and q_2 be extensions of p to B. Choose realisations a_1, a_2 of q_1, q_2, respectively. There is an $\alpha \in \mathrm{Aut}(\mathfrak{C}/A)$ taking a_1 to a_2. Since $\alpha(B) = B$, we have $\alpha(q_1) = q_2$.

EXERCISE 8.4.5. Let $p \in \mathrm{S}(A)$ be algebraic. If p is realised by $b \in \mathrm{dcl}(B)$, then $\mathrm{d}_p\, x\varphi(x, \overline{y}) = \varphi(b, \overline{y})$ is a good definition of p over B. Conversely, if q is an extension of p to M, then q is realised by some b in M and $\mathrm{d}_q\, x(x \doteq y)$ defines the set $\{b\}$. So if q is definable over B, then $b \in \mathrm{dcl}(B)$.

EXERCISE 8.4.6. Let d be a canonical parameter of $\mathbb{D} = \varphi(\mathfrak{C}, d)$. If d' has the same type as d, we have

$$\varphi(\mathfrak{C}, d') = \mathbb{D} \Rightarrow d' = d.$$

By compactness this is true for all d' which satisfy some $\psi(x) \in \mathrm{tp}(d)$. Consider the $L \cup \{P\}$-formula $\theta(x, P) = \psi(x) \wedge \forall y\, (\varphi(y, x) \leftrightarrow P(y))$.

*EXERCISE 8.4.7. 1. Let $\varphi(x, a) \in p$ have the same Morley rank as p and be of degree 1. Then $\ulcorner \mathrm{d}_x\, \varphi(x, y) \urcorner$ is a canonical base of p.

2. Use Part 1 and Exercise 8.4.1.

*EXERCISE 8.4.9. If $\mathrm{stp}(a/A) \neq \mathrm{stp}(b/A)$ there is an $\mathrm{acl}(A)$-definable class $\mathbb{D} = \varphi(x, \overline{a})$ with $\models \varphi(a, \overline{a})$ and $\not\models \varphi(b, \overline{a})$. By Lemma 8.4.4, $\mathbb{D}$ is the union of equivalence classes of an A-definable finite equivalence relation $E_{\overline{a}}$, proving the claim.

EXERCISE 8.4.10. The correspondence is given by $H = \mathrm{Stab}(A)$ and $A = \mathrm{Fix}(H)$. That $A = \mathrm{Fix}(\mathrm{Stab}(A))$ for definably closed A follows from Corollary 6.1.12(1). To see that $H = \mathrm{Stab}(\mathrm{Fix}(H))$ for closed subgroups H, we have to show that every $g \in \mathrm{Stab}(\mathrm{Fix}(H))$ agrees on every finite tuple b with some element h of H. To this end let $\overline{a}$ be a canonical parameter of the finite set Hb. Then $\overline{a}$ is fixed by H and therefore also fixed under g. So $g(Hb) = Hb$, which implies that $gb = hb$ for some $h \in H$.

EXERCISE 8.5.1. Show that the two conditions are equivalent to each of the following

1. $\mathrm{tp}(a/K)$ has a unique extension to K^{sep};
2. $K(a) \cap K^{\mathrm{sep}} = K$.

EXERCISE 8.5.2. Let P be the set of all strong types over A which are consistent with p, and Q be the set of all strong types which are consistent with q. Both P and Q are closed subsets of $\mathrm{S}(\mathrm{acl}^{\mathrm{eq}}(A))$ and are disjoint since strong types are stationary. So they can be separated by a formula $\varphi(x)$ over $\mathrm{acl}^{\mathrm{eq}}(A)$. By Lemma 8.4.4, $\varphi(\mathfrak{C})$ is a union of classes of a finite A-definable equivalence relation $E(x, y)$.

EXERCISE 8.5.4. Use the second part of Exercise 8.3.5. The first claim can now be proved like Theorem 8.5.1.

The second claim is proved like Corollary 8.5.3, but we must be more careful and use the φ-rank introduced in Exercise 8.2.10. We call the minimal φ-rank of a formula in a type p the *φ-rank* of p. Let $A = \mathrm{acl}^{\mathrm{eq}}(A)$, $p \in \mathrm{S}(A)$ a stable type and p' and p'' two non-forking global extensions. Then p' and p'' are definable over A. We want to show that $\varphi(x, b) \in p' \Leftrightarrow \varphi(x, b) \in p''$. We may assume that $\varphi(x, y)$ is stable (containing parameters from A). Let $q(y)$ be a global extension of $\mathrm{tp}(b/A)$ which has the same $\varphi^{\sim}$-rank, where $\varphi(x, y)^{\sim} = \varphi(y, x)$. Since there are only finitely many possibilities for the φ-part of q, the φ-part is definable over A. Now the claim follows from an adapted version of Lemma 8.3.4.

We still have to show that p has a non-forking global extension, i.e., an extension which is definable over A. Choose for every stable $\varphi(x, y)$ a global extension of p with the same φ-rank. By the above the φ-part p^{φ} of this extension is definable over A and as such is uniquely determined. It remains to show that the union of all p^{φ} is consistent. Consider a finite sequence $\varphi_1(x, y), \ldots, \varphi_n(x, y)$ of stable formulas. Choose a stable formula $\varphi(x, y, z)$ such that every instance $\varphi_i(x, b)$ has the form $\varphi(x, b, c)$ for some choice of c. Then p^{φ} contains all p^{φ_i} for $i = 1, \ldots, n$.

*EXERCISE 8.5.5. 1. Argue as in the second part of the proof of Theorem 8.5.10. Replace $p \sqsubset q$ by $\mathrm{MR}(p) = \mathrm{MR}(q)$.

2. By the first part of Exercise 8.3.5 p is stable. Let q be a global extension of p. If $\mathrm{MR}(p) = \mathrm{MR}(q)$, then q has only finitely many conjugates over A. Since q is definable, this implies that q is definable over $\mathrm{acl}^{\mathrm{eq}}(A)$. So by Exercise 8.5.4, q does not fork over A. Now assume that q does not fork over A. Using Exercise 6.2.8 we see that we can assume that $A = \mathrm{acl}^{\mathrm{eq}}(A)$. Let q' be an extension of p with the same Morley rank. Then q' does not fork over A. So by Exercise 8.5.4 $q = q'$.

*EXERCISE 8.5.6. Choose $A_1 \subseteq A$ of cardinality at most $|T|$ over which p does not fork. Let (p^i) be the non-forking extensions of $p \upharpoonright A_1$ to A. For each L-formula $\varphi(y, \overline{y})$ there are only finitely many different p^i_{φ}. Hence there is a finite subset A_{φ} of A such that for all i

$$(p^i \upharpoonright A_{\varphi})_{\varphi} = (p \upharpoonright A_{\varphi})_{\varphi} \Longrightarrow p^i_{\varphi} = p_{\varphi}.$$

Now put $A_0 = A_1 \cup \bigcup_\varphi A_\varphi$.

If p has Morley rank, choose $\varphi \in p$ having the same Morley rank and degree as p. Any set A_0 containing the parameters of φ does the job.

EXERCISE 8.5.7. 1: Easy.

2: Exercise 8.5.6 shows that it is enough to consider types p over a countable set A. The multiplicity is the number of extensions of p to $\mathrm{acl}(A)$. These extensions form a separable compact space. By Exercise 8.4.4, either all or none of them are isolated. In the first case the space is finite; in the second, it has cardinality $2^{\aleph_0}$.

3: Use Exercise 8.5.5.

EXERCISE 8.5.8. 1) Both a and $a \cdot b$ are interalgebraic over b. This implies $\mathrm{MR}(a) = \mathrm{MR}(a/b) = \mathrm{MR}(a \cdot b/b) \leq \mathrm{MR}(a \cdot b)$. If $\mathrm{MR}(a) = \mathrm{MR}(a \cdot b)$, we have $\mathrm{MR}(a \cdot b/b) = \mathrm{MR}(a \cdot b)$ and $a \cdot b$ and b are independent.

2) Let b be independent from a. If a is generic, $\mathrm{MR}(a \cdot b)$ cannot be bigger than $\mathrm{MR}(a)$, so $a \cdot b$ and b are independent by part 1). For the converse we choose b generic. Part 1) (with sides reversed) implies that $a \cdot b$ is also generic. If $a \cdot b$ and b are independent, it follows that $\mathrm{MR}(a) = \mathrm{MR}(a \cdot b)$ and a is generic.

EXERCISE 8.5.9. For 1) notice that each line A_i consists of two elements and their product.

For 2) note that by Exercise 8.5.8, if a, b, c are independent generics, then $a, b, b \cdot c$ are again independent generics. If one applies this rule repeatedly starting with a_1, a_2, a_3, one obtains every non-collinear triple of our diagram.

EXERCISE 8.6.1. It follows from Remark 7.1.3 and Symmetry that a type is algebraic if and only if it has no forking extensions. A type has SU-rank 1 if and only if the algebraic and the forking extensions coincide. So a type is minimal if and only if it has SU-rank 1 and has only one non-forking extension to every set of parameters.

EXERCISE 8.6.2. Use Exercise 7.1.5.

EXERCISE 8.6.4. This follows from Exercise 7.1.2.

EXERCISE 8.6.5. This follows easily from Exercise 7.2.5. Prove by induction on α and γ

$$\mathrm{SU}(a/C) \geq \alpha \Rightarrow \mathrm{SU}(ab/C) \geq \mathrm{SU}(b/aC) + \alpha$$
$$\mathrm{SU}(ab/C) \geq \gamma \Rightarrow \mathrm{SU}(b/aC) \oplus \mathrm{SU}(a/C) \geq \gamma.$$

EXERCISE 8.6.6. The first claim follows from Lemma 7.2.4(2) and the remark thereafter. The second claim is easily proved using the Diamond Lemma (Exercise 7.2.2) and Exercise 7.1.7.

EXERCISE 8.6.7. Totally transcendental theories are superstable by Corollary 8.5.11. It follows also that the multiplicity of a type over arbitrary sets is finite, namely equal to its Morley degree. If T is superstable, one can compute an upper bound for the number of types over a set A of cardinality κ as in the proof of Theorem 8.6.5(2). If T is small, there are only countably many types over a finite set E. If we know also that all $p \in \mathrm{S}(E)$ have finite multiplicity, we have $|\,\mathrm{S}(A)| \leq \kappa \cdot \aleph_0 \cdot \aleph_0 = \kappa$.

EXERCISE 8.6.9. We can assume that all E_i are 0-definable. Choose a sequence $a_0, a_1, \ldots$ such that $\models E_i(a_i, a_{i+1})$ and $\ulcorner a_i/E_i \urcorner$ is not algebraic over $a_0 \ldots a_{i-1}$. Let b be an element in the intersection of all a_i/E_i, $A = \{a_0, a_1, \ldots\}$ and $p = \mathrm{tp}(b/A)$. Then for all i we have $b \not\mathop{\smile\hspace{-0.9em}\mid}_{a_0,\ldots,a_{i-1}} \ulcorner a_i/E_i \urcorner$ by Remark 7.1.3. This shows that p forks over each finite subset of A.

EXERCISE 8.6.10. Define $E_i(x, y)$ as $xy^{-1} \in G_i$. For any imperfect field K of finite characteristic p set $G_i = K^{p^i}$. $x - y \in K^{p^{i+1}}$.

EXERCISE 8.6.11. Half of the claim follows from Remark 6.2.8 and Exercise 8.6.10. Assume that M has the dcc on pp-definable subgroups. Then for every element a and every set B of parameters the positive type $\mathrm{tp}^+(a/B)$ contains a smallest element $\varphi_0(x, b)$. So there are at most $\max(|T|, |A|)$ many types over A. This shows that $M \restriction R_0$ is ω-stable for every countable subring, so M is totally transcendental (see Exercise 5.2.5). Now assume that there is no infinite sequence of pp-subgroups with infinite index in each other. Then $\mathrm{tp}^+(a/B)$ contains a formula $\varphi_0(x, b_0)$ such that $\mathrm{tp}^+(a/B)$ is axiomatised by formulas $\varphi(x, b)$ where $\varphi(M, 0)$ is a subgroup of finite index in $\varphi_0(M, 0)$. There are $\max(|T|, |A|)$ many possibilities for $\varphi_0(x, b_0)$ and for each $\varphi(x, y)$ finitely many possibilities. So the number of types over A is bounded by $\max(2^{|T|}, |A|)$ and M must be superstable. Indeed, the proof of Theorem 8.6.5(3) shows that otherwise for every κ there would be a set A of cardinality κ with $|\,\mathrm{S}(A)| \geq \kappa^{\aleph_0}$.

Chapter 9. Prime extensions

*EXERCISE 9.1.1. Let $p \in \mathrm{S}(A)$ and q a non-forking extension to B. Let $\mathcal{I}$ be a Morley sequence of q, so $\mathcal{I}$ is independent over B. Since every element of $\mathcal{I}$ is independent from B over A, it follows that $\mathcal{I}$ is independent over A as well (see Exercise 7.2.6), hence a Morley sequence of p.

*EXERCISE 9.1.2. a): Since $\mathcal{I} \setminus \mathcal{I}_0$ is independent from B over A, it is independent over B. The elements of $\mathcal{I}$ realise the non-forking extension of p to B.

b): Let $B \supseteq A$ and q the non-forking extension of p. We extend $\mathcal{I}$ to a very long sequence $\mathcal{I}'$ indiscernible over A. Then $\mathcal{I}'$ is still a Morley sequence of p.

If we choose $\mathcal{I}_0 \subseteq \mathcal{I}'$ with $|\mathcal{I}_0| \leq |T| + |B|$ and $B \mathop{\smile\!\!\!\!\!\!\mid}_{A\mathcal{I}_0} \mathcal{I}'$, then $\mathcal{I}' \setminus \mathcal{I}_0$ is an infinite Morley sequence of q having the same average type as $\mathcal{I}$, so $q \subseteq \mathrm{Av}(\mathcal{I})$.

EXERCISE 9.1.3. If $\mathcal{I}_0$ and $\mathcal{I}_1$ are parallel, hence $\mathcal{I}_0\mathcal{J}$ and $\mathcal{I}_1\mathcal{J}$ indiscernible, then

$$\mathrm{Av}(\mathcal{I}_0) = \mathrm{Av}(\mathcal{I}_0\mathcal{J}) = \mathrm{Av}(\mathcal{I}_1\mathcal{J}) = \mathrm{Av}(I_1).$$

If conversely $p = \mathrm{Av}(\mathcal{I}_0) = \mathrm{Av}(\mathcal{I}_1)$, note that by the proof of Theorem 9.1.2 there are sets B_0 and B_1 over which p does not fork and such that $\mathcal{I}_0$ and $\mathcal{I}_1$ are Morley sequences of the stationary types $p \upharpoonright B_0$ and $p \upharpoonright B_1$. Let $\mathcal{J}$ be a Morley sequence of $p \upharpoonright B_0\mathcal{I}_0B_1\mathcal{I}_1$. Then $\mathcal{I}_0\mathcal{J}$ and $\mathcal{I}_1\mathcal{J}$ are Morley sequences of $p \upharpoonright B_0$ and $p \upharpoonright B_1$, respectively.

*EXERCISE 9.1.4. Let p and q be stationary types with infinite Morley sequences $\mathcal{I}$ and $\mathcal{J}$. Then the average types $\mathrm{Av}(\mathcal{I})$ and $\mathrm{Av}(\mathcal{J})$ are the global non-forking extensions of p and q, respectively. Now p and q are parallel if and only if $\mathrm{Av}\,\mathcal{I} = \mathrm{Av}\,\mathcal{J}$, i.e., if and only if $\mathcal{I}$ and $\mathcal{J}$ are parallel.

*EXERCISE 9.2.1. Let $p_1, \ldots, p_n$ be the extensions of p to $\mathrm{acl}^{\mathrm{eq}}(A)$ and let $\mathcal{I}_i$ be the elements of $\mathcal{I}$ which realise p_i.

Chapter 10. The fine structure of $\aleph_1$-categorical theories

EXERCISE 10.1.2. To prove Part 1 either use Remark 6.2.8 to obtain a finite subset of $\mathbb{E}$ which has trivial stabiliser in $\mathrm{Aut}(\mathbb{E}/\mathbb{F})$ or apply Corollary 8.3.3.

EXERCISE 10.1.1. Let N be an elementary submodel which contains $\mathbb{F}(M)$. By Lemma 10.1.4 there is a definable surjection $f : \mathbb{F}^n \to \mathfrak{C}$. Write $f(x) = g(x, a)$ for a 0-definable function g and a parameter tuple a. Since N is an elementary substructure, we may assume that $a \in N$. Then $M = g(\mathbb{F}^n(M), a) = g(\mathbb{F}^n(N), a) = N$.

EXERCISE 10.1.3. If d is in $\mathrm{dcl}^{\mathrm{eq}}(\mathbb{F})$, the set $\{d\}$ is definable over $\mathbb{F}$. The proof of Theorem 8.4.3 shows that $\{d\}$ has a canonical parameter in $\mathbb{F}^{\mathrm{eq}}$.

EXERCISE 10.1.5. a) $\Rightarrow$ b) was implicitly proved in Lemma 10.1.5.
b) $\Rightarrow$ a): This is the proof of Corollary 8.3.3.
b) $\Leftrightarrow$ c): This is Exercise 10.1.4.
b) $\Rightarrow$ d): Same as the proof of Lemma 10.1.5.
d) $\Rightarrow$ a): Assume that $\varphi(a, \mathbb{F})$ is not definable with parameters from $\mathbb{F}$. Let a_i be an enumeration of $\mathfrak{C}$. Construct a sequence φ_i of partial automorphisms of $\mathbb{F}$ so that the domain of φ_i contains some f with $\models \varphi(a, f) \Leftrightarrow\ \models \neg\varphi(a_i, \varphi_i(f))$.

*EXERCISE 10.2.1. Use induction on $\mathrm{MR}(\mathbb{A})$.

EXERCISE 10.2.3. Show first that if p, r are two types over A and A' is an extension of A, then p and q are almost orthogonal if any two non-forking extensions of p and q to A' are almost orthogonal. This implies that q is orthogonal to p if and only if q is orthogonal to all non-forking extensions of p to A'.

EXERCISE 10.2.4. 1. Let $p \in S(A)$ be stationary and $p' \in S(A')$ be a regular non-forking extension. We want to show that p is orthogonal to every forking extension $q \in S(B)$. For this we may assume that $B = \mathrm{acl}^{\mathrm{eq}}(B)$ so that q is stationary. By the Diamond Lemma q has a non-forking extension q' which extends an A-conjugate of p'. So p' and q' are orthogonal and by Exercise 10.2.3 so are p and q.

2. Let us check the four properties in Definition 5.6.5: $A \subseteq \mathrm{cl}(A)$ is true since p is non-algebraic. FINITE CHARACTER follows from the finite character of forking, EXCHANGE from forking symmetry. Regularity is used only for TRANSITIVITY. Show the following: assume that $C \not\downarrow_A d$ and that for all $c \in C$ all extensions of $\mathrm{tp}(c/AB)$ are orthogonal to $\mathrm{tp}(d/A)$. Then $B \not\downarrow_A d$.

3. We assume $A = \emptyset$ to simplify notation. Assume $b \not\downarrow c$, $c \not\downarrow d$ and $b \downarrow d$. Choose an d-independent sequence $(b_\alpha c_\alpha)_{\alpha<|T|^+}$ of realisations of $\mathrm{tp}(bc/d)$. Since $b_\alpha \downarrow d$, it follows that each c_α is independent from $B = \{b_\beta \mid \beta < \alpha\}$. Since all c_β are dependent from B, we conclude by regularity that c_α is independent from $\{c_\beta \mid \beta < \alpha\}$ over A. So $(c_\alpha)_{\alpha<|T|^+}$ is independent. But we have $c_\alpha \not\downarrow d$ for all α, contradicting Exercise 7.2.1.

*EXERCISE 10.4.1. If $A \leq M$ and C is a finite subset of N which contains $A \cap N$ we have $\delta(AC/A) \leq \delta(C/A \cap N)$. This proves $A \leq N$.

Assume $A \leq B \leq M$. Consider a finite extension C of A. Then $B \cap C \leq C$. This implies $\delta(A) \leq \delta(B \cap C) \leq \delta(C)$ and therefore $A \leq C$.

The last implication follows directly from the first two.

For the last part of the exercise chose a finite extension B of A with minimal $\delta(B)$. Then B is strong in M. So we may take for $\mathrm{cl}(A)$ the intersection of all finite extensions of A which are strong in M.

*EXERCISE 10.4.2. Let $E(A)$ denote the set of edges of A. If

$$X = E(A_1 \cup \cdots \cup A_k) \setminus (E(A_1) \cup \cdots \cup E(A_k)),$$

we have

$$\delta(A_1 \cup \cdots \cup A_k) = 2|A_1 \cup \cdots \cup A_k| - |E(A_1) \cup \cdots \cup E(A_k)| - |X|.$$

Now apply Lemma 3.3.10.

EXERCISE 10.4.3. This follows because any path (x_1, a, x_2) is strong in M_μ.

Appendices

EXERCISE B.3.1. Let L be an elementary extension of K. Show first that if S is an integral domain which contains K, then $L \otimes_K S$ is again an integral domain. So $L \otimes_K K^{\text{alg}}$ is an integral domain.

EXERCISE C.1.1. 1. SYMMETRY is clear from the definition. For the other properties show first that a finite set A is independent from B over C if and only if $\dim(A/BC) = \dim(A/C)$. This implies MONOTONICITY and TRANSITIVITY. FINITE CHARACTER and LOCAL CHARACTER follow from the fact that for finite A and any D there is a finite $D_0 \subseteq D$ such that $\dim(A/D) = \dim(A/D_0)$.

2. SYMMETRY, FINITE CHARACTER and WEAK MONOTONICITY are clear. If A is finite and D is any set, let D_0 be a basis of $\mathrm{cl}(A) \cap D$. So D_0 is finite and $A \mathop{\smile\!\!\!\!\!\!|}^{0}_{D_0} D$. This shows LOCAL CHARACTER. Always $A \mathop{\smile\!\!\!\!\!\!|}^{\mathrm{cl}}_{C} B$ implies $A \mathop{\smile\!\!\!\!\!\!|}^{0}_{C} B$. If $\mathop{\smile\!\!\!\!\!\!|}^{0}$ satisfies MONOTONICITY, the converse is true: we may assume that $B = \{b_1, \ldots, b_n\}$ is finite. Then $A \mathop{\smile\!\!\!\!\!\!|}^{0}_{C} B$ implies $A \mathop{\smile\!\!\!\!\!\!|}^{0}_{Cb_1\ldots b_i} b_{i+1}$ for all i. But this is the same as $A \mathop{\smile\!\!\!\!\!\!|}^{\mathrm{cl}}_{Cb_1\ldots b_i} b_{i+1}$, from which follows that $A \mathop{\smile\!\!\!\!\!\!|}^{\mathrm{cl}}_{C} B$.

EXERCISE C.1.3. Choose $a, b, x, c \in K$ p-independent. Set $F_0 = K^p(c)$, $F_1 = K^p(c, a, b)$ and $F_2 = K^p(c, x, ax + b)$. Then F_0 has p-dimension 1, F_1 and F_2 have p-dimension 3 and F_1F_2 has p-dimension 4. To show that $F_0 = F_1 \cap F_2$, prove that $\dim_{F_0} F_1 = \dim_{F_0} F_2 = p^2$ and $\dim_{F_0}(F_1 + F_2) = 2p^2 - 1$.

EXERCISE C.1.5. First show that in any pregeometry for any closed $A \subseteq B$: if $\dim(A/B)$ is finite, then it is the longest length n of a proper chain $B = C_0 \subsetneq C_1 \subsetneq \cdots \subsetneq C_n = B$ of closed sets C_i.

REFERENCES

[1] James Ax, *The elementary theory of finite fields*, ***Annals of Mathematics. Second Series***, vol. 88 (1968), pp. 239–271.

[2] J. T. Baldwin and A. H. Lachlan, *On strongly minimal sets*, ***The Journal of Symbolic Logic***, vol. 36 (1971), pp. 79–96.

[3] John T. Baldwin, *α_T is finite for $\aleph_1$–categorical T*, ***Transactions of the American Mathematical Society***, vol. 181 (1973), pp. 37–51.

[4] ———, ***Fundamentals of Stability Theory***, Perspectives in Mathematical Logic, Springer Verlag; Berlin, Heidelberg, New York, London, Paris, Tokyo, 1988.

[5] ———, *An almost strongly minimal non-Desarguesian projective plane*, ***Transactions of the American Mathematical Society***, vol. 342 (1994), no. 2, pp. 695–711.

[6] A. Baudisch, A. Martin-Pizarro, and M. Ziegler, *Red fields*, ***The Journal of Symbolic Logic***, vol. 72 (2007), no. 1, pp. 207–225.

[7] Andreas Baudisch, *A new uncountably categorical group*, ***Transactions of the American Mathematical Society***, vol. 348 (1996), no. 10, pp. 3889–3940.

[8] Andreas Baudisch, Martin Hils, Amador Martin-Pizarro, and Frank O. Wagner, *Die böse Farbe*, ***Journal of the Institute of Mathematics of Jussieu. JIMJ. Journal de l'Institut de Mathématiques de Jussieu***, vol. 8 (2009), no. 3, pp. 415–443.

[9] Paul Bernays, ***Axiomatic Set Theory*** With a historical introduction by Abraham A. Fraenkel, Dover Publications Inc., New York, 1991, Reprint of the 1968 edition.

[10] N. Bourbaki, ***XI, Algébre, Chapitre 5, Corps Commutatifs***, Hermann, Paris, 1959.

[11] Elisabeth Bouscaren, *The group configuration – after E. Hrushovski*, ***The Model Theory of Groups (Notre Dame, IN, 1985–1987)***, Notre Dame Math. Lectures, vol. 11, Univ. Notre Dame Press, Notre Dame, IN, 1989, pp. 199–209.

[12] Steven Buechler, ***Essential Stability Theory***, Perspectives in Mathematical Logic, Springer-Verlag, Berlin, 1996.

[13] FRANCIS BUEKENHOUT, *An introduction to incidence geometry*, ***Handbook of Incidence Geometry***, North-Holland, Amsterdam, 1995, pp. 1–25.

[14] ENRIQUE CASANOVAS, ***Simple Theories and Hyperimaginaries***, Lecture Notes in Logic, vol. 39, Cambridge University Press, 2011.

[15] C. C. CHANG and H. J. KEISLER, ***Model Theory***, third ed., Studies in Logic and the Foundations of Mathematics, vol. 73, North-Holland Publishing Co., Amsterdam, 1990.

[16] ZOÉ CHATZIDAKIS, *Théorie des modèles des corps finis et pseudo-fini*, Unpublished Lecture Notes, 1996.

[17] ZOÉ CHATZIDAKIS and EHUD HRUSHOVSKI, *Model theory of difference fields*, ***Transactions of the American Mathematical Society***, vol. 351 (1999), no. 8, pp. 2997–3071.

[18] M. M. ERIMBETOV, *Complete theories with 1-cardinal formulas*, ***Akademiya Nauk SSSR. Sibirskoe Otdelenie. Institut Matematiki. Algebra i Logika***, vol. 14 (1975), no. 3, pp. 245–257, 368.

[19] JU. L. ERŠOV, *Fields with a solvable theory*, ***Doklady Akademii Nauk SSSR***, vol. 174 (1967), pp. 19–20, English transl., *Soviet Math. Dokl.*, **8**:575-576, 1967.

[20] ULRICH FELGNER, *Comparison of the axioms of local and universal choice*, ***Polska Akademia Nauk. Fundamenta Mathematicae***, vol. 71 (1971), no. 1, pp. 43–62, (errata insert).

[21] STEVEN GIVANT and PAUL HALMOS, ***Introduction to Boolean Algebras***, Undergraduate Texts in Mathematics, Springer, New York, 2009.

[22] VICTOR HARNIK, *On the existence of saturated models of stable theories*, ***Proceedings of the American Mathematical Society***, vol. 52 (1975), pp. 361–367.

[23] DEIRDRE HASKELL, EHUD HRUSHOVSKI, and DUGALD MACPHERSON, ***Stable Domination and Independence in Algebraically Closed Valued Fields***, Lecture Notes in Logic, vol. 30, Association for Symbolic Logic, Chicago, IL, 2008.

[24] WILFRID HODGES, ***Model Theory***, Encyclopedia of Mathematics and its Applications, vol. 42, Cambridge University Press, Cambridge, 1993.

[25] ———, ***A Shorter Model Theory***, Cambridge University Press, 1997.

[26] EHUD HRUSHOVSKI, *A stable $\aleph_0$-categorical pseudoplane*, Preprint, 1988.

[27] ———, *Unidimensional Theories are Superstable*, ***Annals of Pure and Applied Logic***, vol. 50 (1990), pp. 117–138.

[28] ———, *A new strongly minimal set*, Stability in model theory, III (Trento, 1991), ***Annals of Pure and Applied Logic***, vol. 62 (1993), no. 2, pp. 147–166.

[29] ———, *A non-PAC field whose maximal purely inseparable extension is PAC*, ***Israel Journal of Mathematics***, vol. 85 (1994), no. 1-3, pp. 199–202.

[30] EHUD HRUSHOVSKI and BORIS ZILBER, *Zariski geometries*, ***Journal of the American Mathematical Society***, vol. 9 (1996), no. 1, pp. 1–56.

[31] THOMAS JECH, ***Set Theory***, The third millennium edition, revised and expanded. Springer Monographs in Mathematics, Springer-Verlag, Berlin, 2003.

[32] KLAUS KAISER, *Über eine Verallgemeinerung der Robinsonschen Modellvervollständigung*, ***Zeitschrift für Mathematische Logik und Grundlagen der Mathematik***, vol. 15 (1969), pp. 37–48.

[33] AKIHIRO KANAMORI, ***The Higher Infinite. Large Cardinals in Set Theory from Their Beginnings***, second ed., Springer Monographs in Mathematics, Springer-Verlag, Berlin, 2003.

[34] BYUNGHAN KIM and ANAND PILLAY, *From stability to simplicity*, ***The Bulletin of Symbolic Logic***, vol. 4 (1998), no. 1, pp. 17–36.

[35] SERGE LANG, ***Algebra***, second ed., Addison-Wesley Publishing Company, 1984.

[36] SERGE LANG and ANDRÉ WEIL, *Number of points of varieties in finite fields*, ***American Journal of Mathematics***, vol. 76 (1954), pp. 819–827.

[37] DANIEL LASCAR, ***Stability in Model Theory***, Longman, New York, 1987.

[38] ANGUS MACINTYRE, *On ω_1-categorical theories of fields*, ***Polska Akademia Nauk. Fundamenta Mathematicae***, vol. 71 (1971), no. 1, pp. 1–25, (errata insert).

[39] DAVID MARKER, ***Model Theory***, An introduction, Graduate Texts in Mathematics, vol. 217, Springer-Verlag, New York, 2002.

[40] M. MORLEY, *Categoricity in Power*, ***Transactions of the American Mathematical Society***, vol. 114 (1965), pp. 514–538.

[41] DAVID PIERCE and ANAND PILLAY, *A note on the axioms for differentially closed fields of characteristic zero*, ***Journal of Algebra***, vol. 204 (1998), no. 1, pp. 108–115.

[42] ANAND PILLAY, ***An Introduction to Stability Theory***, Oxford Logic Guides, vol. 8, Oxford University Press, New York, 1983.

[43] ———, *The geometry of forking and groups of finite Morley rank*, ***The Journal of Symbolic Logic***, vol. 60 (1995), pp. 1251–1259.

[44] ———, ***Geometric Stability Theory***, Oxford Logic Guides, vol. 32, Oxford University Press, New York, 1996.

[45] BRUNO POIZAT, ***Cours de Théorie des Modèles***, Nur Al-Mantiq Wal-Ma'rifah, Villeurbanne, 1985.

[46] ———, ***Groupes Stables***, Nur Al-Mantiq Wal-Mari'fah, Villeurbanne, 1987.

[47] MIKE PREST, ***Model Theory and Modules***, London Mathematical Society Lecture Note Series, vol. 130, Cambridge University Press, Cambridge, 1988.

[48] ALEX PRESTEL and CHARLES N. DELZELL, ***Mathematical Logic and Model Theory: A Brief Introduction***, Universitext, Springer, 2011.

[49] V. A. PUNINSKAYA, *Vaught's conjecture*, ***Journal of Mathematical Sciences (New York)***, vol. 109 (2002), no. 3, pp. 1649–1668, Algebra, 16.

[50] GERALD E. SACKS, ***Saturated Model Theory***, Mathematics Lecture Note Series, W. A. Benjamin, Inc., Reading, Mass., 1972.

[51] IGOR R. SHAFAREVICH, ***Basic Algebraic Geometry. 1***, Varieties in projective space, second ed., Springer-Verlag, Berlin, 1994, Translated from the 1988 Russian edition and with notes by Miles Reid.

[52] SAHARON SHELAH, *Every two elementarily equivalent models have isomorphic ultrapowers*, ***Israel Journal of Mathematics***, vol. 10 (1971), pp. 224–233.

[53] ———, *Uniqueness and characterization of prime models over sets for totally transcendental first-order-theories*, ***The Journal of Symbolic Logic***, vol. 37 (1972), pp. 107–113.

[54] ———, ***Classification Theory***, North Holland, Amsterdam, 1978.

[55] ———, *On uniqueness of prime models*, ***The Journal of Symbolic Logic***, vol. 43 (1979), pp. 215–220.

[56] SAHARON SHELAH, *Simple unstable theories*, ***Annals of Mathematical Logic***, vol. 19 (1980), no. 3, pp. 177–203.

[57] JOSEPH R. SHOENFIELD, ***Mathematical Logic***, Association for Symbolic Logic, Urbana, IL, 2001, Reprint of the 1973 second printing.

[58] KATRIN TENT, *Very homogeneous generalized n-gons of finite Morley rank*, ***Journal of the London Mathematical Society. Second Series***, vol. 62 (2000), no. 1, pp. 1–15.

[59] JOUKO VÄÄNÄNEN, *Barwise: abstract model theory and generalized quantifiers*, ***The Bulletin of Symbolic Logic***, vol. 10 (2004), no. 1, pp. 37–53.

[60] FRANK WAGNER, ***Simple Theories***, Kluwer Adacemic Publishers, Dordrecht, 2000.

[61] FRANK O. WAGNER, ***Stable Groups***, London Mathematical Society Lecture Note Series, vol. 240, Cambridge University Press, Cambridge, 1997.

[62] JOHN S. WILSON, ***Profinite Groups***, London Mathematical Society Monographs. New Series, vol. 19, Oxford University Press, New York, 1998.

[63] MARTIN ZIEGLER, *Model theory of modules*, ***Annals of Pure and Applied Logic***, vol. 26 (1984), no. 2, pp. 149–213.

[64] BORIS ZILBER, *Strongly minimal countably categorical theories. II, III*, ***Akademiya Nauk SSSR. Sibirskoe Otdelenie. Sibirskiĭ Matematicheskiĭ Zhurnal***, vol. 25 (1984), no. 4, pp. 63–77.

[65] ———, *Analytic and pseudo-analytic structures*, ***Logic Colloquium 2000***, Lecture Notes in Logic, vol. 19, Association for Symbolic Logic, Urbana, IL, 2005, pp. 392–408.

INDEX